David

2007 Distinguished Instructor Short Course

Distinguished Instructor Series, No. 10

sponsored by the

Society of Exploration Geophysicists

European Association of Geoscientists & Engineers

Concepts and Applications in 3D Seismic Imaging

presented by

Biondo L. Biondi

SEG

EAGE

EUROPEAN ASSOCIATION OF GEOSCIENTISTS & ENGINEERS

Library of Congress Cataloging-in-Publication Data

Biondi, Biondo, 1959-

Concepts and applications in 3D seismic imaging / presented by Biondo L. Biondi.

p. cm. -- (Distinguished instructor series; no. 10)

"Sponsored by the Society of Exploration Geophysicists, European Association of Geoscientists & Engineers."

"Distinguished Instructor Short Course."

Includes bibliographical references and index.

ISBN 1-56080-139-5 (v.) -- ISBN 1-56080-086-0 (series)

1. Imaging systems in seismology. 2. Three-dimensional imaging. 3. Seismic prospecting. I. Society of Exploration Geophysicists. II. European Association of Geoscientists and Engineers. III. Title.

QE538.5.B564 2007

551.22028'7--dc22

2007003947

ISBN 1-56080-086-0 (Series)
ISBN 1-56080-139-5 (Volume)

Printed in the U.S.A.

SEG and EAGE

wish to thank the following
for their generous contributions

SEG Foundation

STANFORD UNIVERSITY
SCHOOL OF EARTH SCIENCES

About the Author

Biondo L. Biondi is associate professor of geophysics at Stanford University in California. He is codirector of the Stanford Exploration Project (SEP) and leads its efforts in 3D seismic imaging. SEP is an academic consortium whose mission is to develop innovative seismic-imaging methodologies and to educate the next generation of leaders in exploration seismology. Biondi has made contributions in several aspects of seismic imaging, ranging from velocity estimation to parallel algorithms for seismic migration. Since the early 1990s, he has been at the forefront of the development of wave-equation 3D prestack imaging methods. In 1994, Biondo cofounded 3DGeo Development, a seismic-services company for practical applications of seismic imaging. It brings innovative technologies to the exploration industry, such as wave-equation imaging and Internet-based seismic processing.

Biondi graduated from Politecnico di Milano in 1984 and received an M.S. (1988) and a Ph.D. (1990) in geophysics from Stanford. He is a member of the Society of Exploration Geophysicists, the European Association of Geoscientists and Engineers, and the American Geophysical Union. In 2004, SEG honored him with its Reginald Fessenden Award for his development of azimuthal moveout.

EAGE

EUROPEAN
ASSOCIATION OF
GEOSCIENTISTS &
ENGINEERS

Dear DISC Participant:

It is a great pleasure to welcome you to the tenth annual SEG/EAGE Distinguished Instructor Short Course (DISC), "Concepts and Applications in 3D Seismic Imaging," by Dr. Biondo L. Biondi. SEG, EAGE, and your local society are proud to provide this premier course in geophysics education.

With rapidly changing technologies, geoscientists around the world have an increasing need to acquire expert technical knowledge. SEG's Professional Development program — the SEG/EAGE DISC, the SEG Distinguished Lecturer, the SEG/AAPG Distinguished Lecturer, the new SEG Regional Lecturers, and other live and online short courses — aids in the promotion of technologies that will have a significant impact on geophysics and geology. Likewise, EAGE supports Geoscience Education through its Distinguished Lectures and Short Courses and its Learning Geoscience e-learning initiative.

Previous DISC programs were:

- "Time-Lapse Seismic in Reservoir Management," by Ian Jack, in 1998
- "The Seismic Velocity Model as an Interpretation Asset," by Phil Schultz, in 1999
- "Shear Waves from Acquisition to Interpretation," by Robert Garotta, in 2000
- "Seismic Amplitude Interpretation," by Fred J. Hilterman, in 2001
- "Understanding Seismic Anisotropy in Exploration and Exploitation," by Leon Thomsen, in 2002
- "Geostatistics for Seismic Data Integration in Earth Models," by Olivier Dubrule, in 2003
- "Petroleum Systems of Deepwater Settings," by Paul Weimer, in 2004
- "Insights and Methods for 4D Reservoir Monitoring and Characterization," by Rodney Calvert, in 2005
- "Seismic Attribute Mapping of Structure and Stratigraphy," by Kurt Marfurt, in 2006

Education in geophysics and geoscience is one of the top priorities for both SEG and EAGE. The DISC affords important opportunities for local organizations to provide first-rate geophysical education at modest cost. The program is truly a cooperative effort of many people dedicated to the promotion and advancement of geophysics. Your participation is key to the continued success of this annually renewed program.

We are honored to have Biondo L. Biondi for our 2007 DISC program. This is a great opportunity to learn from one of our profession's recognized experts in the use of seismic imaging to improve the quality of interpretations. We encourage you to take full advantage of this opportunity to broaden your perspectives through participation in the 2007 DISC.

Sincerely,

Leon Thomsen
SEG President

Theo Kortekaas
EAGE President

Table of Contents

Acknowledgments

This book benefits in endless ways from the contributions of faculty and students at the Stanford Exploration Project (SEP) at Stanford University in California. Jon Claerbout, the inspiring founder of SEP, has influenced every aspect of my work. I am particularly grateful to him for demonstrating how much fun and how insightful the "exploration" of field seismic data can be. This book would be much poorer in data examples without his influence.

All the data examples have been produced using the SEPlib software library, particularly its 3D extension named SEPlib3d. SEPlib and its related graphic software were built and refined by several generations of SEP students. Jon Claerbout and Stew Levin started SEPlib itself. Rob Clayton, Dave Hale, Joe Dellinger, and Rick Ottolini developed most of the graphic software, and Dave Nichols and Steve Cole continued their work. SEPlib3d owes its existence to Bob Clapp. He and I collaborated in the design and initial implementation of SEPlib3d, which was based on early work by Martin Karrenbach. Clapp deserves all the credit if SEPlib3d is now an excellent tool for 3D seismic research, still unmatched in flexibility and capabilities.

Students of my 3D seismic-imaging class at Stanford University have provided innumerable original suggestions to make the text clearer and more readable. SEP students produced several figures used in this book, and I acknowledge these important contributions in the captions.

I am also grateful to professional geophysicists who read early versions of the book when I posted them on the Web and have contributed to the correction of several errors in the text and in the equations. Among them, Bill Schneider and Zheng-Zheng (Joe) Zhou have been particularly helpful. Dawn Burgess and Fannie Gray, SEP technical editors, carefully edited earlier versions of the manuscript and patiently expunged the Latinisms in my English writing. Anne H. Thomas and Gary F. Stewart, who copy-edited the manuscript for SEG, did a great job of forging it into its final shape.

I am grateful to SEG, the SEG Foundation, and the European Association of Geoscientists and Engineers (EAGE) for giving me the wonderful opportunity to teach around the world the one-day 2007 Distinguished Instructor Short Course (DISC) on seismic imaging and for publishing this book that accompanies the course. I am particularly grateful to the following SEG staff members: Rowena Mills (manager of GEOPHYSICS and books) for patiently driving the process of publishing the book, Julie Colley and Sarah Tyler for assisting with proofreading, William D. Underwood for helping the proofreading process when geophysical issues surfaced, and Ted Bakamjian (director of publications) for his support. I also thank Candice Chinsethagid of SEG and Alejandra Reynaldos Rojas of EAGE for effectively and patiently working with me on the organization of my DISC tour.

The sponsors of SEP supported my research financially. I am grateful to all of them for making the writing of this book possible. Several of these companies and a few other organizations have provided the field data sets and the images used in the data examples. I thank the following people and organizations (in alphabetical order of the organization's name) for providing the data sets used in this book:

Mike Howard of BHP Billiton and Nick Moldoveanu of WesternGeco for the sketch shown in Chapter 1 and the field-data images shown in Chapter 9; Jerry Beaudoin and Allan Ross at BP, and BHP Billiton for the 3D rendering of the OBS grid shown in Chapter 1; BP (ARCO) for the button-patch data geometry used in Chapter 1; Carl Regone at BP for the sketch shown in Chapter 1 and the synthetic-data images shown in Chapter 9; Frederic Billette and John Etgen, along with BP and ExxonMobil, for the deepwater Gulf of Mexico data set used in Chapter 9; Tim Summers and Scott Michell of BP for the wide-azimuth field-data images shown in Chapter 9; Bob Brown and Chevron for the Gulf of Mexico data set used in the introduction; John Toldi and Chevron for the OBC data geometry used in Chapter 1; Ross Hill and Chevron for the Gaussian-beam migrations shown in Chapter 5; Doug Hanson with ConocoPhillips and ExxonMobil for the North Sea data set used in the introduction and in Chapters 4 and 9; Luca Cazzola of ENI for the field-data example of phase-encoding migration shown in Chapter 5; Anatoly Baumstein and Mohamed Hadidi at ExxonMobil for the data and images shown in Chapter 4; Joe Reilly, Karen Dittert, and Jerry Krebs of ExxonMobil for the Gulf of Mexico anisotropic parameters and data set shown in Chapters 2 and 7; Husky Oil for the cross-swath data geometry used in Chapter 1; IFP for the Marmousi data set used in Chapter 8; SEG and EAGE for the salt data set used throughout the book; Josef Paffenholz (now at Fairfield Industries) and the SMART JV for the Sigsbee data set used in Chapter 9; Henri Calandra and Total for the North Sea data set used in the introduction and in Chapters 2, 3, 5, 7, and 9; Wook Lee (now at SeisLink Corporation) and Unocal for the Gulf of Mexico OBC data set used in the introduction and in Chapter 8; Phil Schultz and Unocal for the deepwater Gulf of Mexico data set used in Chapter 5 and 7.

The images of this data set shown in Chapter 5 were produced by Sean Crawley and Moritz Fliedner (of 3DGeo) and kindly were provided by 3DGeo Development. Scott Cheadle at Veritas DGC provided the field-data images shown in Chapter 2, with the generous permission of Nexen Inc., Devon Canada Corp., and Burlington Resources Canada. Moritz Fliender of 3DGeo provided the example of automatic MVA shown in Chapter 7.

Introduction

The introduction of 3D seismology had a dramatic impact on petroleum exploration and production. Thus it has contributed substantially to the world's proven reserves of energy. Companies have been enabled to explore and produce in areas of the world (e.g. deepwater subsalt) that until recently were believed to have no commercial accumulation of hydrocarbons, and previously uneconomical, or marginally economical, plays have become viable.

These benefits have been achieved because of the dramatic increase in information and accuracy that 3D seismic images yield, in comparison with images obtained by 2D seismic. That increase is substantial not only for structures that are obviously three-dimensional — such as the Gulf of Mexico salt dome described in this introduction — but also for areas where the geology has a predominant direction of dip, such as the North Sea province. Furthermore, at the reservoir scale, the earth is not two-dimensional. Only 3D imaging can provide the necessary detailed detailed information about small-scale reservoir features, such as small faults and thin channel-fill deposits. This information often is crucial to an effective exploration and production strategy.

The economic value provided by the quality and quantity of available information is compounded with substantial reduction in costs related to acquisition and processing of 3D seismic data. This technological trend is continuing unabated, and it is likely to continue for the foreseeable future, because the acquisition and processing of 3D seismic data is becoming more widespread and achieving even larger economies of scale.

This introduction starts by illustrating how data and images obtained by 3D seismology can be visualized. Once the reader has become acquainted with the displays of 3D images, the differences between images obtained from 2D seismology and from 3D seismology will be illustrated, and causes of these differences will be explained intuitively. Finally, the new challenges presented by 3D seismic imaging will be discussed briefly.

Exploring 3D Seismic Data and Images

When one works with 3D seismic data, the multidimensional nature of the data — and of images produced from the data — are the first obvious challenge. As will be discussed in more detail in Chapter 1, 3D seismic prestack data are defined in 5D space, compared with the 3D space of 2D prestack data. The five dimensions are recording time (t), two components of the source position (x_s, y_s), and two components of the receiver position (x_g, y_g). Similarly, in Chapters 2 and 3 the images produced from 3D data are also defined in 5D space. In addition to coordinates defining the location in the subsurface — depth (z), and horizontal coordinates (x, y) — in the general case a seismic image is defined also by reflection angles at the reflectors; that is, the reflection opening angle (γ) and reflection azimuth angle (ϕ).

Visualization of 5D fields that are full of small-scale details and that represent large geologic structures is a challenge in itself. Throughout this book images and data are displayed by rendering the value of the field along hyperplanes (2D sections) that cut the 5D hypercube orthogonally with respect to one or more of its coordinate axes. To start with the simplest case, Figure 1 depicts a 3D cube. Images on exposed faces of the cube render the value of the field (in this case, results of prestack migration) on those faces by use of a gray-scale display. Blue lines superimposed on the images identify locations of sections that are displayed. In this case they are at the edge of the cube, because values on outer faces of the cube are rendered. Usually one is more interested in analyzing the interior of the image cube rather than its boundaries. Figure 2 displays the interior of the seismic cube shown in Figure 1. As in Figure 1, blue lines identify locations of sections that are displayed on the outer faces. Now one can see informative geology instead of the migration boundary artifacts shown in Figure 1. Whereas the prospective view of the image rendered in Figure 2 provides an intuitive understanding of the geometry, rendering of images on slanted planes hampers analysis of their details. Therefore, orthogonal projections (Figure 3) are usually preferred to the prospective 3D views, such as those shown in Figure 2. To facilitate analysis of details in the sections, Figure 3 displays the same images shown in Figure 2, after the seismic cube was unfolded.

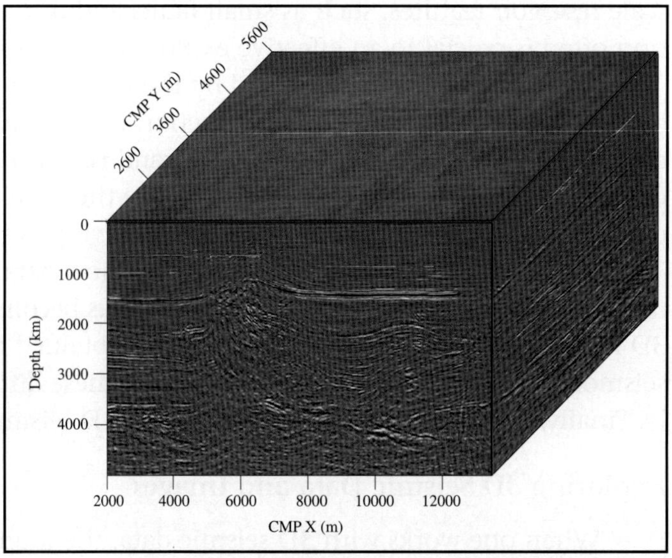

Figure 1. Three-dimensional display of a seismic cube. Images on exposed faces of the cube render the value of the field (in this case, results of prestack migration) by gray-scale display. Blue lines superimposed onto images identify locations of the sections that are displayed.

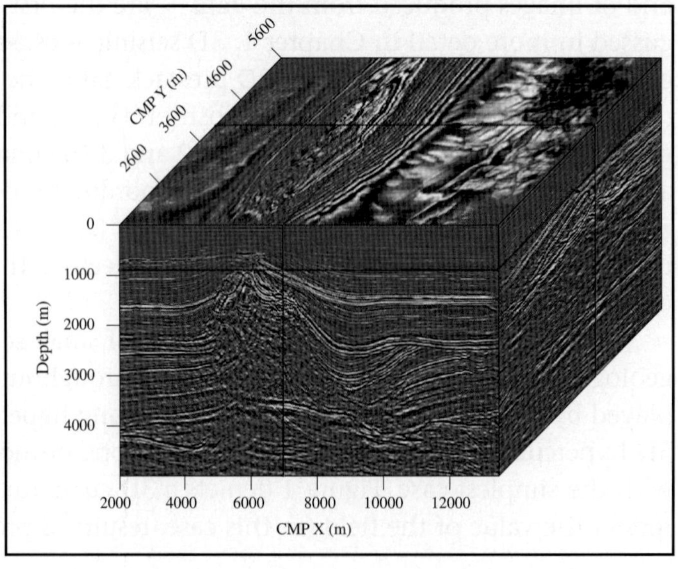

Figure 2. Three-dimensional display of the interior of the seismic cube shown in Figure 1. Blue lines identify locations of the inner sections displayed on the outer faces.

Figures 1, 2 and 3 display 3D image cubes. However, seismic data and images are in general five-dimensional. To display 5D cubes, the methodology used to visualize 3D cubes can be generalized by displaying 2D sections cut at constant value of the other three coordinates. Figure 4 is an example of the visualization of a 5D image obtained by prestack migration of a synthetic data set recorded over five dipping planes. The three sections on the left display values of the image as a function of the subsurface coordinates (z,x,y), for constant values of the aperture angle (γ) and of the reflection azimuth (ϕ). The section on the right shows values of the image as a function of aperture angles at the reflector, at fixed location (i.e., fixed z,x,y). The reader can notice that the image contains energy only in a narrow azimuthal range, because the data were computed with common-azimuth acquisition geometry.

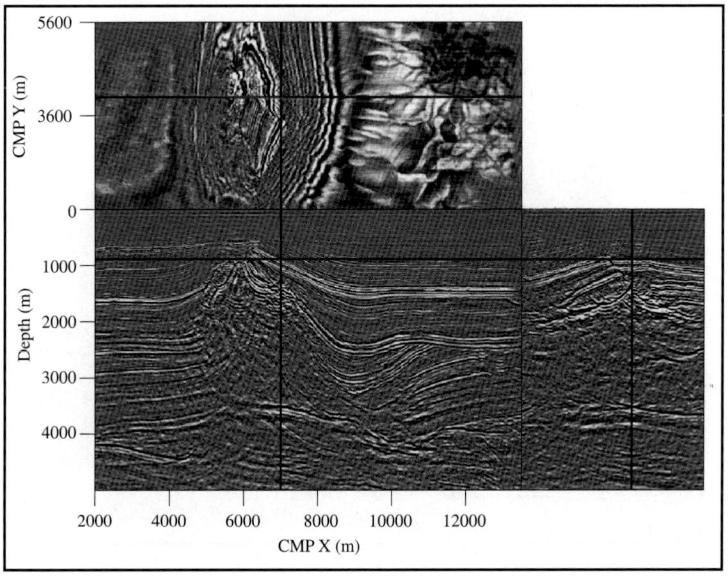

Figure 3. Display of the images shown in Figure 2. The seismic cube was unfolded, to facilitate analysis of details in the images. This kind of orthogonal-projections display is the one used most commonly in this book.

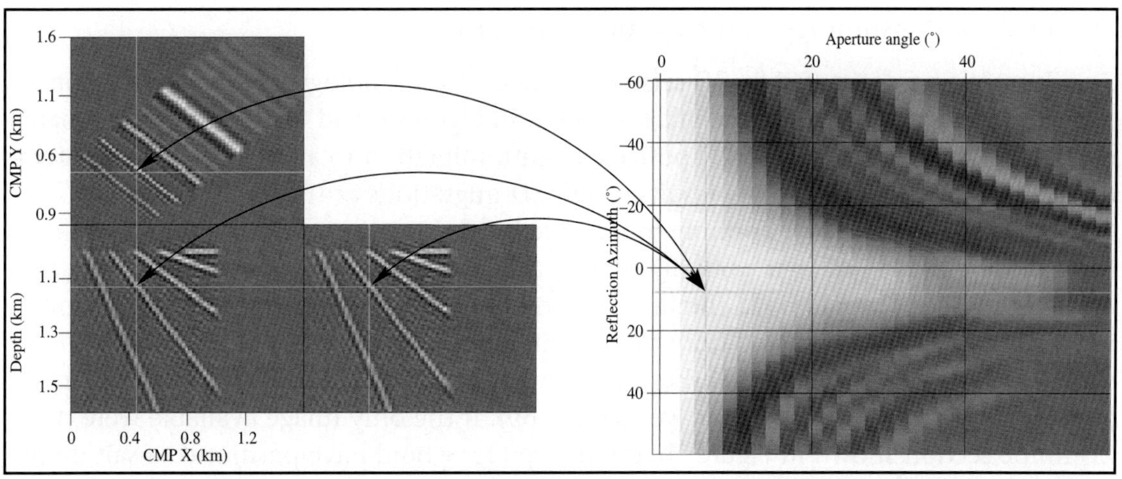

Figure 4. Display of a five-dimensional seismic image. The three sections on the left display the values of the image for a constant value of the aperture angle (γ) and of the reflection azimuth (ϕ), and as a function of the subsurface coordinates (z,x,y). The section on the right shows the values of the image at fixed location, and as a function of the aperture angles at the reflector.

With respect to the visualization of seismic images, visualization of seismic data is further complicated by the irregular acquisition geometries commonly used to record field data (Chapter 1). However, after the data geometry is regularized by interpolation or by other means, seismic data can be visualized in a fashion similar to that used in previous figures. Figure 5 displays a 3D cube of interpolated near-offset traces. The vertical axis records time instead of depth; otherwise the display is similar to Figures 1 through 4. This data cube is a subset of the input used to generate the image cube shown in Figure 3. Notice the numerous unfocused diffractions visible in Figure 5, as compared to the sharp geological structures interpretable from Figure 3. Comparing the sections displayed in Figure 3 with the sections shown in Figure 5 provides an intuitive understanding of the focusing process that is at the heart of 3D seismic imaging.

Figure 5. Display of the three-dimensional cube formed by the near-offset traces extracted from the seismic data. These data correspond to the image cube shown in Figure 3. The comparison of the sections shown in Figure 3 with the sections shown in this figure provides an intuitive understanding of the focusing process that is at the heart of 3D seismic imaging.

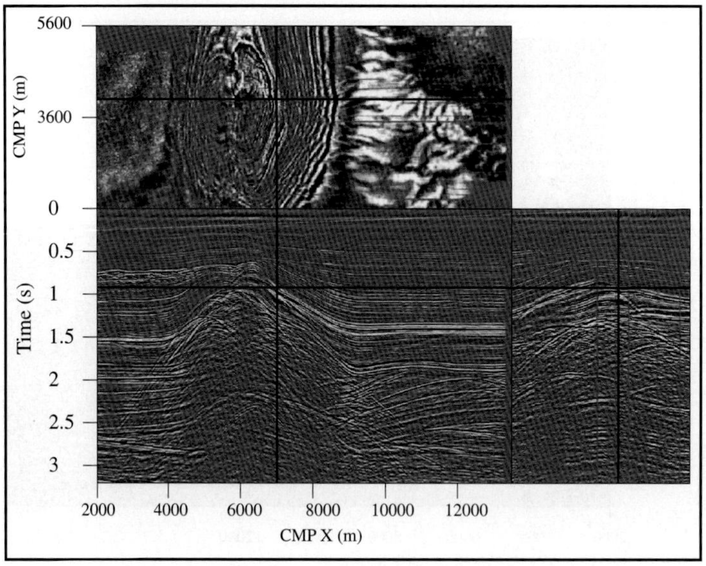

2D versus 3D Seismology: More Accurate Imaging

Improvements assured by 3D seismology over 2D seismology in imaging the subsurface are illustrated well by the images shown in Figures 6 and 7. The figures compare the results of applying 3D and 2D poststack depth migration to a zero-offset cube (Chapter 2). Sections on the left show results of full 3D migrations of the data; sections on the right show results of performing independent 2D migrations on each inline section (horizontal direction in Figure 6b). If one had available only 2D data collected along any of the inlines, one would have obtained distorted images of the subsurface corresponding to inline sections extracted from the 2D migrated cube (e.g., Figure 7b). Depth slices shown in Figure 6 show how the shape and extent of the salt dome would have been grossly misinterpreted from 2D images (Figure 6b). If the only image available were the 2D inline section shown in Figure 7b, the interpreter would have positioned a salt dome in the subsurface at a location where only sediments are present (Figure 7a).

When a 2D seismic survey is acquired, we assume implicitly that the earth is a cylinder,the axis of which is orthogonal to the direction of the survey. If this underlying assumption is fulfilled, the image obtained by 2D seismic interpretation is an accurate rep-

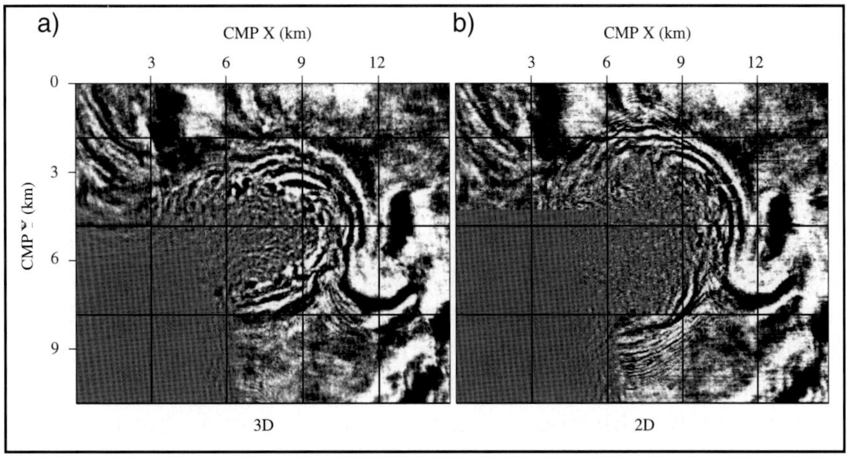

Figure 6. Depth slices of a salt dome, Gulf of Mexico: (a) 3D versus (b) 2D. The slice is cut at depth = 3.63 km.

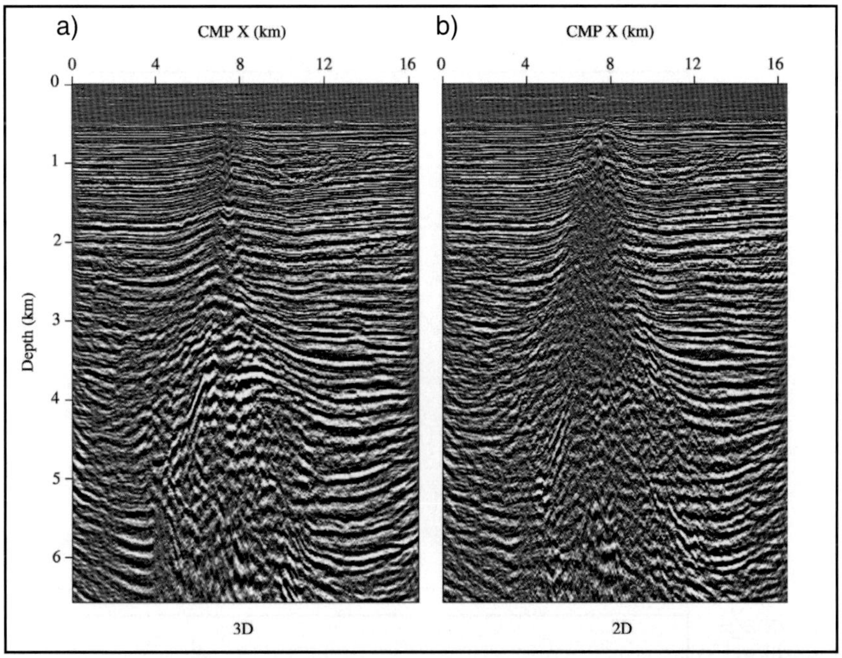

Figure 7. Inline sections of a salt dome, Gulf of Mexico: (a) 3D versus (b) 2D. The section is cut at CMP Y = 3.25 km.

resentation of a vertical section of the subsurface. However, where this assumption is not fulfilled, interpretation of 2D seismic produces a distorted image. The schematic in Figure 8 shows the main reasons why 2D seismic interpretation produces distorted images when one is imaging 3D structures. The point diffractor R_{3D} is out-of-the-plane with respect to the 2D-acquisition direction. Reflections are incorrectly back-propagated in the earth along the vertical plane passing through the acquisition direction, and are imaged at the wrong location, R_{2D}. To image the diffractor at its correct location, R_{3D}, one must apply 3D imaging for back-propagating the recorded reflection along an oblique plane. However, even after applying 3D imaging methods to a single 2D line, one would be left with the ambiguity of where to place the diffractor along the semicircular curve perpendicular to the acquisition direction, marked as "crossline migration" in the Figure 8.

To resolve this ambiguity, one estimates the crossline component of the propagation direction. This estimate is possible only by measuring the reflection stepout along the crossline direction — that is, by recording 3D data. Figures 4 and 5 illustrate this concept schematically. To estimate the direction of event propagation,seismic imaging algorithms rely on the difference in arrival times of the same event recorded by nearby traces (time stepout). The nearby traces may correspond to different receiver locations (G in Figures 8, 9 and 10) or to different source locations (S in Figures 8, 9 and 10). In most instances, this estimate usually is implicit in the imaging algorithms, but nevertheless, it is crucial to the imaging process. If data are recorded along one direction only (Figure 9), the only information available is about the angle that the propagation direction makes with the recording direction. No information on the crossline angle is available. Conversely, if data are recorded along both horizontal directions (Figure 10), the propagation direction can be resolved fully. Notice that at least in principle, for the imaging problem to be fully determined, the wavefield should be sampled properly at both the source and the receiver locations. Unfortunately, with practical acquisition geometries that is rarely the case.

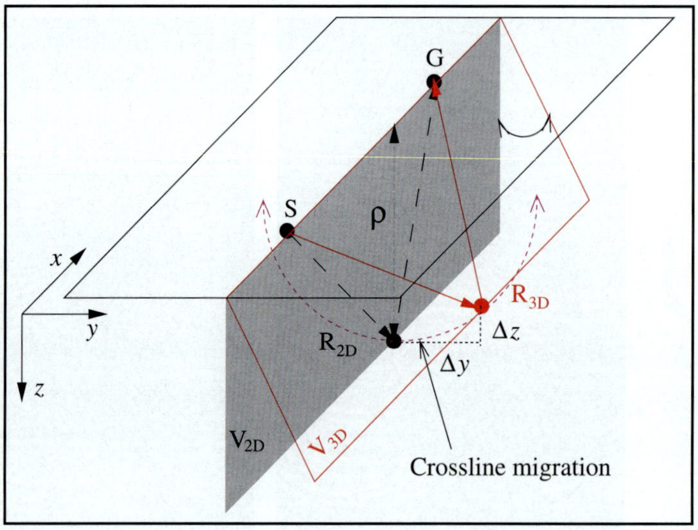

Figure 8. Imaging of an out-of-plane diffractor, 2D versus 3D.

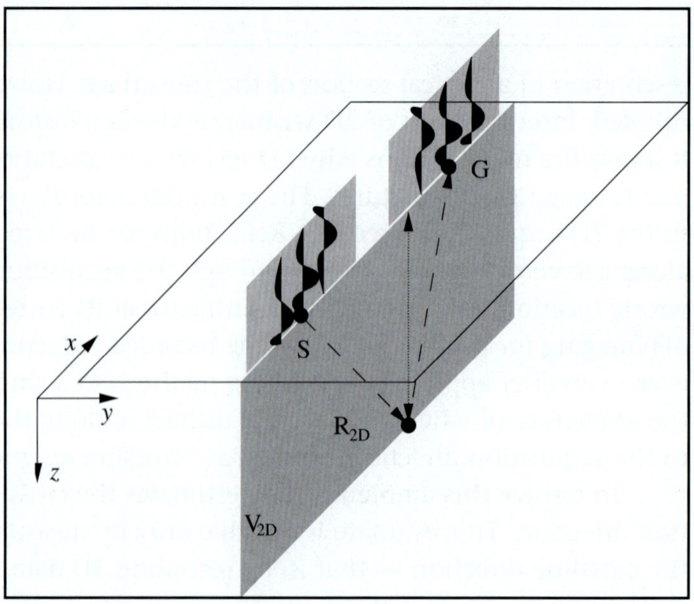

Figure 9. 2D acquisition and imaging. Only the in-plane propagation direction can be determined from 2D data.

When 2D imaging is performed, the error in positioning of the diffractor has both a crossline direction component, Δy and a depth component, Δz. The reflector R_{2D} should be moved along a semicircle in the crossline direction to be placed at its correct position R_{3D}. In the following, we will see that at least where the velocity is constant, this semicircular trajectory corresponds to the semicircle of a zero-offset crossline migration. The analytical expression of this semicircle is independent of velocity and depends only on the apparent depth ρ of the reflector; that is,

$$\Delta y^2 + (\rho - \Delta z)^2 = \rho^2. \tag{I.1}$$

Therefore in this simple case the imaging of 3D data can be performed as two steps: (1) imaging along 2D lines, followed by (2) crossline migration. However, where the propagation velocity is not constant, this simple decomposition of 3D imaging is not correct, and full 3D migration is required. In the general case, the inline 2D migration cannot focus all the data properly because the velocity function V_{2D} along the vertical plane is different from the true velocity function V_{3D} along the oblique plane. Where variations in velocity are large, propagation paths do not even lie on planes; they lie along complex surfaces.

Figures 11 through 14 show concrete examples of the types of error introduced by 2D imaging when it is applied to geologic structures that present 3D features. The figures permit comparison of the sections of prestack depth-migrated cubes. Sections on the left (a) show results of full 3D migrations of the data; sections on the right (b) show results of independent 2D migrations on parallel seismic lines. In a crossline section, Figure 11 shows the mispositioning of reflectors caused by 2D imaging. Arrows point to two dipping reflectors that are grossly mispositioned by 2D imaging. From the 2D image, width of the channel between the two reflectors would be underestimated by about 300 m.

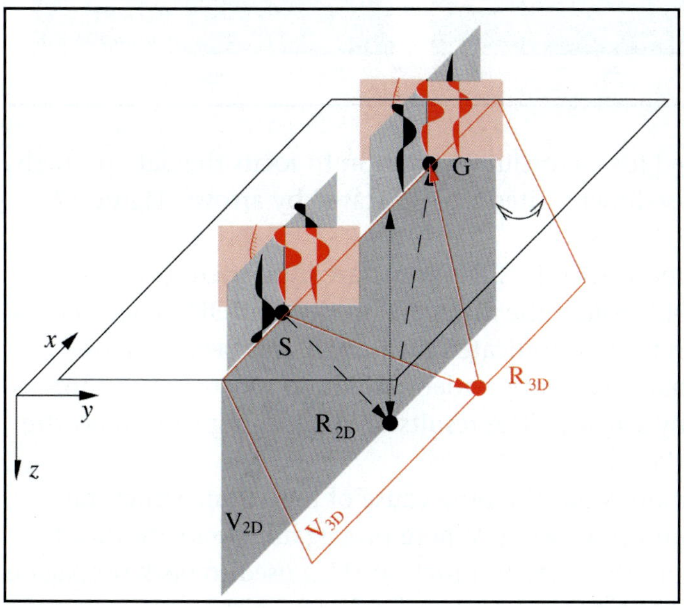

Figure 10. 3D acquisition and imaging. Data contained in the red traces allow unambiguous determination of the propagation direction and positioning of the diffractor.

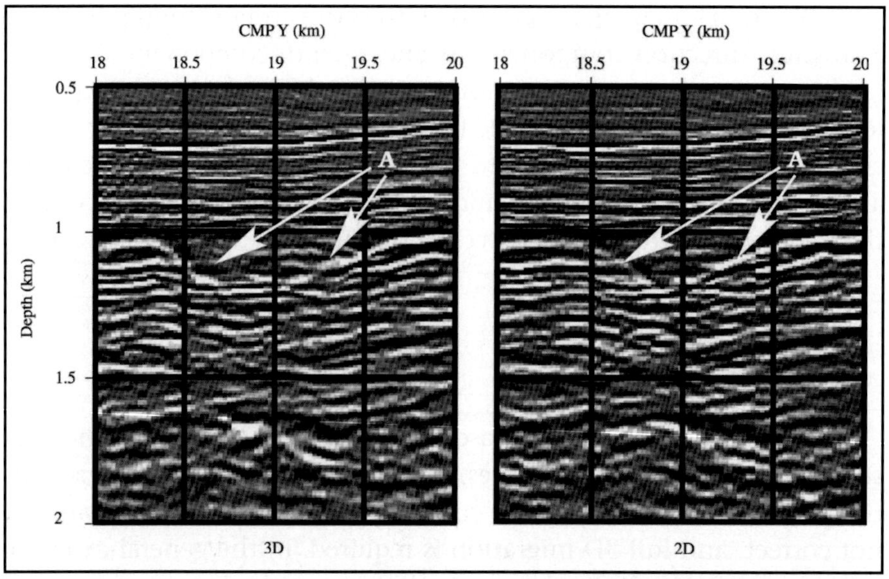

Figure 11. Crossline sections showing that 2D imaging causes lateral mispositioning of dipping reflectors.

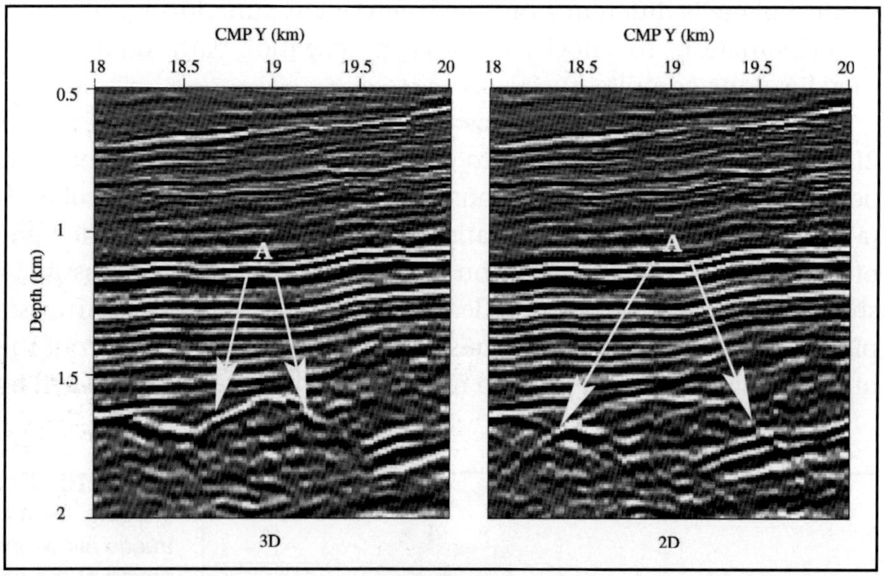

Figure 12. Crossline sections showing that 2D imaging does not focus diffractions properly.

Figure 12 shows evidence of the need for a crossline migration to focus the data properly. Diffractions created by a rough salt-sediment interface (indicated by arrows, Figure 12) are collapsed correctly only in the 3D image.

Errors in structural interpretation caused by 2D seismic are illustrated further in Figure 13. It shows depth slices that cut through the same 2D-imaged and 3D-imaged cubes as in Figures 11 and 12. The broad reflection indicated by arrow B has been mispositioned by several hundred meters, but even more evident are the misfocusing and mispositioning of the reflector indicated by arrow A. The results of 2D imaging show a bowtie shape instead of the correct sinusoidal shape.

Lack of proper crossline migration is not the only cause of inaccurate structural imaging by 2D seismic acquisition and processing. Where propagation velocity varies significantly in the crossline direction, the velocity function (V_{2D}) used to back-propagate reflections along the vertical plane generally is different from the velocity function (V_{3D}) that should be used for propagating reflections along the correct oblique plane. Application of an incorrect velocity function causes both mispositioning and defocusing of the imaged reflectors. Figure 14 shows comparison of two inline sections from the image cubes previously presented; this is an example of the misfocusing introduced by 2D seismology. Reflections from flanks of the salt (arrow A) and for the bottom of the salt (arrow B) are significantly better focused in the 3D image. Notice also the better focusing of the reservoir interval (Arrow C).

2D versus 3D Seismology: Extra Information

Three-dimensional seismic information not only can increase the accuracy of structural images of the subsurface, it also can provide a wealth of stratigraphic information that is not present in 2D seismic data.

Figure 15 shows an example of geologic information that would be impossible to recover from a set of 2D lines. The time slice in Figure 15a illustrates a complex channel system embedded in sediments near a salt dome, in the Gulf of Mexico. These

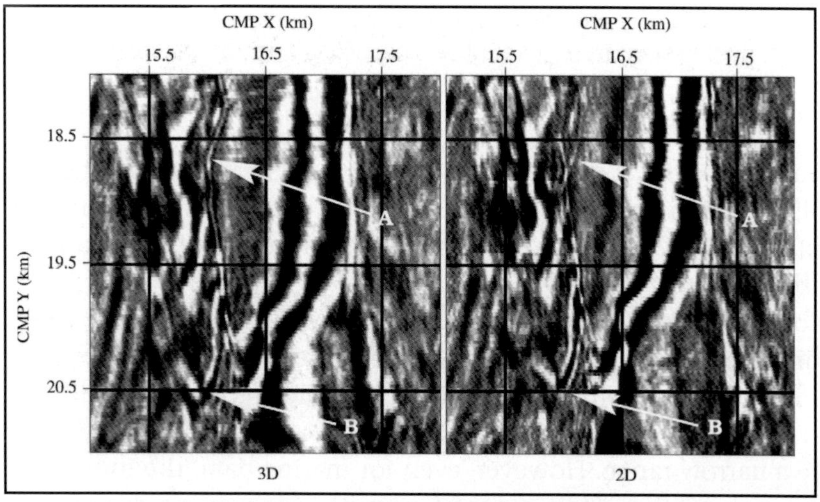

Figure 13. Depth slices showing that 2D imaging causes misfocusing and lateral mispositioning.

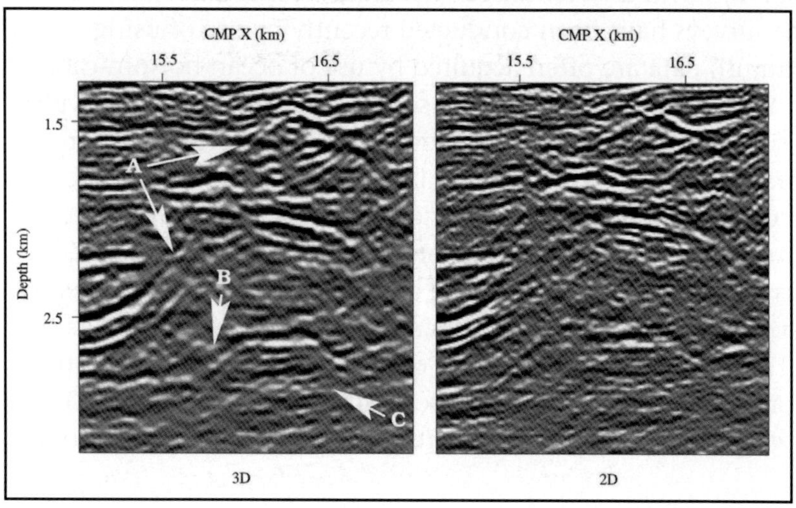

Figure 14. Inline sections showing that 2D images are misfocused because of lateral velocity variations.

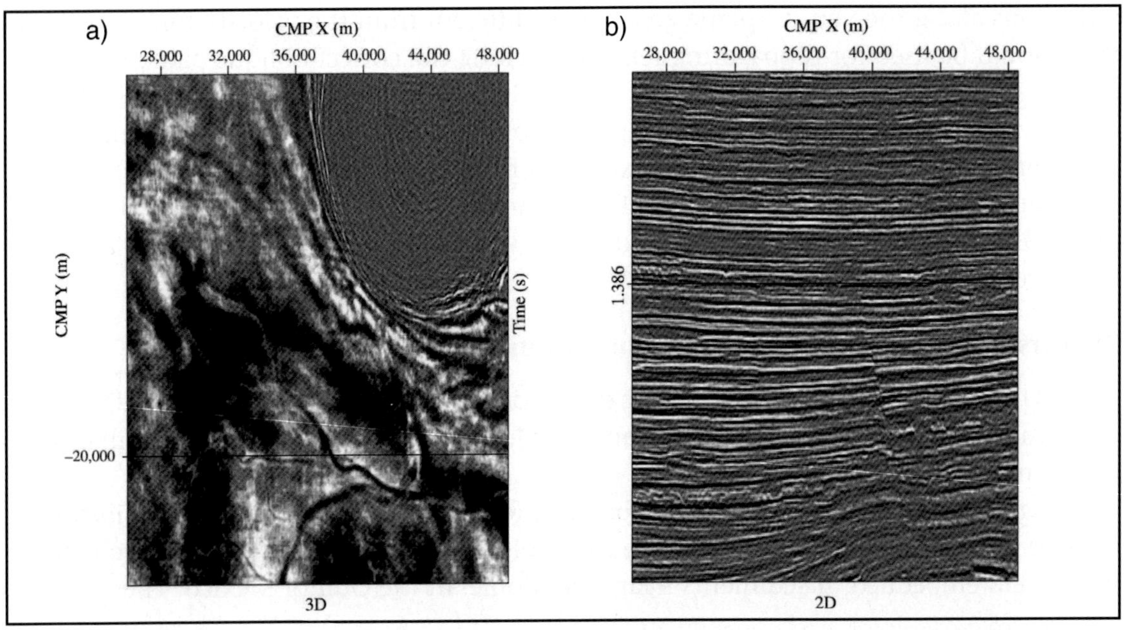

Figure 15. Time slice (t = 1.386 s) and inline section (CMP X = –20,000 m) of a 3D image from data recorded in the Gulf of Mexico. Observe that the complex channel system is visible in the time slice.To reconstruct the channel system from a set of vertical slices such as the inline section would be impossible.

paleochannels are thin features, barely above the threshold of seismic resolution. If one had available only the inline section shown in Figure 15b, to detect even the presence of the channels would be difficult. To reconstruct the complex 3D structure of the channel system would be impossible

Even when the geologic structure is two-dimensional or one-dimensional, 3D seismic data can provide additional information on the rock's parameters. That additional information is contained in the data recorded along different source-receiver azimuthal directions. Typically, land data are acquired with a wide range of azimuthal directions, whereas marine data have a narrow range. However, even for marine data, the industry's trend is toward acquiring surveys with a wider range of azimuths for source-receiver pairs. Wide-azimuth marine surveys have been conducted recently by use of using multiple boats and wide-azimuth data are often acquired by use of ocean-bottom-cable (OBC) surveys (Chapter 1). Recording data with a wide source-receiver azimuth provides additional illumination of subsurface targets. This additional illumination may be crucial to generation of interpretable images where targets lie under complex overburden (Chapter 9). Wide-azimuth data also may provide additional information on properties of subsurface materials. For example, measurement of propagation velocity and reflectivity along different azimuths can be useful in estimating anisotropic properties caused by "vertical" fractures in reservoir rocks. This book focuses mainly on imaging methods that assume an isotropic earth. However, the basic concepts of 3D imaging still are essential when one is extracting information about anisotropy from 3D data. Tsvankin (2001) and Thomsen (2002) provide a good introduction to issues related to imaging of anisotropic materials.

The availability of high-resolution 3D images also has paved the way for time-lapse seismic data (Jack, 1998; Calvert 2005), which often is referred to as 4D seismic data. Time-lapse seismic data are used to monitor reservoirs during production and to contribute to development of optimal strategies for management of reservoirs. In this book, I do not focus on application of 3D imaging to 4D seismic data; nevertheless, the success of time-lapse studies depends on the quality of 3D imaging. In many cases, the changes over time in a reservoir's seismic properties are small and close to the seismic resolution; only high-fidelity imaging can reveal them properly.

2D versus 3D Seismology: New Challenges

Many of basic concepts developed for 2D seismic imaging are still valid for 3D imaging. However, 3D imaging presents several new problems that are caused by the higher dimensionality of 3D data. An obvious problem related to high dimensionality of 3D seismic data is the amount of data that is recorded in the field and that should be processed. Many modern surveys produce several terabytes of data. Such large quantities are difficult to handle, visualize, and process. An important aspect of processing a large 3D data set is to devise the right strategy for reducing the time and resources necessary for obtaining an image of the subsurface, without compromising accuracy of the results.

The two most common methods for reducing the cost of are aimed at reducing the dimensionality of the computational domain. These methods are useful in 2D imaging, but they are essential in 3D imaging, because of the higher dimensionality of the data space. The first method, commonly called stacking, reduces the size of the data set by averaging recorded traces along appropriate ranges of the four spatial axes. The most common example of stacking is offset stacking, in which recorded traces are averaged over offset after the application of normal moveout (NMO) (Chapter 4). Offset stacking reduces the cost of computation greatly, because in many instances, the cost imaging operators is directly proportional to the size of the input data. Common-azimuth migration (Chapter 5) is another example of an imaging algorithm that achieves high computational efficiency by exploiting reduction in data dimensionality (from five dimensions to four). In this case, a common-azimuth data set is generated from marine data that are averaged over azimuths after applying azimuth moveout (AMO) (Chapter 4). Then, common-azimuth data are imaged by a wavefield-continuation migration method.

The second methodology used to reduce computational complexity is based on full separation — splitting of the computations along different data axes. Two-pass migrations are typical examples of application of full-separation methods. In general, decoupling the computations along an axis x of length n_x and an axis y of length n_y reduces the computational cost from order $O(n_x \text{ x } n_y)$ to order $O(n_x + n_y)$. Cost advantages of stacking and of separation are achieved at the expense of generality; the derived imaging methods are accurate only for particular classes of data sets. Because selection of the appropriate processing strategy is a crucial decision, I will discuss the available choices as a function of their trade-offs between accuracy and speed.

The other major challenge of 3D seismic data collection and processing is that 3D prestack seismic data are sampled sparsely and irregularly along the spatial coordinates (x_s, y_s, x_g, y_g). Because of practical and economic considerations, actual 3D surveys recorded in the field (Chapter 1) do not record seismic traces at every possible location for the source and the receiver; therefore the data never fill the 5D space that is available. On the contrary, much effort is spent to optimize survey design, so that the desired accuracy in the image is achieved and the data-recording efforts are minimized. The result is that the seismic wavefield almost never is well sampled along both the source axes and the receiver axes. Thus, the condition for unambiguous imaging that was introduced in the discussion of Figure 10 is seldom fulfilled. To generate high-fidelity images, one must safeguard the imaging algorithm from generating aliasing artifacts (Chapter 8). Crucial to this goal is exploitation of the redundant information in the recorded data; this is done by making assumptions on both the reflectivity model and the complexity of wave propagation.

Irregularity in spatial sampling may cause severe variation in spatial illumination of the subsurface. Depending on the degree of variability in data density and the complexity of the imaging algorithm, irregular sampling may cause distortions in the image amplitudes and it may create coherent artifacts in the image. Chapter 9 introduces the basic concepts for compensating for variable data density by normalizing imaging operators. It also introduces a method that exploits data redundancy to reduce both aliasing and irregular sampling artifacts.

References

Calvert, R., 2005, Insights and methods for 4D reservoir monitoring and characterization: Society of Exploration Geophysicists, Distinguished Instructor Series, No. 8, 219 p.

Jack, I. G., 1998, Time-lapse seismic in reservoir management: SEG/EAGE Distinguished Instructor Series, no. 1.

Thomsen, L., 2002, Understanding seismic anisotropy in exploration and exploitation: SEG/EAGE Sistinguished Instructor Short Course, 1-27/1-42, 2-1/2-43.

Tsvankin, I., 2001, Seismic signatures and analysis of reflection data in anisotropic media: Elsevier Science, 456 p.

Chapter 1 3D Data Geometries

Introduction

Because of practical and economical considerations, 3D surveys are not acquired with complete and regular sampling of the spatial axes. The design of 3D surveys presents many more degrees of freedom than does the design of 2D surveys, and it has no standard or unique solution. Design of 3D acquisition geometries is the result of many trade-offs among data quality, logistics, and cost. Furthermore, nominal designs commonly must be modified to accommodate operational obstacles encountered in the field.

Data-acquisition design and processing are becoming ever more connected, because characteristics of the acquisition geometry strongly influence the data processing. One must understand the principles of acquisition design if one is to understand many data-processing issues.

The main goal of conventional acquisition design is to obtain an adequately and regularly sampled stacked cube that can be imaged accurately by poststack migration. Other important parameters are the minimum and maximum offsets. Minimum offset must be small enough to guarantee adequate coverage of shallow targets. Maximum offset must be large enough to allow accurate velocity estimates, which are necessary for both stacking and poststack imaging. However, in common acquisition geometries, sampling of the offset axes may be inadequate when the data require prestack processing that is more sophisticated prestack processing than simple stacking. Even application of standard dip moveout (DMO) (Chapter 4) may be problematic with some commonly used acquisition geometries. In such cases, the requirement that midpoint axes be adequately sampled is not sufficient, because offset and azimuth sampling also are important.

Another important issue emerging in modern 3D survey design is subsurface illumination, in which targets are imaged under increasingly complex overburden and where more control is required on amplitudes of the images. Complex wave propagation associated with large contrasts in velocity (e.g., salt or basalt bodies) may cause data to illuminate the target only partially, even if the wavefield is sampled sufficiently at the surface. Because target illumination is related strongly to data offset and azimuth, a careful survey design can improve image quality dramatically. Survey-modeling tools that analyze target illumination by ray tracing through an a priori model of the subsurface geology help guide the design process toward successful acquisition geometries. However, the usefulness of these tools is limited by how accurately the subsurface geology and the velocity model are known.

Although there are no preset solutions to the problem of acquisition design, a few acquisition schemes commonly are used as templates to be adapted to individual surveys. Common marine- and land-acquisition geometries and their characteristics are discussed in this chapter. More details and deeper analysis are presented in two excellent books: Stone (1994) presented a comprehensive introduction to seismic-data acquisition design; Vermeer (1990) discussed many issues related to wavefield sampling and seismic processing, although the analysis was limited to 2D data.

Data Coordinates

In any survey, each seismic trace is characterized by the corresponding positions of its source and receiver. The coordinates of source and receiver are defined relative to an orthogonal coordinate system, with x and y as the horizontal axes and z as the vertical axis. In this book, a flat recording surface is assumed, with the origin of the coordinates at the surface — unless topography of the recording surface is mentioned explicitly. Therefore, the source position is a 2D vector $\mathbf{s} = (x_s, y_s)$, and the receiver position is a 2D vector $\mathbf{g} = (x_g, y_g)$. If the receivers are aligned preferentially along one direction, as they are for marine surveys and most land surveys, the x-axis is assumed to be aligned with this preferential direction. In that case, the x-axis is called the inline axis and the y-axis the crossline axis. Given these definitions, the geometry of a 3D prestack data set is defined in a 4D continuum (x_s, y_s, x_g, y_g), and the data are defined in a 5D continuum (t, x_s, y_s, x_g, y_g), where t is the recording time. Each trace represents a discrete sample in the 4D geometry space. Usually the geometry space is sampled irregularly along all axes, and also, often it also is sampled inadequately along some or all axes.

Data coordinates expressed in terms of source-receiver coordinates commonly are called the field coordinates, as opposed to the term midpoint-offset coordinates. Midpoint-offset coordinates are defined by transformation of the field coordinates:

$$\mathbf{m} = \frac{\mathbf{g}+\mathbf{s}}{2}, \qquad \mathbf{h} = \frac{\mathbf{g}-\mathbf{s}}{2}, \tag{1.1}$$

where the vector $\mathbf{m} = (x_m, y_m)$ is the midpoint and the vector $\mathbf{h} = (x_h, y_h)$ is the half-offset. In many instances, the half-offset vector is defined more conveniently in terms of the absolute half-offset h and the source-receiver azimuth θ_h, so that $\mathbf{h} = h(\cos\theta_h, \sin\theta_h)$. Because the source-receiver azimuth is defined uniquely for each trace, it commonly is referred to as simply the trace azimuth or data azimuth. This angle must be distinguished from the midpoint azimuth ϕ_m, which is related to geologic dips in the image; this will be discussed in following chapters. Figure 1 shows the geometric relationships between the field coordinate vectors (**g** and **s**) and the offset-midpoint coordinate vectors (**m** and **h**).

Acquisition geometries vary greatly with variations in the physical environment (e.g., whether it is land or marine), with the subsurface structural complexity, and with the overall goals of the survey. In the following section of this chapter, several templates of acquisition geometries are presented; these templates are used commonly in survey design. All geometry templates share the characteristic that the midpoint axes are sampled fairly well, and in many instances, regularly. Good sampling of the midpoint axes is necessary if one is to accomplish the paramount goal of obtaining accurate poststack-migration results. In contrast, sampling of offset axes may vary according to the choice of an acquisition template. In the analysis that follows, the focus will be on sampling of the source and receiver axes, and of the offset axes, for each geometry template.

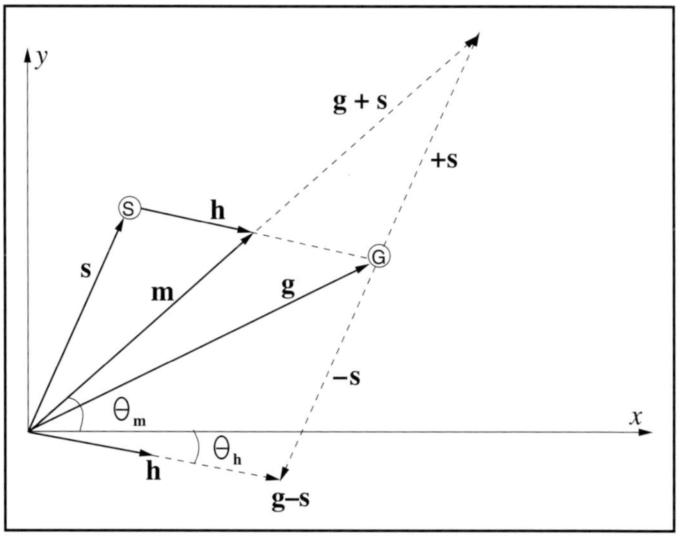

Figure 1. Geometric relationships between the field coordinate vectors (**g** and **s**) and the offset-midpoint coordinate vectors (**m** and **h**).

Marine Data Geometries

Conventional streamer geometries

Modern marine data are acquired by towing several streamers behind a boat. Each streamer incorporates hundreds of hydrophones, located at regular intervals. Some streamers are as long as 6 to 10 km. At least one source of seismic energy is positioned close to the stern of the boat. Most sources are arrays of air guns, but experiments with marine vibroseis are yielding promising results. As the cartoon in Figure 2 shows, the boat moves over the target on multiple passes along (ideally) straight lines (called sail lines), which are spaced equally and are parallel. The sail-line direction is considered the inline direction of the surveys. Because of currents and other factors, the sail lines often are not straight. Even more commonly, currents cause the streamers to follow trajectories that are not straight. Lateral displacement of the streamers is often called cable feathering.

Another difficulty encountered in areas already developed for hydrocarbon production is the fact that platforms or other permanent obstacles can be in the desired path of the boat. A technique called undershooting often is employed to avoid gaps in the coverage just below obstacles that are just above targets of interest. As Figure 2 shows, undershooting requires two boats — one carrying the sources and the other towing the hydrophone streamers. In comparison with conventionally recorded data, the data recorded by undershooting commonly have wider and irregular azimuthal coverage and larger minimum offset. The lack of near-offset data and the irregular azimuthal sampling may cause problems during processing of data.

Figure 3 shows the source and receiver locations for two consecutive shot gathers recorded in the North Sea. (A shot gather is the collection of all the traces recorded with a particular source location.) The shooting configuration had three streamers and two sources. The sources were activated in alternation. Dots in the figure represent receiver locations along the streamers; the two asterisks represent the source locations, at either side of the central streamer. The streamers were 100 m apart in the crossline direction and were not along straight lines, because of strong cable feathering.

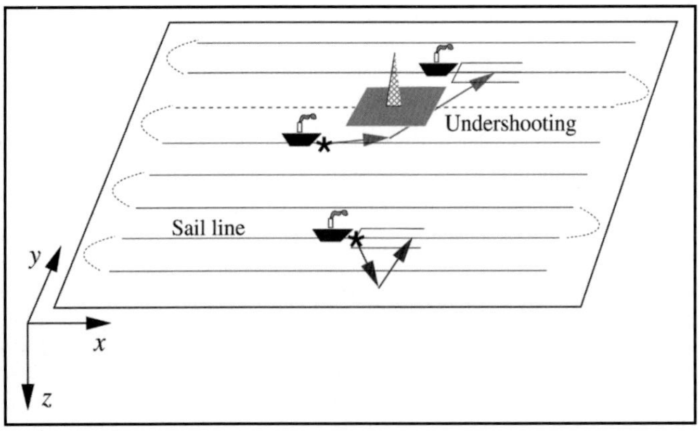

Figure 2. Cartoon representing typical marine streamer acquisitions.

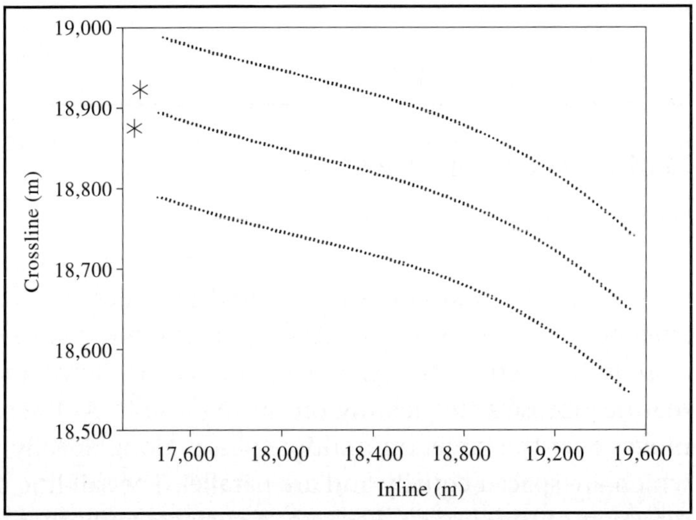

Figure 3. Locations of sources and receivers for two consecutive shot gathers of a three-streamer double-source acquisition in the North Sea. Sources are represented by asterisks and receivers by dots.

Figure 4 shows the offsets of 1000 traces selected randomly from the data set. The inline component of the offset x_h covers a much wider range (200 m to 2200 m) than does the crossline component y_h (– 400 m to 400 m). The crossline range is wider at far offset, because cable feathering is more pronounced at the far ends of the streamers. The crossline-offset range does not drop to zero at near offset because of the use of multiple streamers. As is measurable from Figure 3, the crossline distance between each source and the outer streamer on the opposite side is 120 m. Because of this narrow range of azimuths, the inline component of the offset is sampled densely and prestack imaging is robust.

The trend in streamer surveys is to increase the number of streamers and consequently, to increase the azimuth range of the data. Increase in the number of streamers is motivated by cost considerations: The more streamers towed, the wider the subsurface swath a boat can cover with a single pass, and thus reduce total acquisition time.

Multiazimuth and wide-azimuth streamer geometries

Conventional single-boat streamer acquisition is very efficient in covering large areas of the subsurface and provides well-sampled data in the offset plane, though over a limited range of azimuths. Furthermore, the processing of narrow-azimuth data has substantial advantages both in simplicity and computational cost, as is discussed in the

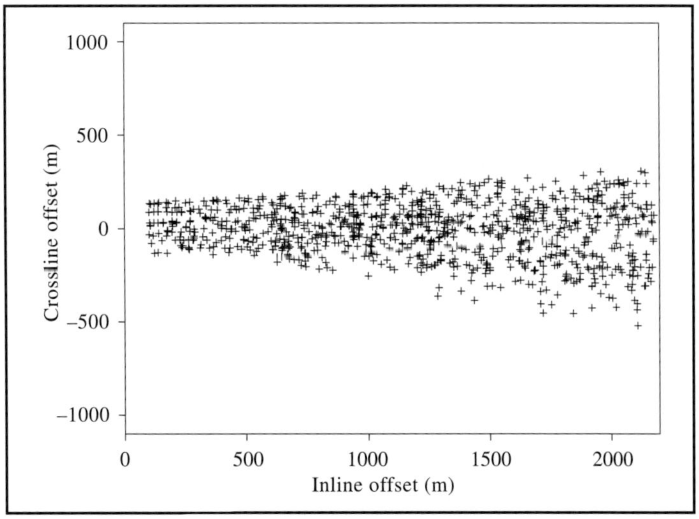

Figure 4. Offsets of 1000 traces selected randomly from the same streamer data set as shot gathers shown in Figure 3.

section titled "Narrow-azimuth versus wide-azimuth geometries." However, where the overburden is so complex that the propagating wavefield is severely distorted, narrow-azimuth data may provide information on targets in the subsurface that may be insufficient.

A practical and economical solution is to acquire several conventional narrow-azimuth data sets, each with its sailing direction at an angle with respect to the others. Data sets acquired by this technique are called multiazimuth surveys. Although the first experiment with "planned" multiazimuth acquisition was reported by O'Connell et al. (1993), in the recent past, multiazimuth surveys were "patched" together by the combining of several surveys acquired over the years. A few surveys have been planned and recorded as multiazimuth using two (Dewey et al., 2006; Riou et al., 2006), three (Rekssnes et al., 2002) and even six sailing directions (Keggin et al., 2006). Results of imaging obtained from these data sets indicate that substantial improvement can be achieved with this simple solution; these results are signs to a potential broader use of genuine multiazimuth surveys.

One operational advantage of multiazimuth data sets is that the data acquired along each direction can be processed independently and imaged as if they were from a conventional single-boat data set. The partial images from each of the directions are combined to form the final image. At its simplest, this final combination can be straight averaging of the images. Further enhancement in the image can be achieved by a weighted average, where the time-spatial variable weights are calculated according to the relative coherency of the each partial image.

The introduction of multi-boat acquisitions represents a more drastic departure from conventional streamer acquisition. This innovation enables the recording of data with wide and dense azimuthal coverage, and thus potentially more complete illumination of the subsurface. Two kinds of acquisition design have been proposed so far: the wide-azimuth towed streamer (WATS) (Regone, 2006) and the rich-azimuth towed streamer design (Howard and Moldoveanu, 2006). The WATS relies on multiple parallel passes of the recording boats to achieve an effective rectangular areal array of receivers with the crossline maximum offset (4 km) only half of the inline maximum offset (8 km). The rich-azimuth design is a hybrid of multiazimuth acquisition and multi-boat acquisition. Data are acquired on multiple passes directed along three directions. At each pass, as many as three boats are used to acquire a fairly wide azimuth data set.

Figure 5 illustrates the WATS shooting configuration. Two shooting-only vessels and one recording-only vessel are deployed. The recording vessel is equipped with several cables. In this particular case eight cables, each 8 km long, and spaced 125 meters apart. The shooting vessels sail on the side of the cable array: one at the front and one at the back. During the first pass the shooting vessels sail as close as possible to the outer cable. For the second pass the sail line of the shooting vessels is 1 km apart from the outer cables, for the third pass it is 2 km apart, and so on. Four passes are sketched in this figure. The four passes are repeated for the next sail line. In this case the distance between sail lines was 250 m; that is, double the crossline-receiver spacing. Geometry of the WATS data can be visualized as an extension along the cross-line direction of the one-sided (along the inline direction) geometry of conventional single-boat acquisition. Only half the offset plane is filled by actual data; the other half can be synthesized by use of the reciprocity principle (Knopoff and Gangi, 1959).

The WATS design described above was adopted to record data for guidance of the development of a large subsalt reservoir, in deep waters of the Gulf of Mexico. By reducing the number of passes and increasing the ratio between the sail line and the intra-cable distances, the data-acquisition effort could be reduced, and thereby wide-azimuth data became economical for exploration — in addition to being useful for development of the subsalt reservoir.

Figure 6 illustrates the RAZ shooting geometry (Howard and Moldoveanu, 2006). As in the WATS case, three vessels are used for the data acquisition. However, in this case all three vessels are shooting and only one is recording. The recording vessel tows 10 cables 120 m apart and 7 km long. The blue patches in the top panel in Figure 6 represent the CMP coverage corresponding to each source and are 600 m wide. The rollover interval between the sailing lines is 600 meters. Each patch is illuminated by all three

Figure 5. Sketch illustrating the shooting geometry of wide-azimuth towed streamer (WATS) data. One recording-only vessel and two shooting-only vessels are deployed. (This sketch provided by Carl Regone, of BP.)

sources and thus it has wide-azimuthal coverage. Furthermore, the vessels pass over the same area three times with the sailing directions rotated by 60 degrees with respect to each other, as illustrated by the sketch at the bottom of Figure 6. This acquisition geometry achieves wide azimuthal coverage without sacrificing data density. The shot grid is hexagonal, and thus it benefits from the excellent sampling properties of hexagonal grids (see discussion below on OBS recording). RAZ acquisition is particularly cost effective when the target area has limited extent and several surface obstructions are present, as it is often the case in development projects, because it allows extensive undershooting of the obstacles.

Parallel-swath geometries (OBC and OBS)

Because seismic data are used increasingly to optimize development and production strategies, more and more surveys are acquired in highly developed areas where towing kilometer-long streamers would be difficult or even hazardous. Therefore, an increasing number of surveys is acquired with an array of receiver cables at the bottom of the ocean. The sources are at the surface and are carried on boats, as in standard streamer surveys. This kind of acquisition technology is called OBC, for ocean-bottom-cable surveys.

A particular kind of OBC survey deploys three-component geophones to record shear waves in addition to pressure waves. This kind of survey is often called OBS, for ocean-bottom-seismometer surveys. Because the sources are at the ocean surface, and

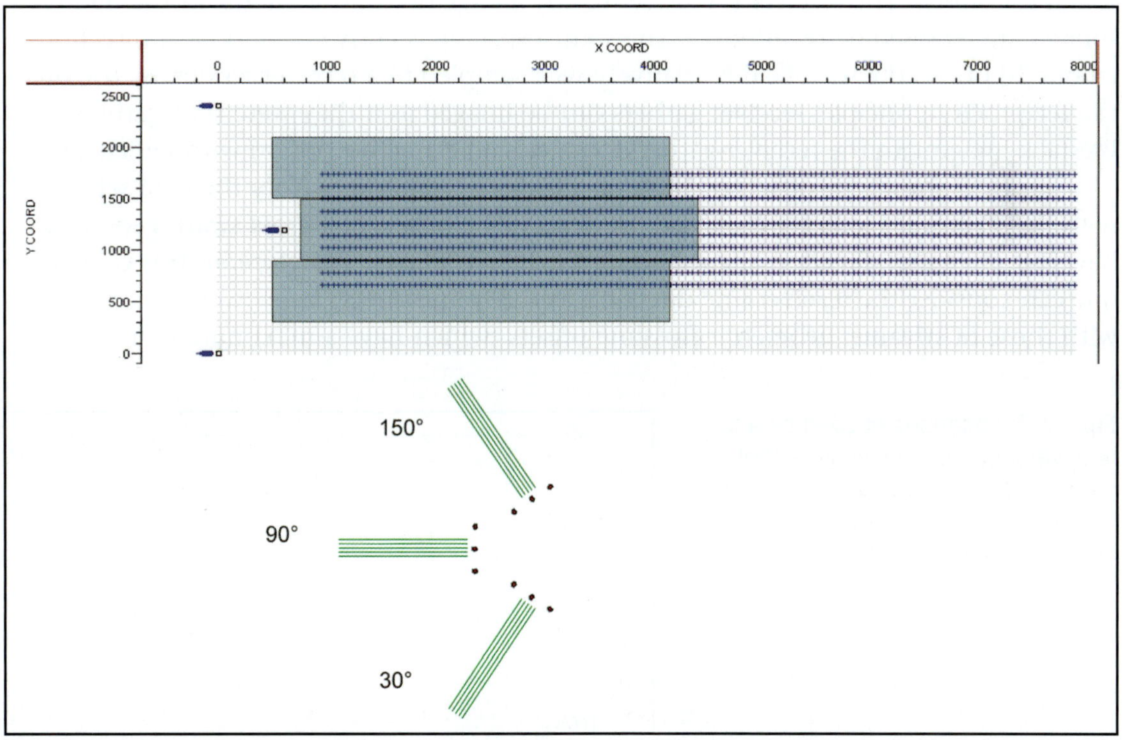

Figure 6. Sketches illustrating the shooting geometry of rich-azimuth (RAZ) towed streamer data. Three vessels are deployed; all three vessels are shooting whereas only one is recording. To record the whole data set the vessels complete three passes over the target area. The sailing directions of these three passes are rotated by 60 degrees with respect to each other (bottom panel). (This sketch was kindly provided by Mike Howard with BHP-Billiton.)

because shear waves do not propagate in liquids, the shear waves must be created by mode conversion either at the water bottom or at the reflectors. Shear waves carry information on rock and fluid properties that complements the information carried by pressure waves. Furthermore, the propagation of shear waves is less affected by slow and attenuating areas in the subsurface (e.g., by gas clouds). Beneath these obstacles, shear-wave data may provide better structural images than pressure-wave data provide.

Both OBC and OBS geometries resemble land geometries more than they resemble classical marine geometries. A popular acquisition template is parallel-swath shooting, which has an advantage over cross-swath shooting (analyzed in the next section), in that it allows deployment of long receiver cables and requires fewer turns of the shooting boat.

Figure 7 shows the source-and-receiver positions for two consecutive shot gathers of an OBS survey recorded in the North Sea. As in previous figures, receivers are represented by dots and sources by asterisks. The receivers were laid along parallel lines; the nominal separation between receiver lines was 400 m. The sources moved parallel to the receiver lines, hence the name *parallel swath*. Notice that, in contrast to streamer surveys (Figure 3), the receiver locations are exactly the same for the two shot gathers; this is because the receiver cables are stationary at the ocean bottom.

Figure 8 shows the geometry of a common receiver gather for the same OBS data set. (A receiver gather is the collection of all traces recorded at the same receiver location, for all the shots.) The receiver is represented by a large dot in the middle of the figure; sources are represented by asterisks. The source spacing is fairly small in both directions (about 50 m), so the wavefield is relatively well sampled in a receiver gather. By use of the reciprocity principle, receiver gathers can be imaged as if they were shot gathers. Prestack imaging in the receiver-gather domain is an attractive option with OBS data.

Figure 9 shows the offsets of 1000 traces selected randomly from the data set. As with the streamer geometry discussed previously, the offsets of a cross-swath geometry have a preferential direction aligned along the receiver lines. Therefore, in this case also, one can consider the receiver-line direction as the inline direction. However, compared with those of streamer geometry, crossline offsets (which have a maximum of 1.4 km)

Figure 7. Locations of sources and receivers for two consecutive shot gathers of an OBS data set from the North Sea. Sources are represented by asterisks and receivers by dots.

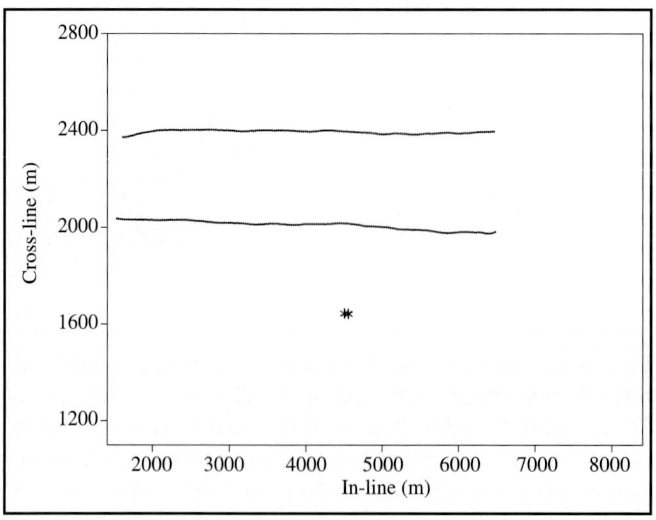

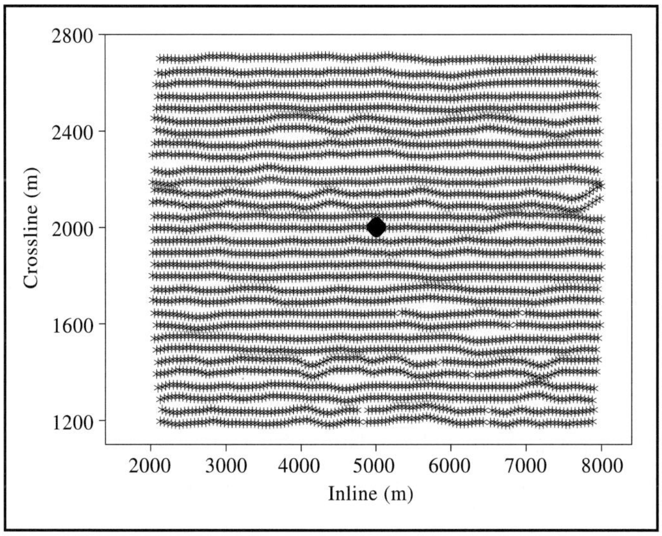

Figure 8. Locations of sources and receivers for one-receiver gather of an OBS data set. The sources are represented by asterisks; the receiver is represented by the large dot.

are not negligible; they create potential problems for some prestack processing algorithms (e.g., DMO, discussed in Chapter 4).

Notice that for OBC and OBS data, the elevation of the sources (ocean surface) differs from the elevation of the receivers (ocean bottom). This difference may create some practical problems; it may require adaptation of algorithms developed on the assumption that sources and receivers are at the same elevation. However, the difference in elevation rarely creates new conceptual challenges.

Arrays of autonomous OBS

Another recent advancement in marine acquisition technology is the development of economical autonomous OBS (often called nodes) that can be deployed in deep waters (>2000 m). An array of these instruments can be laid on the ocean floor by remotely operated vehicles, the desired seismic signal can be recorded, and then the array can be recovered . This acquisition technology is expensive when deployed in large areas, but it provides the advantage of recording wide-azimuth multi-component data even where production facilities inhibit acquisition by surface streamers.

Figure 10 shows a 3D rendering of the grid that was used in the Gulf of Mexico for recording a 3D rendering with autonomous OBSs (Ross and Beaudoin, 2006). Red dots in the image correspond to OBS locations. The 3D image conveys the challenge of the rugged seafloor topography; depth of water ranges from approximately 1300 m to approximately 2300 m. The grid was filled in two stages. The OBSs were deployed in the northern half of the grid (upper part of the figure). Then the OBSs were recovered, data were retrieved, and the same OBSs were deployed in the southern half of the grid. Altogether the grid has 22 x 74 locations, with the distance between adjacent OBS of 426 m. As for the ocean-bottom cable data analyzed above, the receiver gathers are well sampled because of the low additional cost of shooting on a dense grid; thus they can be imaged effectively as if they were shot gathers. Notice that the OBSs were laid on a hexagonal grid instead of a Cartesian grid. A hexagonal array samples 3D wavefields more across incident azimuths in a more nearly uniform fashion than does a Cartesian array (Mersereau, 1979).

Land-data Geometries

Land-data geometries vary more widely than marine geometries because receivers on land are not constrained to being attached to a towed streamer. Therefore, several templates can be adopted when a land survey is designed. Two templates will be analyzed that illustrate some of the issues encountered in land-geometry design and their respective advantages and disadvantages will be discussed. Despite its theoretical advantages, one important template that will not be discussed is the cross-spread template (Vermeer, 1998), which seldom is applied because of its cost.

Cross-swath geometries

The first template to be examined commonly is called cross-swath shooting. An example of this template's geometry is given in Figure 11, which shows the source and receiver positions for two consecutive shot gathers recorded in the Canadian Rockies. As in previous figures, receivers are represented by dots and sources by asterisks. The

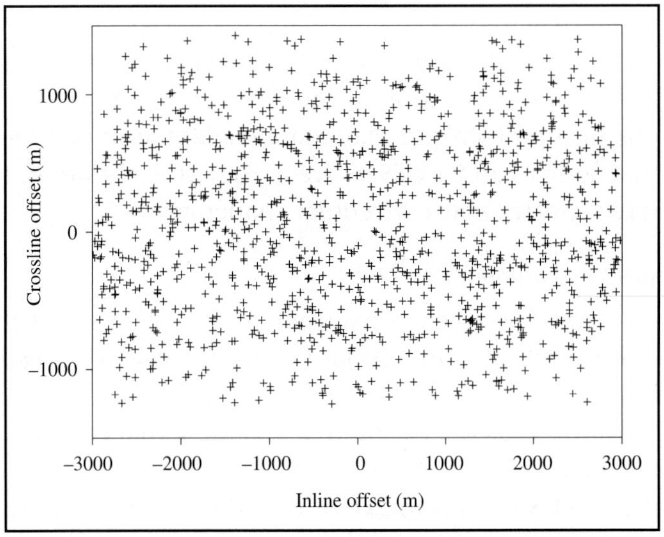

Figure 9. Offsets of 1000 traces selected randomly from an OBS data set.

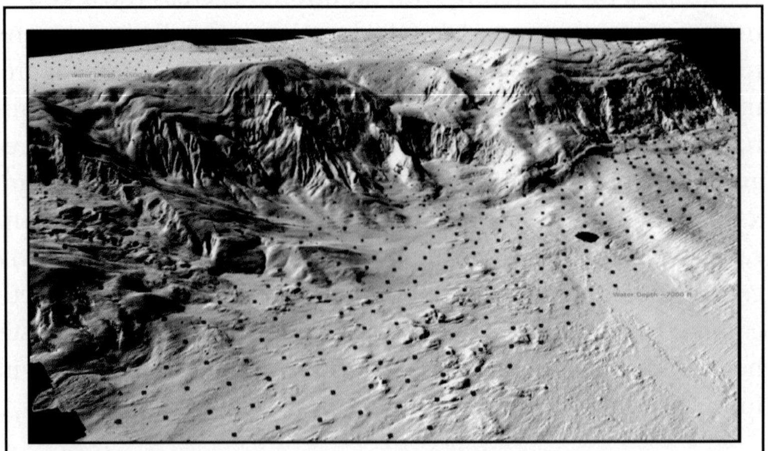

Figure 10. 3D rendering of the grid that was used for recording an actual survey with autonomous OBSs, Gulf of Mexico. Red dots correspond to OBS locations. Water depth ranged from approximately 1300 m to approximately 2300 m. The area is approximately 16 km by 16 km. (This image provided by Jerry Beaudoin, of BP.)

receivers were laid along parallel lines; separation between receiver lines was 250 m. The receiver lines are crooked because of obstacles in the terrain. Sources were moved in a direction approximately orthogonal to receiver lines, hence the name *cross swath*. When the source is moved outside the swath covered by the receiver array, the farthest receiver line is rolled over to the side of the new shots. Such configuration assures that the receiver arrays of individual shot gathers are well sampled along the receiver-line direction, but they are aliased in the orthogonal direction. Conversely, common-receiver gathers are well sampled along the crossline direction but poorly sampled along the inline directions. Only the midpoint axes are well sampled along both directions.

Figure 12 shows the offsets of 1000 traces selected randomly from the data set. As was true of the marine geometries discussed previously, the offsets of a cross-swath geometry have a preferential direction aligned along the receiver lines. Therefore, in this case also, one can consider the receiver-line direction to be the inline direction, with the sources moving along the crossline direction. However, compared with those of streamer geometries, the trace azimuths are not so clustered around the inline azimuth, and the offset plane is sampled densely only at near offsets.

Button-patch geometries

Button-patch acquisition was developed to sample data azimuths over the whole range between 0° and 360°. Wide-azimuth data are necessary for gathering information on how seismic properties, such as velocity and reflectivity, vary with propagation direction. Such detailed information can be useful where the subsurface is anisotropic.

The acquisition is called *button patch* because receivers are deployed in a checkered pattern (buttons), as shown in Figure 13. An ensemble of several buttons makes up a patch that is rolled along to cover contiguous areas. Receiver locations (dots) and the source location (asterisk) from a shot gather using the button-patch technique are shown in Figure 14; this shot gather was recorded in Wyoming. The irregularity in location of buttons, visible on the bottom right corner of the receiver array, was caused by obstacles on the surface. As expected, the offsets of button-patch acquisition are well distributed

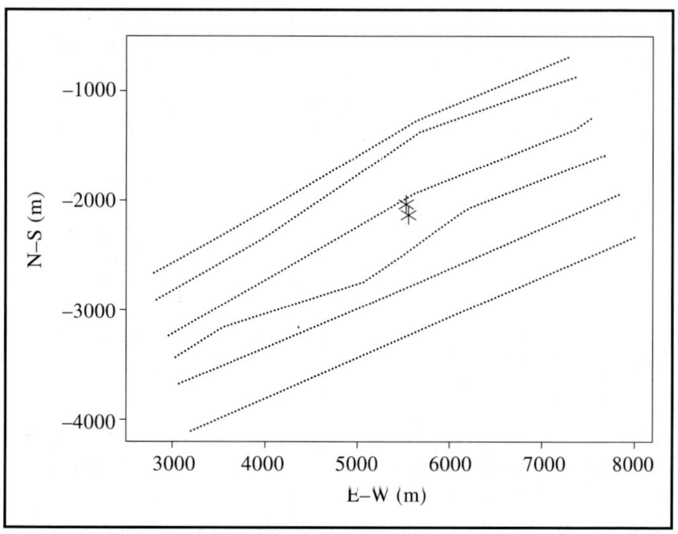

Figure 11. Locations of sources and receivers for two consecutive shot gathers of a cross-swath land-acquisition geometry, Canadian Rockies. Sources are represented by asterisks and receivers by dots.

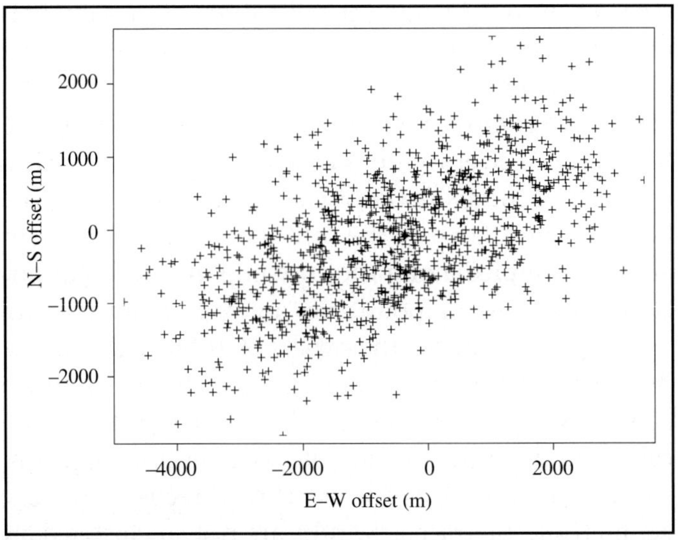

Figure 12. Offsets of 1000 traces selected randomly from a cross-swath data set.

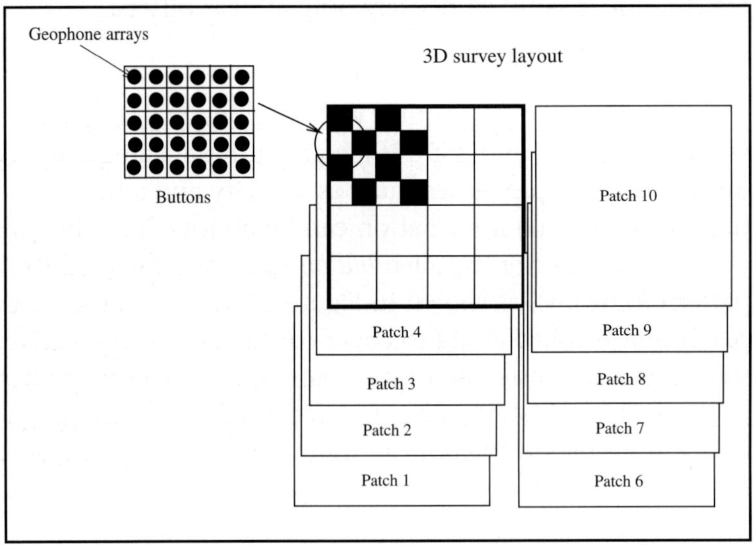

Figure 13. Button-patch acquisition geometry. (Diagram provided by J. M. Darnal, Arco Exploration and Production Technology.)

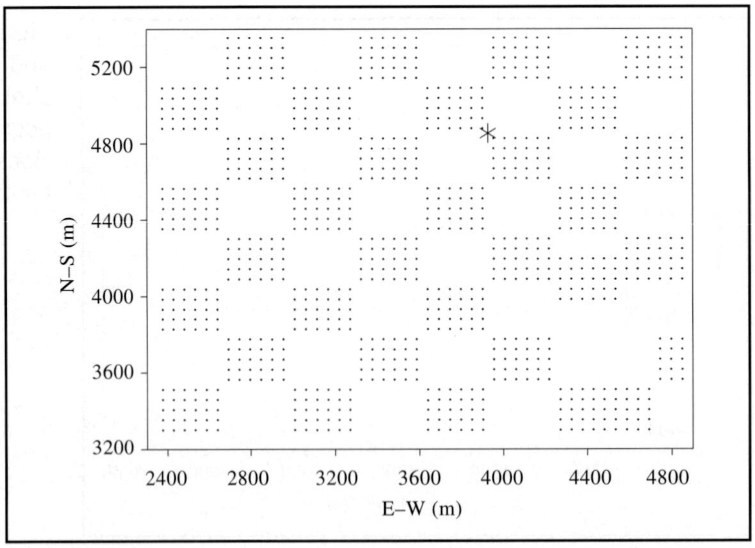

Figure 14. Locations of source and receivers for one shot gather of a button-patch land-acquisition geometry, Wyoming. The source is represented by an asterisk and the receivers are represented by dots.

on the entire offset plane, as shown in Figure 15. The figure shows the offsets of 1000 traces selected randomly from the data set.

Narrow-azimuth versus Wide-azimuth Geometries

The main distinction among the four acquisition geometries I have discussed in detail above is in their sampling of the offset plane. To understand the different sampling configurations of the offset plane achieved by different geometries, Figures 4, 9, and 12 were replotted together and on the same scale as Figure 15. The top left panel in Figure 16 shows Figure 4 rescaled, the top right panel shows Figure 9 rescaled, the bottom left panel shows Figure 12 rescaled, and the bottom right panel shows Figure 15. For the streamer geometry, only the positive half of the inline offset component was plotted, because reciprocal traces (a trace pair with the respective source and receiver locations exchanged) provide equivalent sampling for prestack imaging operators. Such a comparison is only qualitative; when one determines the sampling criteria of seismic reflections one should consider other factors, such as propagation velocity, target depth, and signal frequency. However, the difference in offset coverage and trace density among the four geometries is quite dramatic. Azimuthal coverage increases as one moves from streamer geometry to parallel-swath to cross-swath and finally to button-patch geometry.

Wide-azimuth geometries have the obvious advantage that data with different azimuths can provide useful information about the anisotropy of the subsurface, as well as about the heterogeneity of the velocity field. On the other hand, if one considers the number of recorded traces to be fixed, wide-azimuth geometries have this drawback: they sample the offset plane less densely, and thus are more subject to aliasing problems of the offset axes. This usually is not a problem if no imaging operator is applied in the prestack domain and if only trace-to-trace transformations, such as NMO or stacking (Chapter 4), are applied to prestack data. However, if prestack imaging is necessary because of structural complexity (see the discussion of DMO in Chapter 4) or strong velocity variations (see discussion of full prestack migration in Chapter 2) or both, the sparse sampling of the offset axes may cause artifacts in the imaging results (Canning

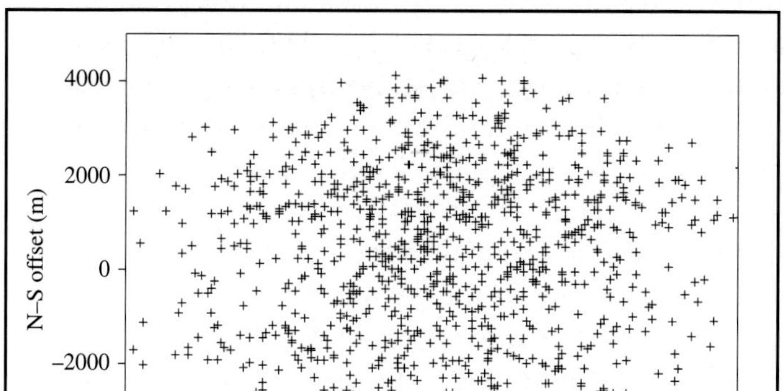

Figure 15. Offsets of 1000 traces selected randomly from a button-patch data set.

and Gardner, 1996). Such a trade-off between the width of the azimuth coverage and data density can be avoided either by recording more traces or by improving the methods used to image wide-azimuth prestack data.

The trade-off between narrow- and wide-azimuth data may be influenced further by considerations of subsurface illumination and multiple-reflections attenuation. Under complex overburden, target illumination varies with the data azimuth; therefore, a wide-azimuth survey has illumination holes that are more "uniformly" distributed in the subsurface. Therefore, the survey has better probability of providing at least partial information on all the geologic structures than does a narrow-azimuth survey. Recent numerical-modeling experiments (Regone, 2006) and field-data results (some shown in Chapter 9) support this intuition. Furthermore, images obtained from wide-azimuth data seem to be more nearly immune to strong artifacts caused by diffracted multiples, or in general, 3D multiple reflections not propagating along the vertical plane but passing through the source and receiver locations. When modern methods for attenuating surface multiples in presence of complex geologic structure are applied to conventional narrow-azimuth marine data, multiple events of these kinds are difficult to model, and thus to attenuate (Verschuur et al., 1992).

Sorting and Binning

To facilitate data processing, often the order of the traces is changed according to their geometry. Sorting and binning are two commonly used operations that reorder the traces.

Sorting reorders traces according to specific trace coordinates, called sorting indexes. Typical sorting operations are reordering of the data in shot gathers or receiver gathers. Such sorting operations are conceptually simple, because all traces that share the same source (or receiver) location have exactly the same source (or receiver) coordinates. On the other hand, sorting by common midpoints is complicated by the fact that because of irregularities in data acquisition, traces do not share exactly the same midpoint coordinates. Therefore, common-midpoint binning is usually required before common-midpoint sorting.

Common-midpoint binning is necessary not only to sort the data, but because it also can be useful in performing basic operations, such as stacking. The data are binned by superimposition of a regular grid onto the midpoint plane. Next, all traces that fall within each grid cell, or bin, are assigned effective midpoint coordinates equal to the central point of the cell. The nominal bin size — that is, the length of each side of the cells — is determined by the acquisition parameters; it can vary between the inline direction and the crossline direction. The nominal fold is equal to the number of traces that would fall in each grid if there were no irregularities in the geometry.

Figure 17 shows a small part of the midpoint plane of the streamer geometry discussed in the previous section titled "streamer geometries." The grid superimposed onto the traces represents the binning grid. The nominal bin sizes are 9.375 m in the inline direction and 25 m in the crossline direction. Because of geometric irregularities, caused mostly by cable feathering, the effective fold varies from 24 to 30 for this subset of the data; the nominal fold of the acquisition was 28.

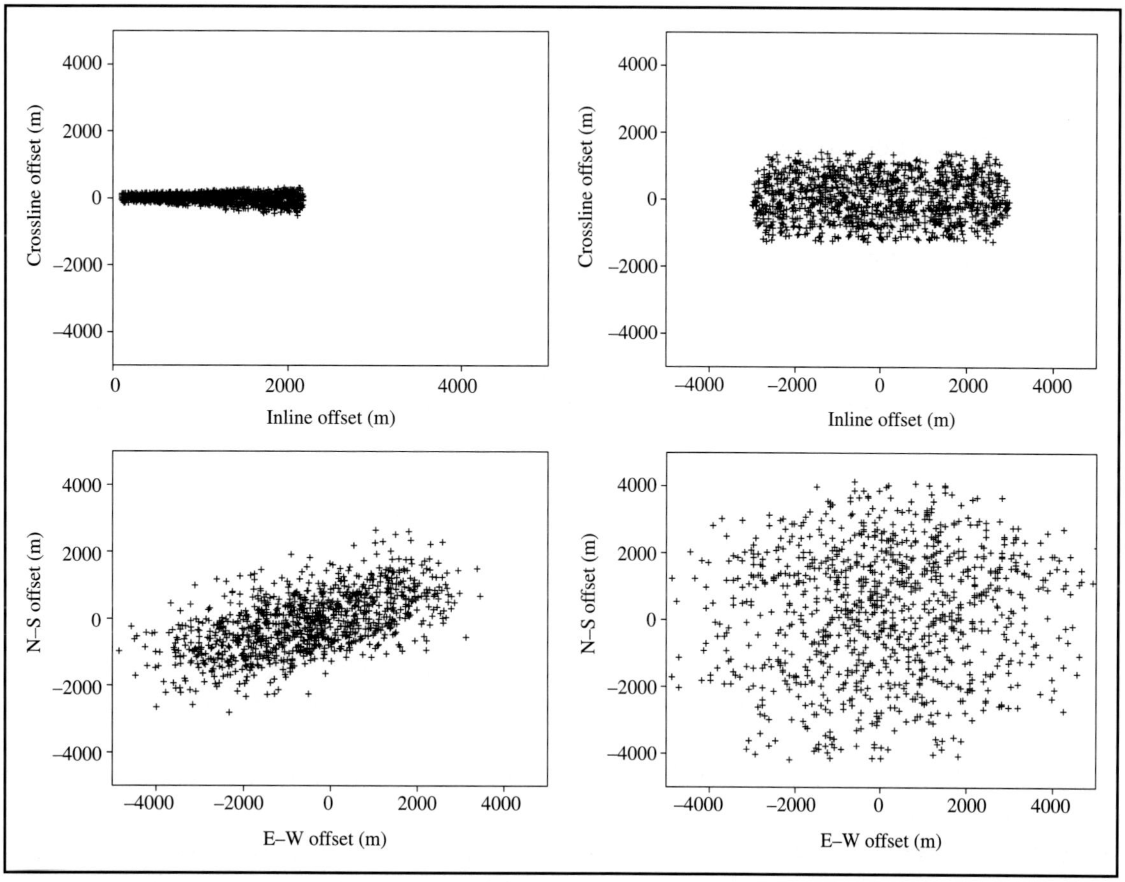

Figure 16. Rescaled plots of the offset planes shown in Figure 4 (top-left), Figure 9 (top-right), Figure 12 (bottom-left), and Figure 15 (bottom-right). Observe the substantial differences in offset coverage and trace density among the four geometries.

Figure 17. Midpoint plane for the streamer-data set. The grid superimposed onto the plot represents the binning grid.

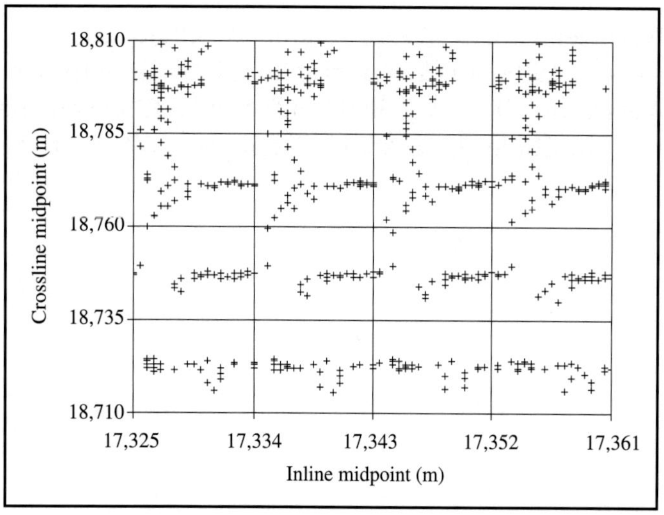

References

Canning, A., and G. H. F. Gardner, 1996, Another look at the question of azimuth: The Leading Edge, **15**, 821–823.

Dewey, F., M. van der Meulen, and P. Gardner, 2006, Using dual-azimuth data to image below salt domes: First Break, **24**, 55–60.

Howard, M. S., and N. Moldoveanu, 2006, Marine survey design for rich-azimuth seismic using surface streamers: 76th Annual International Meeting, SEG, Expanded Abstracts, SS 2.7.

Keggin, J., M. Benson, W. Rietveld, T. Manning, B. Barley, P. Cook, E. Jones, M. Widmaier, T. Wolden, and C. Page, 2006, Multiazimuth towed streamer 3D seismic in the Nile delta, Egypt: 76th Annual International Meeting, SEG, Expanded Abstracts, SS 2.2.

Knopoff, L., and A. F. Gangi, 1959, Seismic reciprocity: Geophysics, **24**, 681–691.

Mersereau, R.M., 1979, The processing of hexagonally sampled two-dimensional signals: Proceedings IEEE, **67**, 930–949.

O'Connell, J. K., M. Kohli, and S. Amos, 1993, Bullwinkle: A unique 3-D experiment: Geophysics, **58**, 167-176.

Regone, C., 2006, Using 3D finite-difference modeling to design wide azimuth surveys for improved subsalt: 76th Annual International Meeting, SEG, Expanded Abstracts, SS2.3.

Reksnes, A., E. Haugane, and S. Hegna, 2002, How PGS created a new image for the Varg field: First Break, **20**, 773–777.

Riou, A., K. Kravik, P. Sexton, L. Lemaistre, V. Aubin, and F. G. Bertini, 2006, Hild structure multiazimuth seismic experiment: First Break, **24**, 61–65.

Ross, A.A. and Beaudoin, G., 2006, Design and operation of a deep water, wide azimuth node seismic survey: 76th Annual International Meeting, Society of Exploration Geophysicists, Expanded Abstracts, SS 2.8.

Stone, D. G., 1994, Designing seismic surveys in two and three dimensions: SEG, Geophysical References Series, 244 p..

Vermeer, G. J. O., 1990, Seismic wavefield sampling: SEG, 120 p.

———, 1998, 3-D symmetric sampling: Geophysics, 63, 1629–1647.

Verschuur, D. J., A. J. Berkhout, and C. P. A. Wapenaar, 1992, Adaptive surface-related multiple elimination: Geophysics, **57**, 1166–1177.

Chapter 2 Principles of Migration Using Kirchhoff's Methods

Introduction

The ultimate goal of recording seismic data is to recover an image of the geologic structure in the subsurface. Imaging is the most computationally demanding and data-intensive component of seismic processing. Therefore, researchers have spent considerable effort in devising imaging strategies that are effective and that yield accurate images, but that are computationally affordable. The most accurate and most expensive types of migration are those that operate directly on the entire prestack data set; they are called full-prestack migrations. All other imaging methods are designed to approximate full-prestack migration; they aim to achieve the same image accuracy but at a fraction of the computational cost. In this chapter, the main characteristics of 3D prestack migration are analyzed. An understanding of these features is important if one is to determine when full-prestack migration is required and to evaluate its several available approximations.

Imaging of the subsurface is the result of two processing steps that are tightly connected — velocity estimation, and migration. Migration focuses the data and produces an image of the subsurface. However, adequate information on propagation velocity is needed for migration to focus the data properly and to produce the correct image of strata at depth. This chapter discusses the strong dependence of prestack migration on accuracy of the velocity function. In particular, these topics are discussed: principles of time migration and depth migration, and their differences with respect to reliability of the velocity-function estimates that each of them requires.

The link between migration and velocity estimation is provided by computation of partial prestack images that are defined (in addition to the subsurface coordinates) as functions of either the data offset and azimuth or the reflection aperture angle and reflection azimuth. Introduced in the last section of this chapter are the general concept of partial images, and of common image gathers (CIGs) that are a specific section of the partial prestack image cube and that are widely used for velocity estimation. The use of CIGs for velocity estimation is discussed in Chapter 7.

Imaging as a Data-focusing Process

To image reflectors in the subsurface, the effects of wave propagation from the surface to reflectors and back to the surface must be removed. This is accomplished by migration, which is founded in wave-propagation theory. Different types of migration are based on different ways to solve the wave equation. Kirchhoff migration is based on Green's function theory and on an integral solution of the wave equation. Therefore, Kirchhoff migration methods are often called integral migration methods.

In practice, Kirchhoff migration focuses reflections by multidimensional summations (integrals) over the recorded data. These summations can be described easily by use of simple geometrical constructs; thus Kirchhoff migration methods are a natural choice for introduction of basic imaging concepts. In addition to didactical advantages, Kirchhoff migration has the characteristic of being flexible, computationally efficient and easy to adapt for irregularly sampled prestack data. Kirchhoff migration often is the method of choice for 3D prestack migration, except when the complexity of the velocity creates severe multipathing in the propagating wavefields (see Chapter 3).

Rigorous theoretical derivation of Kirchhoff migration is rather involved, but the final result is extraordinarily simple (Schneider, 1978). The general form is given by the following integral expression:

$$I(\xi)=\int_{\Omega_\xi} W(\xi,\mathbf{m},\mathbf{h})D\left[t=t_D(\xi,\mathbf{m},\mathbf{h}),\mathbf{m},\mathbf{h}\right]d\mathbf{m}\,d\mathbf{h}\,. \tag{2.1}$$

The image $I(\xi)$, which is defined in a 3D space $\xi = (z_\xi, x_\xi, y_\xi)$, is equal to the integral of the data values $D(t, \mathbf{m}, \mathbf{h})$ evaluated at the time $t_D(\xi, \mathbf{m}, \mathbf{h})$ and weighted by an appropriate factor $W(\xi, \mathbf{m}, \mathbf{h})$. The amplitude factor also should include a 1D convolution by a derivative filter. The order of the derivative depends on the dimensionality of the integration. A first-derivative filter should be applied for 3D migration. The domain of integration in equation 2.1 does not span the whole input space; it is limited to a region Ω_ξ that is centered on the image point location. This region Ω_ξ often is called the migration aperture; it is important in determining the dip limitation and the cost of the migration procedure.

In practice, 3D prestack data are recorded on a discrete set of surface points, and the integral in equation 2.1 is approximated with a finite sum. Equation 2.1 can be rewritten as

$$I(\xi)\approx\sum_{i\in\Omega_\xi} W_i(\xi,\mathbf{m}_i,\mathbf{h}_i)D\left[t=t_D(\xi,\mathbf{m}_i,\mathbf{h}_i),\mathbf{m}_i,\mathbf{h}_i\right]. \tag{2.2}$$

In the equation above, the subscript is a reminder that a finite number of data traces are recoded and that trace locations are defined on a discrete but irregular spatial grid.

The time shift $t_D(\xi, \mathbf{m}, \mathbf{h})$ is given by the total time delay accumulated when the reflections propagate from the source position **s** to the image point $I(\xi)$, come back to the surface, and are recorded at the receiver position **g**. The time delays t_D define multidimensional summation surfaces, which form the basis of Kirchhoff migration. Summation surfaces have the important physical interpretation of being diffraction surfaces for scatterpoints located in the subsurface. The diffraction from an isolated scatterpoint in the earth's subsurface would create events with the same shape as that of the summation surfaces.

Figures 1 and 2 provide information for an intuitive understanding of the focusing effect of the summation described in equation 2.2. Figure 1 shows a near-offset cube from a 3D marine data set recorded in the North Sea. The top of a salt diapir and of the overlying chalk layer create several diffractions in the data. These diffractions are focused in the image cube (Figure 2), where the tip of the salt diapir and complex faulting in the chalk layer are clearly visible. Dashed red lines superimposed onto the images in Figure 1 outline the shape of one of the summation surfaces used by Kirchhoff migration. After data summation using equation 2.2, this diffraction collapses into the tip of a fault block, which is circled in green in Figure 2.

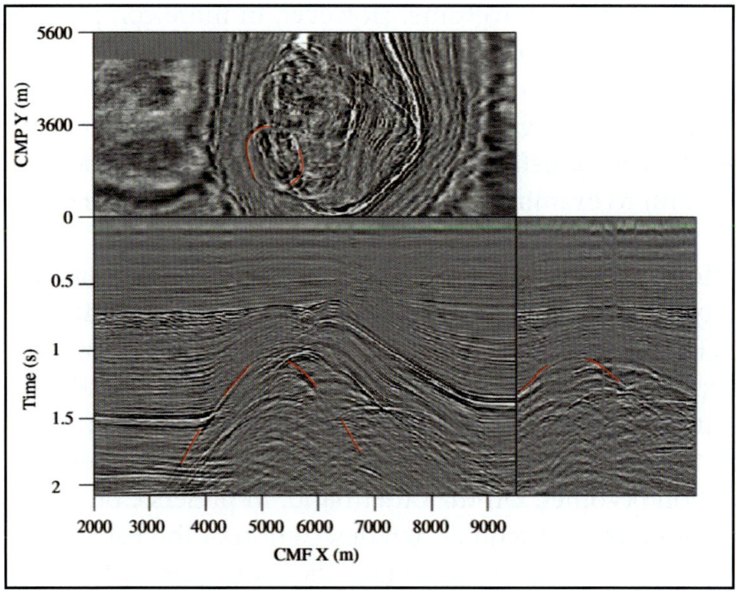

Figure 1. Near-offset cube from a 3D marine data set recorded in the North Sea. The top of a salt diapir and the top of the overlying chalk create several diffractions in the data. Dashed red lines superimposed on the images outline the shape of one of the summation surfaces used by Kirchhoff migration.

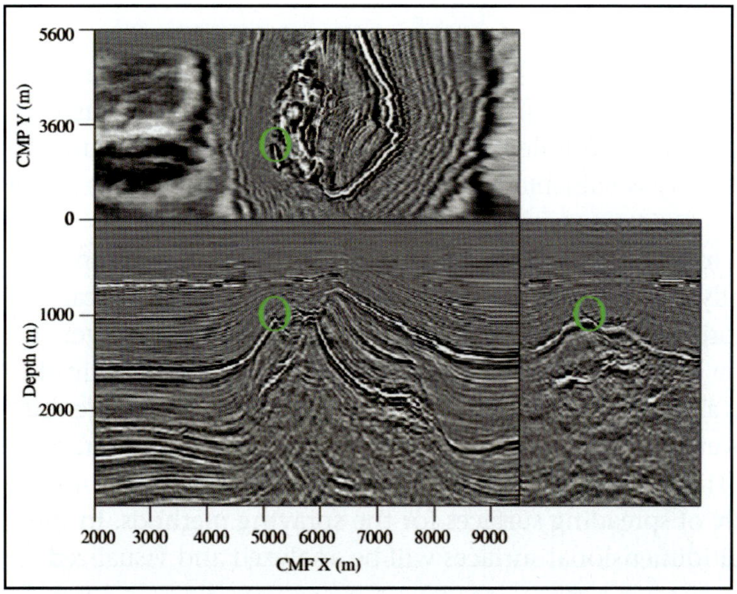

Figure 2. Image cube corresponding to the data cube shown in Figure 1.Diffractions in Figure 1 are well focused in the image cube. In particular, the diffractions outlined by red dashed lines in Figure 1 are focused, to image the tip of a fault block in the chalk that overlies the salt diapir (green circle).

Accurate evaluation of the time shift t_D and an appropriate definition of the migration aperture Ω_ξ are essential to proper focusing of the data by the migration operator defined in equation 2.2. In contrast, accurate definition of the weights in equation 2.2 is seldom important for producing a structural image, but it is crucial when reflection amplitudes are used to determine relative strengths of the reflectors and to estimate petrophysical parameters. An entire body of research is dedicated to defining the appropriate integration weights, so this issue will not be analyzed in detail here. Bleistein and Gray (2001) provided a clear, intuitive introduction to the subject.

Integration weights obviously are dependent on the kinematics of wave propagation (shape of the summation surfaces) but also on dynamics — that is, on defocusing and focusing of the seismic wave as it propagates through the subsurface. In simple velocity media, this effect often is called geometric spreading; it can be approximated well by a factor proportional to the inverse of traveltime. However, in more complex media, it must be computed numerically together with the traveltime functions.

Because the definitions of the time shifts t_D and of the migration aperture Ω_ξ have a first-order effect on quality of the migration results, the remainder of this chapter is dedicated to (a) discussion of the choices for defining the time shifts and their respective advantages and disadvantages, and (b) to examination of the effects on migration results of varying the migration-aperture size.

The shape of the summation surfaces is strictly linked to the choice of parameterization of the propagation velocity. A fundamental choice is whether to parameterize the velocity function with average velocities (e.g. $V_{\rm rms}$), or with interval velocities. Another important choice is whether to assume an isotropic velocity function (e.g. independent from the propagation angle), or an anisotropic velocity function. In general, the more degrees of freedom one allows in the description of the velocity function, the more challenging and uncertain the estimation becomes. On the other hand, in presence of complex geology only the migration operators that require rich description of the velocity function can focus the data — which leaves the seismic imager no choice but to tackle a complex estimation problem.

Constant-velocity Migration

For constant velocity, the time delays t_D can be evaluated analytically. Obviously, a constant-velocity assumption is seldom fulfilled in practice, but the analysis of time delays computed in constant velocity is valuable to understanding the basic properties of the imaging process.

The summation expressed in equation 2.2 can be implemented numerically in two ways that are mathematically equivalent, but that may have different numerical properties. The first class of algorithms includes gathering methods, which loop over each image point and gather the contributions of all the input traces that are within the migration aperture. In contrast, algorithms belonging to the second class, often referred to as spraying methods, loop over each input trace and spray data over all image traces within the migration aperture. Time delays define a family of summation surfaces for the gathering methods, and a family of spreading surfaces for the spraying methods. In the next two subsections, these multidimensional surfaces will be analyzed and visualized.

Summation surfaces

In constant velocity, time delays can be evaluated analytically as

$$t_D = t_s + t_g = \frac{\sqrt{z_\xi^2 + \left|\overline{\mathbf{xy}}_\xi - \mathbf{s}\right|^2}}{V} + \frac{\sqrt{z_\xi^2 + \left|\overline{\mathbf{xy}}_\xi - \mathbf{g}\right|^2}}{V}, \tag{2.3}$$

or, in midpoint-offset coordinates, as

$$t_D = \frac{\sqrt{z_\xi^2 + \left|\overline{\mathbf{xy}}_\xi - \mathbf{m} + \mathbf{h}\right|^2}}{V} + \frac{\sqrt{z_\xi^2 + \left|\overline{\mathbf{xy}}_\xi - \mathbf{m} - \mathbf{h}\right|^2}}{V}, \tag{2.4}$$

where $\overline{\mathbf{xy}}_\xi = (x_\xi, y_\xi)$ represents the horizontal projection of the image-coordinates vector ξ.

Summation surfaces are defined in the 5D data space. However, visualizing 5D surfaces is still beyond our capabilities; thus, we are limited to showing 3D sections that represent interesting and meaningful special cases. The simplest of those cases is the summation surface for zero-offset data. In this case, the double square root of equation 2.4 simplifies into a single square root, and the summation surface becomes the hyperboloid of rotation given by the equation

$$t_D = 2\frac{\sqrt{z_\xi^2 + \left|\overline{\mathbf{xy}}_\xi - \mathbf{m}\right|^2}}{V}. \tag{2.5}$$

Figure 3 shows the zero-offset hyperboloid in a 3D perspective. The vertical axis (identified by Z in the axis triplet at the lower left corner of the figure) is the time axis, whereas the horizontal axes are the midpoint axes. Contour lines identify circles with equal time delays. The cross at the apex of the hyperboloid identifies the image point that corresponds to the summation surface. The time of the cross is equivalent to the two-way traveltime, $\tau_\xi = 2z_\xi/V$.

A constant-offset summation surface is another meaningful 3D section of the whole 5D summation surface. Figure 4 shows a constant-offset hyperboloid in a 3D perspective. For simplicity, but without loss of generality, the offset vector is assumed to be aligned along the inline direction, which is identified in the figure as "X." The constant-offset hyperboloid is flat at the top and is not circularly symmetric, as was the zero-offset one. Instead, it is squeezed on the top along the vertical direction and squeezed horizontally

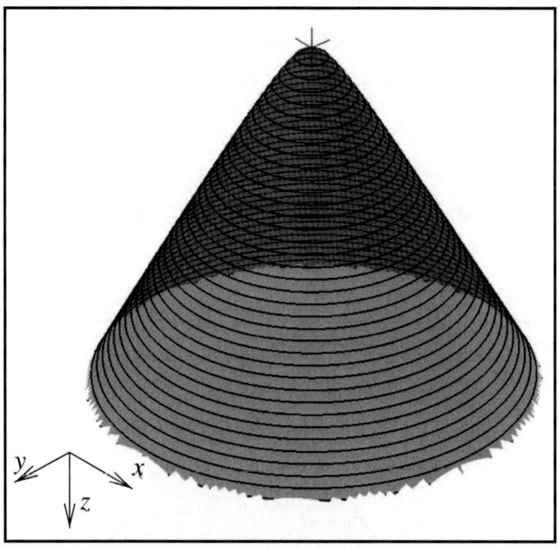

Figure 3. Zero-offset summation surface defined in the data space. The vertical axis is the time axis; horizontal axes are the midpoint axes. Contour lines identify circles with equal time delays. The cross at the apex of the hyperboloid identifies the image point corresponding to the summation surface. The time of the cross is equivalent to the two-way traveltime $\tau_\xi = 2z_\xi/V$.

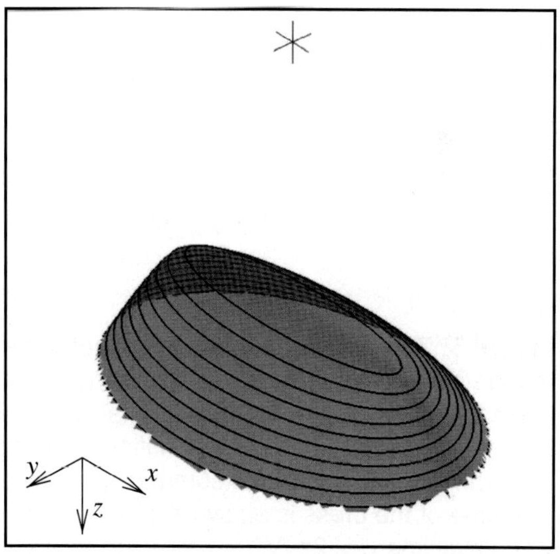

Figure 4. Constant-offset summation surface defined in the data space. The vertical axis is the time axis; horizontal axes are the midpoint axes. Contour lines identify ellipses with equal time delays. The cross identifies the image point corresponding to the summation surface. The time of the cross is equivalent to the two-way traveltime $\tau_\xi = 2z_\xi/V$.

along the crossline direction, which is identified as in Figure 4 as "Y." Contour lines identify points with equal time delays; the zero-offset circles have become constant-offset ellipses. As before, the cross identifies the image point that corresponds to the summation surface, but in this case, the image point is detached and is above the summation surface. Varying the data azimuth while keeping the absolute offset constant would be equivalent to rotating the hyperboloid shown in Figure 4 around the time axis. In contrast, varying the absolute offset while keeping the azimuth constant would result in different amounts of both vertical and horizontal squeezing.

To gain better intuition for the whole 5D surface, Figure 5 shows simultaneously the zero-offset hyperboloid and the constant-offset hyperboloid that correspond to the same image point. For large time delays, the two surfaces approach the same asymptote and at the limit, they become tangent. Inline edges of the constant-offset surfaces approach the zero-offset surface faster than do the crossline edges.

Spreading surfaces

Spreading surfaces are dual to the summation surfaces and belong to a family of ellipsoids defined in the 3D image space. The analytical representation of such ellipsoids can be derived with a little algebraic manipulation from equation 2.4. The canonical form for this family of ellipsoids is

$$\frac{4\left(x_\xi - x_m\right)^2}{t_D^2 V^2} + \frac{4\left(y_\xi - y_m\right)^2}{t_D^2 V^2 - 4x_h^2} + \frac{4z_\xi^2}{t_D^2 V^2 - 4x_h^2} = 1, \tag{2.6}$$

where x_m and y_m are the midpoints and x_h is the offset of the input data trace. It can be verified immediately that a zero-offset trace is sprayed along a spherical surface defined by the equation

$$\frac{4\left(x_\xi - x_m\right)^2}{t_D^2 V^2} + \frac{4\left(y_\xi - y_m\right)^2}{t_D^2 V^2} + \frac{4z_\xi^2}{t_D^2 V^2} = 1. \tag{2.7}$$

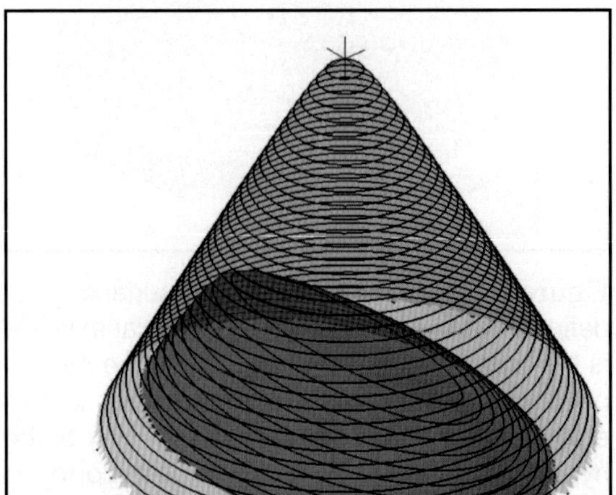

Figure 5. Zero-offset and constant-offset summation surfaces defined in the data space. The vertical axis is the time axis; horizontal axes are the midpoint axes. The cross identifies the image point corresponding to both summation surfaces. The time of the cross is equivalent to the two-way traveltime $\tau_\xi = 2z_\xi/V$.

Figure 6 shows a 3D rendering of the zero-offset hemisphere. The spraying surfaces are defined in the 3D image space; therefore, in Figure 6, 7 and 8, the vertical axis is depth. Contour lines identify circles with equal depth. The cross barely visible at the bottom of the hemisphere represents an impulse in the input trace. The equivalent depth of the impulse is given by $z_\xi = V\tau_\xi/2$.

A constant-offset trace is sprayed along the ellipsoid shown in Figure 7. Compared with the zero-offset hemisphere, the constant-offset ellipsoid is squeezed both vertically and along the crossline direction; thus, the contour lines become elliptical. The bottom of the ellipsoid is above the equivalent input impulse, which is identified by the cross in Figure 7. Figure 6 shows simultaneously the zero-offset hemisphere and the constant-offset ellipsoid. Observe that the two surfaces become tangent along the inline edges at the $z_\xi = 0$ plane.

Migration in Heterogeneous Media

Where velocity is not constant, the time-delay function between the image point in the subsurface and the source/receiver point at the surface cannot be computed by use of the simple relationship expressed in equation 2.3. Consequently, the summation surfaces have shapes that are more complex than the one represented in Figures 3 through 5.

Generally, time delays used for migration are computed numerically. Several methods exist for computing time delays through a complex velocity function. All of them solve the Eikonal equation, a high-frequency approximation of the wave equation (Bleistein, 1984). Most numerical methods are either ray-tracing methods (Červeny and

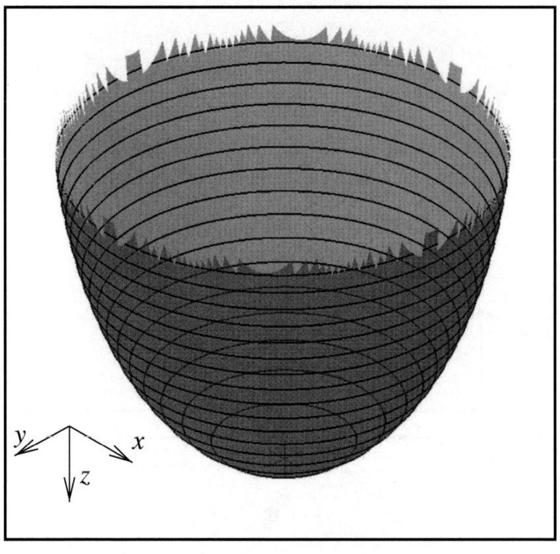

Figure 6. Zero-offset spreading surface defined in the 3D image space. The vertical axis is the depth axis. Contour lines identify circles at equal depth. The cross at the bottom of the hemisphere identifies the input impulse that is sprayed along the surface. The equivalent depth of the impulse is given by $\tau_\xi = 2z_\xi/V$.

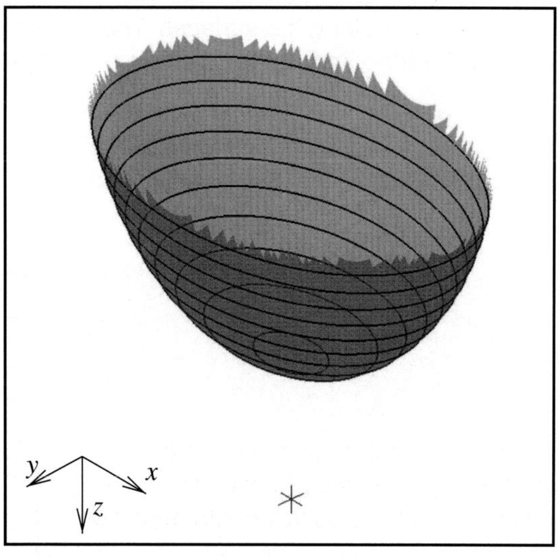

Figure 7. Constant-offset spreading surface defined in the 3D image space. The vertical axis is the depth axis. Contour lines identify ellipses at equal depth. The cross at the bottom of the figure identifies the input impulse that is sprayed along the surface. The equivalent depth of the impulse is given by $\tau_\xi = 2z_\xi/V$.

Pšenčík, 1983; Sava and Fomel, 2001), which solve the Eikonal equation along its characteristics, or finite-difference methods (Vidale, 1990; van Trier and Symes, 1991; Popovici and Sethian, 2002).

Figure 9 illustrates how downgoing rays and upgoing rays are traced through an interval velocity function $v(z, x, y)$. The resulting time delays can be expressed analytically in the expression

$$t_D = t_s\left[\xi, \mathbf{s}, v(z,x,y)\right] + t_g\left[\xi, \mathbf{g}, v(z,x,y)\right], \tag{2.8}$$

where t_s is the time delay from the source location **s** to the image point ξ, and t_g is the time delay from the same image point ξ to the receiver location **g**.

When reflection amplitudes are of interest, the geometric spreading (a_s, a_g) along the rays connecting the image points and the source/receiver points also can be computed numerically, by solving the associated transport equation. The combination of traveltimes (t_s, t_g) and amplitudes (a_s, a_g) defines the asymptotic Green's functions for the wave-propagation phenomenon in the given complex media. According to Green's theorem, these functions can be used to compute the wavefield $P(\omega, \xi)$ generated by a source $P_s(\omega)$ as

$$P(\omega,\xi) \approx P_s(\omega)\, a_s \exp(i\omega t_s). \tag{2.9}$$

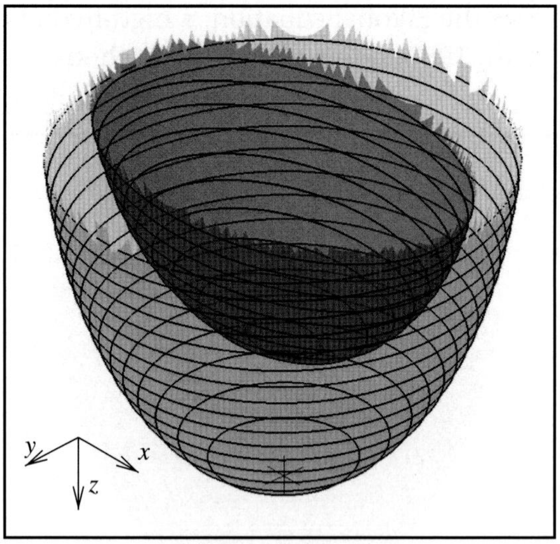

Figure 8. Zero-offset and constant-offset spreading surfaces defined in the 3D image space. The vertical axis is the depth axis. The cross at the bottom of the figure identifies the input impulse that is sprayed along the surfaces. The equivalent depth of the impulse is given by $\tau_\xi = 2z_\xi/V$.

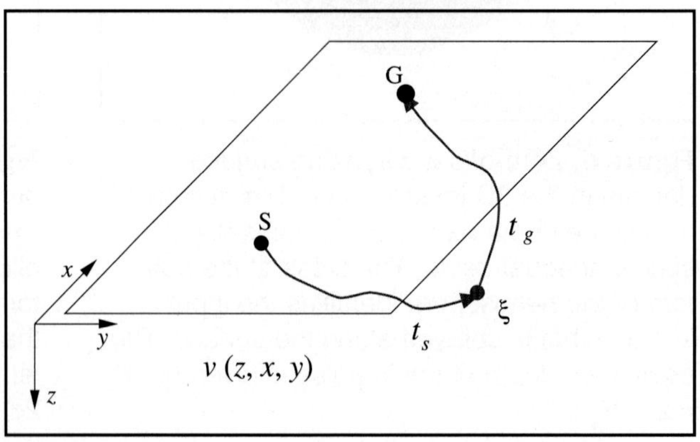

Figure 9. Graphic illustration of the ray tracing necessary to compute time delays used for prestack depth migration.

By assuming that both t_s and a_s are solutions of asymptotic approximations of the wave equation (respectively the Eikonal and the transport equation) and that they are independent from the temporal frequency ω, one introduces an asymptotic approximation in the imaging process. That approximation is the basis of Kirchhoff migration and constitutes a fundamental theoretical limitation of the method.

When the velocity function is very complex and the wavefronts triplicate, the Green's functions become multivalued; consequently, the summation surfaces become multibranched. In theory, computation of the migration integral (summation) is possible for arbitrary complexity of the summation surfaces. In practice, when multipathing is severe, computation of the Green's functions and numerical evaluation of the migration integral become expensive and inaccurate. That is a practical limitation of Kirchhoff migration.

Time imaging versus depth imaging

One does not always need to compute the time-delay functions t_s and t_g numerically. When velocity is changing slowly in the horizontal direction, the time-delay functions often can be approximated by equation 2.3, but with the constant velocity V replaced by the root-mean-square (rms) velocity function V_{rms}. The V_{rms} function is related to the interval velocity function by the Dix formula (Chapter 6). In such situations, the image often is computed in the time domain, and the migration is called time migration, in contrast to depth migration.

Their names notwithstanding, the main difference between time migration and depth migration is not related to definition of the vertical axis of the image (i.e., time or depth). Depth migration uses time-delay functions that are computed numerically by equation 2.8. Time migration uses time-delay functions that are computed analytically as follows:

$$t_D = \sqrt{\frac{\tau_\xi^2}{4} + \frac{\left|\overline{\mathbf{xy}}_\xi - \mathbf{s}\right|^2}{V_{\mathrm{rms}}\left(\tau_\xi, x_\xi, y_\xi\right)^2}} + \sqrt{\frac{\tau_\xi^2}{4} + \frac{\left|\overline{\mathbf{xy}}_\xi - \mathbf{g}\right|^2}{V_{\mathrm{rms}}\left(\tau_\xi, x_\xi, y_\xi\right)^2}}. \tag{2.10}$$

Notice that to compute the time delays for time migration, an estimate of average velocities at each image location is required. In contrast, for depth migration, an estimate of interval velocity at every point in the subsurface is required.

As is demonstrated in Chapter 7, the quality of data focusing can be measured from the image obtained by prestack migration. For time migration, average velocities can be estimated directly, by measuring the focusing quality on a point-to-point basis at each image location. This direct estimate is possible because the average velocity (V_{rms}) enters directly in expressions of the corresponding summation surfaces. In contrast, there is no point-to-point direct correspondence between time delays and interval velocity. Therefore, estimating interval velocity from measures of focusing quality in the image is a complex inverse problem known as migration tomography.

Another crucial distinction between time migration and depth migration is that time migration only focuses the data and leaves the task of time-to-depth mapping to a subsequent independent step that often is called map migration. When velocity is constant, the depth coordinate z_ξ of the image and the corresponding two-way traveltime τ_ξ are connected uniquely by the simple relationship $z_\xi = V\tau_\xi/2$. When velocity is not constant, the relationship between two-way traveltime and depth is complex. However, if velocity is a function only of depth, the relationship is still univocal, and one can use a simple 1D stretching to transform an image defined in time into an image defined in depth, and vice versa. If velocity varies laterally also, the relationship becomes equivocal. In many cases, time still can be mapped into depth, but use of image rays is required (Hubral, 1977). Image rays are convenient conceptual constructs, but they are not real physical entities.

Such splitting of the focusing and mapping steps is an important advantage of time migration. In many cases, velocity parameters necessary to focus the data properly (average velocities) are easier to estimate than are velocity parameters necessary for an accurate time-to-depth conversion (interval velocities). In contrast, depth migration relies on interval-velocity estimates for both focusing and mapping. When interval-velocity estimates are inaccurate, depth migration mispositions the reflectors in depth and also misfocuses the data.

The robustness of time migration makes it an essential imaging tool. However, it must be applied within limitations imposed by its underlying assumptions. When velocity variations are strong enough to create nonhyperbolic diffraction surfaces, time migration cannot focus the data properly, and depth migration becomes necessary. The following example illustrates this issue.

- Theoretical imitations of time imaging

To illustrate the differences between time migration and depth migration, an example from the SEG-EAGE salt data set (Aminzadeh et al., 1996) will be discussed. The differences in results of migration will be explained by analyzing the Green's functions that correspond to two representative image-point locations for an inline section (locations above and below a salt body). Figure 10 shows the velocity model for the section of interest.

Figure 10. Inline section of the interval-velocity model for the SEG-EAGE salt data set.

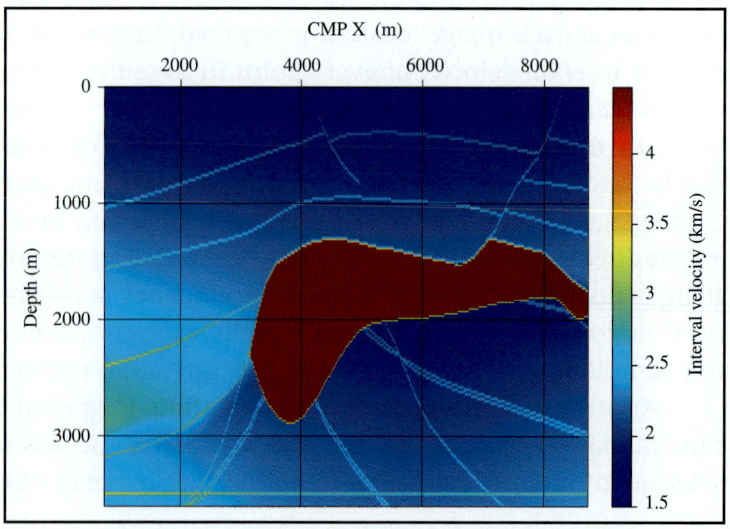

Figure 11 shows a vertical inline section of the 3D prestack-migrated cube using time migration. The true V_{rms} function was used, as computed from the true interval-velocity function by the Dix formula. Figure 12 shows a vertical inline section of the 3D prestack-migrated cube using depth migration. The true interval-velocity function was employed for depth migration. The two sections are similar in regions above the salt body. In regions below the salt body, they differ substantially — beyond the obvious laterally variant stretching of the vertical axis. In particular, the bottom of the salt and the horizontal basement reflector ($z = 3.6$ km) are imaged much better by use of depth migration. However, even depth migration does not image all subsalt reflectors perfectly, because the marine-like acquisition geometry of the data illuminates some of reflectors insufficiently. The differences between Figure 11 and Figure 12 can be explained easily by analyzing how well the time-delay functions computed by equation 2.10 approximate time delays of the true Green's functions computed by a wavefield modeling program.

Figure 13 shows the superposition of two snapshots of the wavefield, taken at times $t = 0$ s and $t = 0.6$ s, when the source was above the salt. Notice the circular shape of the

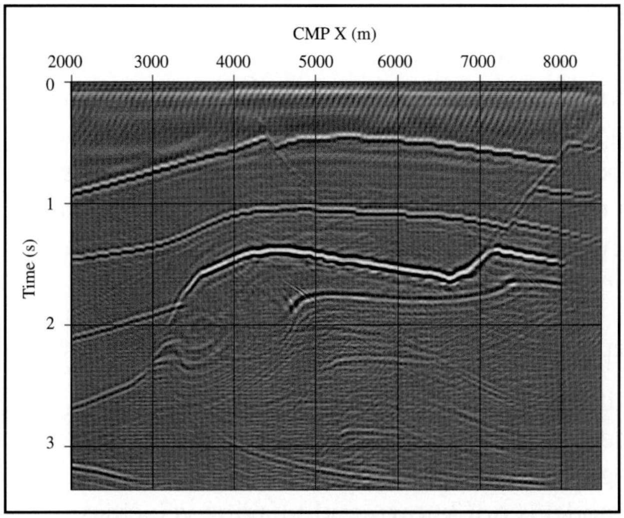

Figure 11. 3D time migration of the inline section shown in Figure 10, computed using the true V_{rms} function.

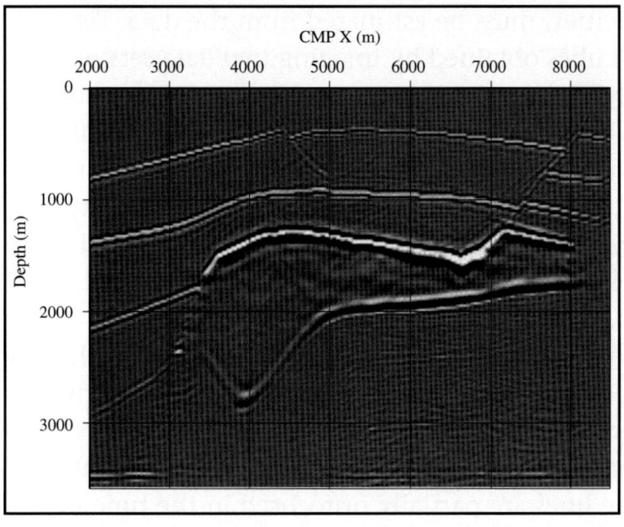

Figure 12. 3D depth migration of the inline section shown in Figure 10, computed using the true interval-velocity function.

Figure 13. Snapshots of the wavefield when the source is located above the salt (t = 0 s and t = 0.6 s).

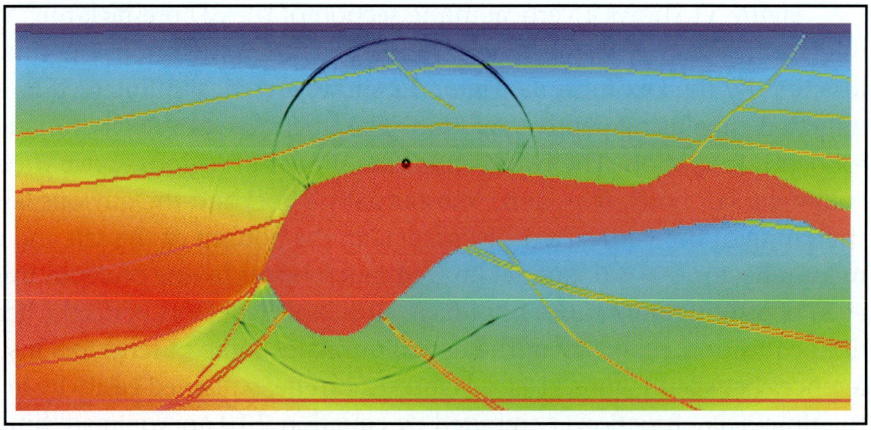

upgoing wavefield. Figure 14 shows the wavefield recorded at the surface, as a function of recording location and recording time. It represents the full-wavefield Green's function in the time-space domain corresponding to a reflector point above the salt. A hyperbolic trajectory, which represents the approximate time-delay function employed by time migration, is superimposed onto the wavefield. In this case, the hyperbolic trajectory overlaps the true Green's function almost perfectly.

Figures 15 and 16 show the corresponding results when the image point is below the salt. Notice that the shape of the upgoing wavefield shown in Figure 15 is far from circular. In Figure 16, the analytic hyperbolic trajectory (yellow-on-blue line) differs substantially from the true Green's function computed by wavefield modeling. In contrast, the time-delay function computed by solving the Eikonal equation (yellow-on-red line) overlaps the true Green's function almost perfectly.

- Field-data comparisons of time and depth imaging

The previous figures illustrate the differences between time imaging and depth imaging by a synthetic-data example where the velocity model was known perfectly. Fairer comparison of the two methodologies is achieved by analyzing migration results from field data, where the true velocity model is not known and imaging velocities (either average velocities or interval velocities) must be estimated from the data. Figures 17 to 20 show such a comparison, from cubes obtained by imaging two data sets recorded in the foothills of the Canadian Rocky Mountains, in western Alberta. These examples are from a region where time migration has been used successfully, and where it still is widely used — because the theoretical advantages of depth migration are challenging to achieve in practice. Velocity contrasts between layers are relatively small and quality of the data may hamper accurate estimation of interval velocities. However, even in this difficult area a careful building of the velocity model for depth migration can yield significant improvements.

Figure 17 shows the image cube produced by Kirchhoff prestack migration, whereas Figure 18 shows the image cube produced by Kirchhoff prestack anisotropic depth migration (see the next section for a discussion of anisotropic migration). The differences between the two image cubes in both focusing and reflector positioning are evident. The structures around the crossing of the blue lines are partially unfocused in the time migration results, along both the inline and the crossline directions. In contrast, the same structures are imaged nicely by depth migration. The depth slice also shows more continuity than the corresponding time slice. Because of lateral velocity variations, dip of the deeper reflectors is shown as being substantially different, as depicted in the two images. In the crossline section (CMP Y) of the time-migration results (Figure 17), reflectors dip

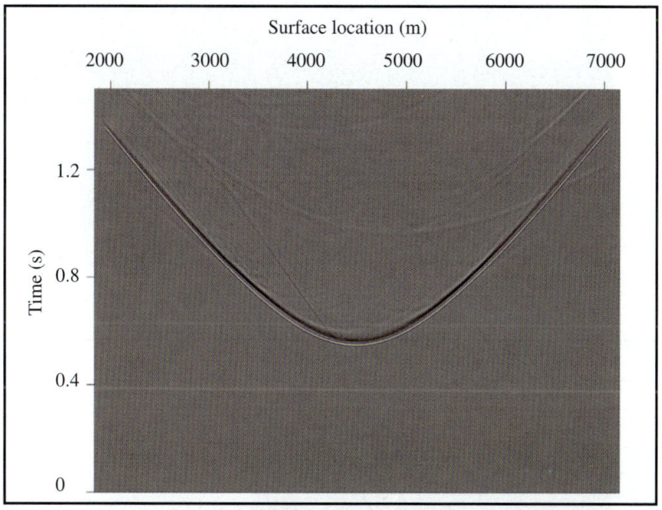

Figure 14. Wavefield recorded at the surface, corresponding to the wave modeling shown in Figure 13. The hyperbolic trajectory superimposed onto the wavefield represents the approximate time-delay function employed by time migration.

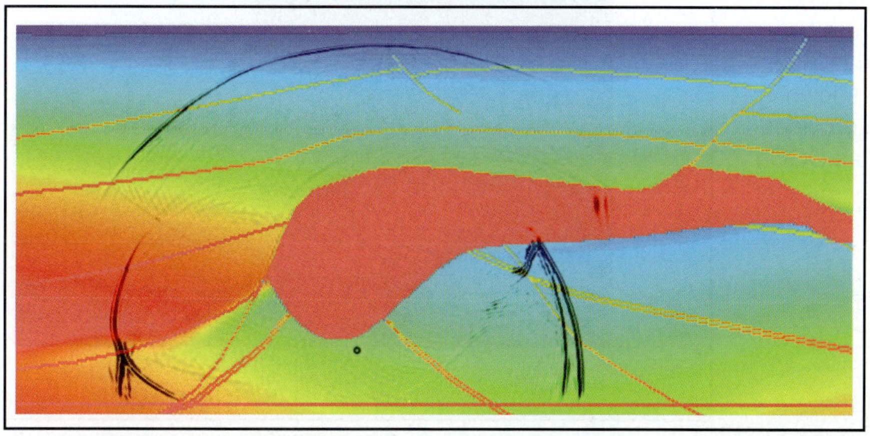

Figure 15. Snapshots of the wavefield when the source is located below the salt (t = 0 s and t = 1 s).

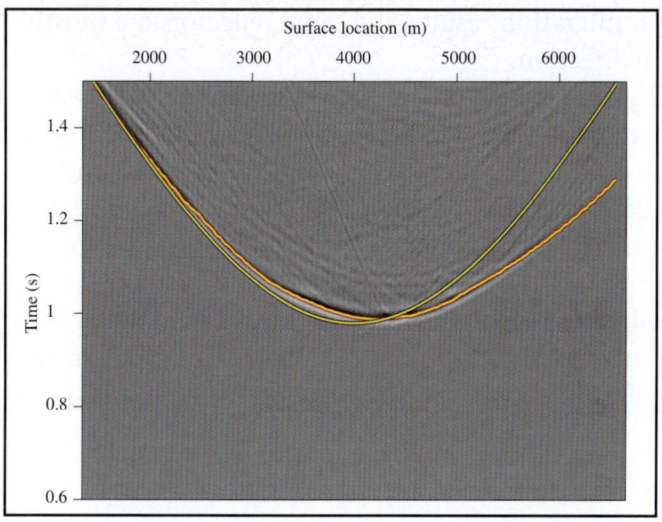

Figure 16. Wavefield recorded at the surface, corresponding to the wave modeling shown in Figure 15. Hyperbolic trajectories superimposed onto the wavefield represent the approximate time-delay function employed by time migration (blue line) and the time-delay function computed by numerically solving the Eikonal equation (red line).

Figure 17. Image cube obtained by prestack time migration of a data set recorded in the foothills of the Canadian Rocky Mountains. Structures around the crossing blue lines are incoherently imaged in both the inline and crossline sections. (This image provided by Scott Cheadle, Veritas DGC.)

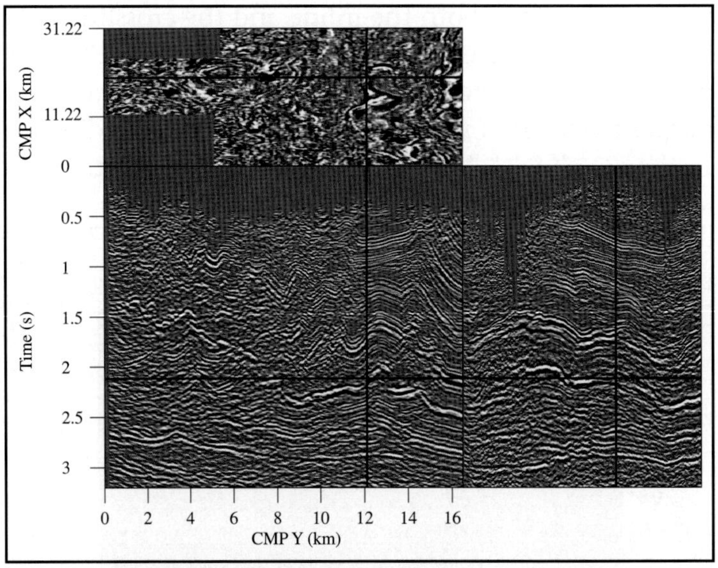

Figure 18. Image cube obtained by prestack depth migration of a data set recorded in the foothills of the Canadian Rocky Mountains. Structures around the crossing blue lines are better focused and more continuous than in the time image shown in Figure 17. (This image provided by Scott Cheadle, Veritas DGC.)

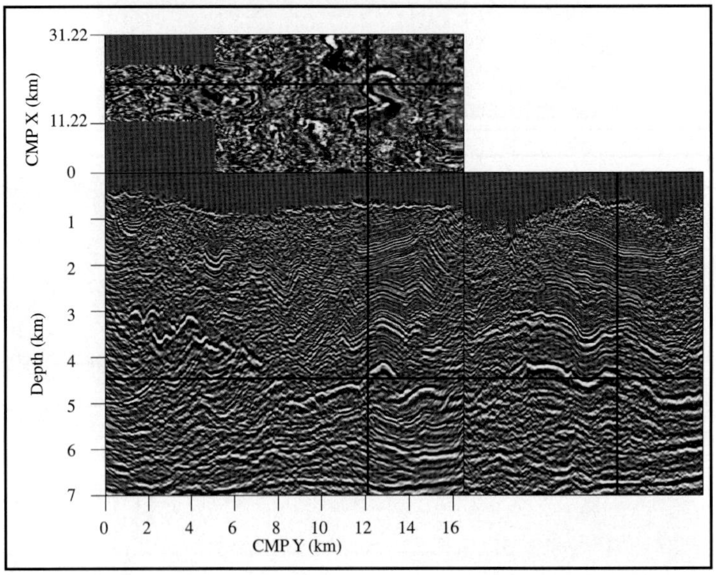

from left to right, whereas in the depth migration results, the same reflectors are mostly flat or they dip slightly in the opposite direction.

The second field-data example shows evidence of simpler subsurface geology, and differences between the time-migration image cube (Figure 19) and the depth-migration image cube (Figure 20) are more subtle than in the previous example (Figures 17 and 18). However, the depth migration shows better continuity, and crisper images of the deep structures. Furthermore, close to the crossing point between the blue lines the top of the structure is shifted towards lower CMP Ys in the depth-migration results by about 200 m (Figure 20), with respect to the time migration results. This shift clearly is visible by comparing the time slice (Figure 19) with the depth slice (Figure 20). Such a shift may make the difference between a successful well that reaches the structural top and an unsuccessful one, which misses the target.

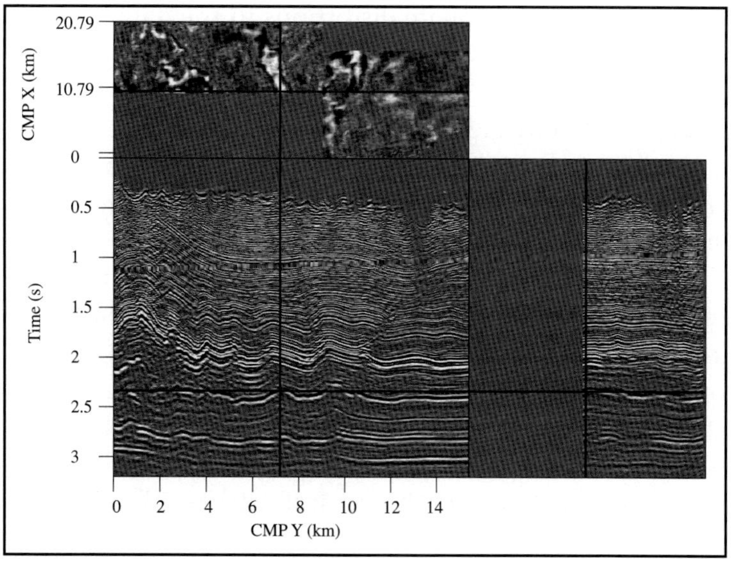

Figure 19. Image cube obtained by prestack time migration of a data set recorded in the foothills of the Canadian Rocky Mountains. Deep structures are partially misfocused. (This image provided by Scott Cheadle, Veritas DGC.)

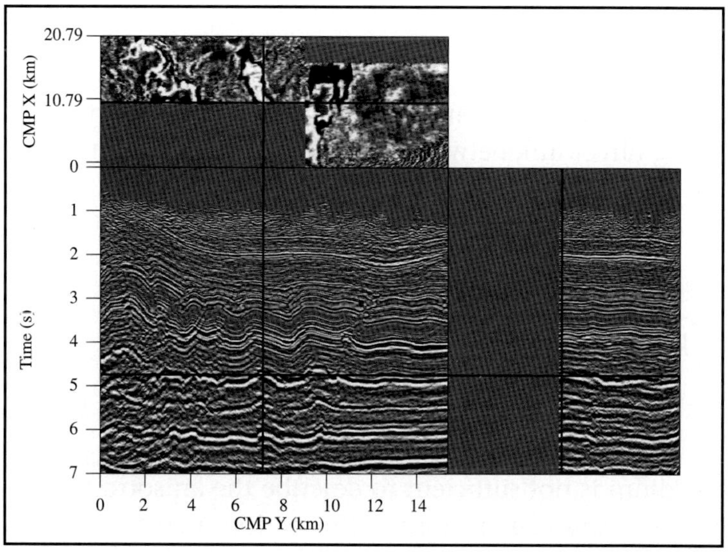

Figure 20. Image cube obtained by prestack depth migration of a data set recorded in the foothills of the Canadian Rocky Mountains. Deep structures are crisply imaged. With respect to images shown in Figure 19, notice the lateral shift along CMP Y of the crest of the structure close to the crossing point between the blue lines. (This image provided by Scott Cheadle, Veritas DGC.)

Anisotropic Migration

The previous examples demonstrate that data acquired over complex geology can be focused better by depth migration than by time migration. The price to pay for the additional focusing power of depth migration is this: estimation of the interval velocity model is more difficult than estimation of average velocities. However, when the propagating wavefields are distorted by velocity heterogeneities, the migration summation surfaces must adapt to the complexity of wave propagation, so as to sum coherently over the recorded events. An analogous situation is when propagation velocity in the subsurface is anisotropic and propagation velocity of seismic waves depends on the propagation angle. There are many physical reasons for geological formations to show anisotropic behavior (Tsvankin, 2001; Thomsen, 2002). To produce an optimally focused image when imaging under sufficiently thick anisotropic layers, summation surfaces of the migration operator must be adapted to accommodate distortions in the data kinematics, which are caused by propagation in the anisotropic layers.

Both time migration and depth migration can be adapted to be anisotropic migrations. For time migration, a parameter related to the anellipticity of the velocity function is added to the usual V_{rms} parameter, to describe the summation surfaces. This parameter can be estimated from seismic data if the data offsets are sufficiently long, or if steeply dipping reflectors are in the subsurface (Alkhalifah and Tsvankin, 1995).

When applying anisotropic depth migration, the anisotropic interval velocity is usually described by using some simplifying assumptions over the description of a general anisotropic medium. The most common assumption is to characterize the medium as being vertical transverse isotropic (VTI), and the Thomsen parameters ε and δ are used (Thomsen, 1986). These parameters define the relationships between the vertical velocity V_V, the horizontal velocity V_H, and the NMO velocity V_N (the physical interpretation of V_N is the paraxial curvature of the wavefront around the vertical axis) as follows:

$$\begin{aligned} V_H^2 &= V_V^2(1+2\varepsilon), \\ V_N^2 &= V_V^2(1+2\delta). \end{aligned} \tag{2.11}$$

Often the anellipticity parameter η (Alkhalifah and Tsvankin, 1995) is used instead of ε, because it provides the following direct link between the two velocities that might be independently measurable from surface seismic data (i.e. V_N and V_H):

$$V_H^2 = V_N^2(1+2\eta), \tag{2.12}$$

where η is related to the Thomsen parameters as

$$\eta = \frac{\varepsilon - \delta}{1+2\delta}. \tag{2.13}$$

In some cases a simple VTI medium is not sufficient to describe the anisotropic velocity function, because geologic processes (e.g., tectonic deformation) may have folded the VTI layers after their deposition. In these cases, another degree of freedom is introduced in description of the anisotropic velocity, as the tilting angle of the axis of symmetry with respect to the vertical direction. These anisotropic layers are defined as being characterized by tilted transverse anisotropy (TTI). The estimation of TTI parameters is even more challenging than estimation of VTI parameters, and usually it is based on interpretation of dips of strata that are known to be characterized by VTI.

When only surface seismic data are available, estimation of the vertical velocity (or δ) is practically impossible because depth of the reflectors is unknown a priori, and the data are recorded with the source and receivers at the same depth; in other words, there is no vertical component in the data offset. However, if borehole seismic data or well logs are available, a joint inversion of borehole data and surface seismic data may yield reliable estimates (Bear et al., 2003).

Unfortunately, in use of surface seismic data only, it is often impossible to distinguish the effects of anisotropy from the effects of simple velocity heterogeneities of strata in the subsurface. Because of ray bending, even a simple vertical velocity gradient can induce distortions in the kinematics that can easily be mistaken as evidence of anisotropic velocity. Geologic information and borehole data can help to resolve this ambiguity; thus, at least one of these two elements should be an integral part of the estimation process of anisotropic velocity functions.

When the anisotropy is a property of strata in the subsurface, ignoring it may lead to mispositioning of reflectors. Or, in some cases, it may lead to incomplete focusing of the reflectors. In particular, it might be impossible to completely focus both flat reflectors and steeply dipping reflectors, without the introduction of unrealistic heterogeneities into the velocity function. Focusing of flat reflectors is dependent mostly on the NMO velocity, whereas focusing of dipping reflectors is more sensitive to horizontal velocity, as the propagation angle of events increases.

Figures 21 and 22 compare the results of isotropic and anisotropic depth migration, applied to a field-data set recorded in the Gulf of Mexico. The three sections in Figure 21 show the anisotropic velocity parameters estimated by using both surface and borehole data (Krebs et al., 2003). The top panel displays the vertical velocity function, the middle panel displays the δ function, and the bottom panel displays the η function. Observe the strong difference between vertical and the horizontal velocity on the right-hand side of the salt body, and the strong anelliptic configuration of the shallow layers on the left-hand side. Figure 22 shows the result of anisotropic migration (top panel) using the

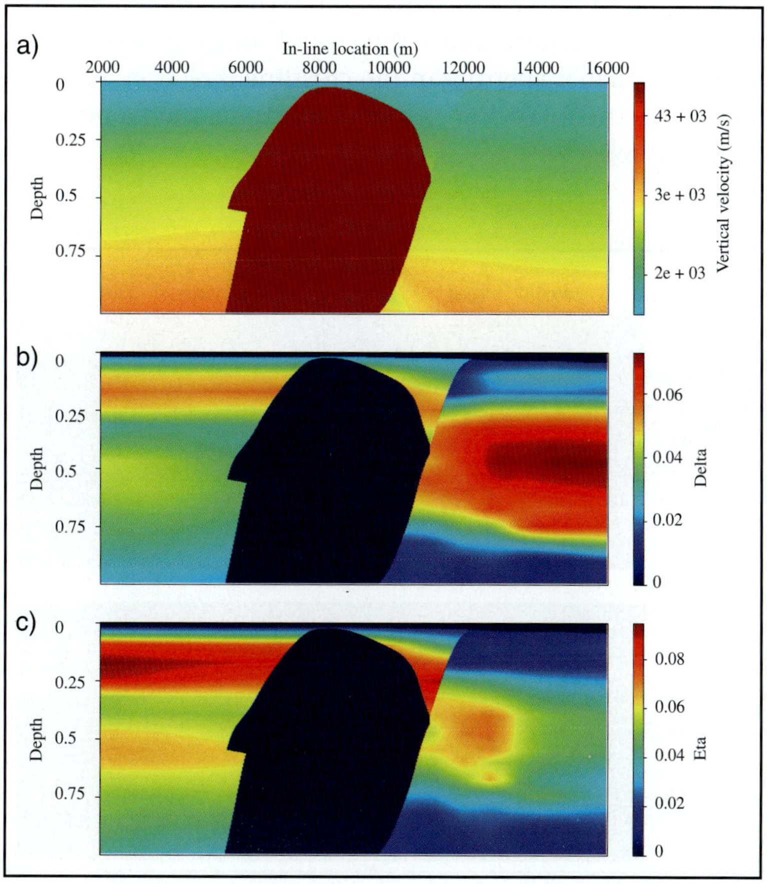

Figure 21. Anisotropic parameters describing a VTI velocity function: (a) vertical velocity, (b) δ parameter, and (c) η parameter.

parameters illustrated in Figure 21, and the result of isotropic migration (panel) using the vertical velocity shown in Figure 21a. The steep flanks on both the left-hand and right-hand sides of the salt body are focused incompletely in the isotropic migration, because the vertical velocity is too low for obliquely propagating events (both δ and η are positive). The reflection of steeply dipping strata on the right-hand side of the salt body also is not as coherent in the isotropic migration as it is in the anisotropic migration, because of poor stacking over reflection angles.

Effects of Migration Aperture on Images

Correct determination of the migration aperture Ω_ξ has a strong influence on the quality of migration results — almost as much as correct determination of the shape of the summation surfaces. When determining migration aperture, the cost of computation, the desired range of dip in the image, and data sampling (i.e., aliasing in the image) must be considered. Typically, the aperture is cone-like, such as the one illustrated in Figure 23. The input midpoints that contribute to a single image point are limited to a circular area centered on the surface location $\overline{\mathbf{xy}}_\xi$ of the image point. The diameter of this circle is directly proportional to ($z_\xi \tan\alpha$), where α is the aperture of the cone, as shown in Figure 23. Because of the growth in migration aperture with depth, the cost of 3D prestack Kirchhoff migration grows with the cube of the target depth; this makes the full-volume imaging of deep targets extremely expensive.

The migration aperture in the inline direction seldom is the same as in the crossline direction. Commonly, strata in the subsurface dip predominantly in one direction; a priori knowledge of this dip can be exploited to reduce the cost of imaging. To do so, the data must be migrated by use of a smaller aperture in the crossline direction (α_y) than in the inline direction (α_x). However, as the following examples illustrate, such a practice may lead to incomplete imaging of all reflectors.

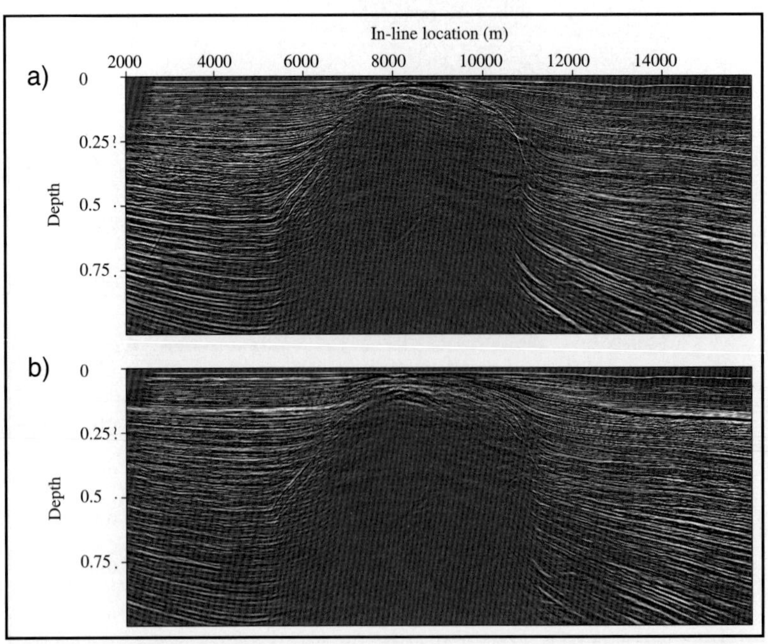

Figure 22. Comparison of anisotropic prestack depth migration (panel a) using the parameters displayed in Figure 21 with isotropic prestack migration (panel b) using the vertical velocity function displayed in Figure 21a. Notice the incomplete focusing of steeply dipping salt-flank reflections in panel (b) and the more coherent image of steeply dipping strata reflections on the right-hand side of the salt body in panel (a).

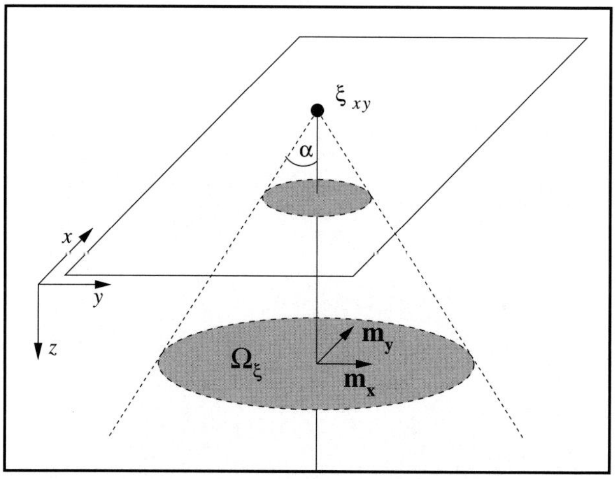

Figure 23. Illustration of the cone-shaped migration aperture for Kirchhoff 3D prestack migration.

When one is setting the migration aperture, certain trade-offs are made between cost and accuracy; these trade-offs are illustrated by the following 3D migration examples (Figures 24, 25, and 26). The examples show images for the same line as that of images presented in Figures 11 and 12. However, observe that these examples are near-offset (0–0.5-km) migrations, whereas migrations shown in Figures 11 and 12 are full-offset ones (0–2.3-km). Imaging of the two normal faults above the salt body — which are visible in the velocity model (Figure 10) — depends strongly on width of the migration aperture.

Figure 24 shows the result of time-migrating the data with a narrow aperture in both the inline and crossline directions ($\alpha_x = 30°$ and $\alpha_y = 5°$). Only the flat reflectors in the strata and the flat parts of the top and bottom of the salt are imaged. When the data are migrated with a wide aperture in the inline direction ($\alpha_x = 70°$) and a slightly wider aperture in the crossline direction ($\alpha_y = 30°$), the image of the top and flanks of the salt improves considerably. However, the two fault planes in strata above the salt are missing (Figure 25). These fault planes have a strong dip component in the crossline direction, so they require a full crossline aperture as well. Only when $\alpha_y = 60°$ are the fault planes imaged nicely (Figure 26). The wide-aperture image (Figure 26) is not uniformly better than the narrow-aperture images. The subsalt reflectors are better focused in Figure 24 and 25 than in Figure 26. When the summation surfaces do not track the kinematics in the data accurately, summation on larger surfaces adds more of coherent noise than of signal.

Common-image Gathers by Kirchhoff Migration

The migration method presented at the beginning of this chapter produces a migrated cube that is a function of the three spatial coordinates (z, x, y). To analyze velocity and amplitude, often one should exploit the redundancy of seismic data and produce images with more dimensions than the three coordinates of physical space. When Kirchhoff migration is applied, it is straightforward to image the data as a function of their recording offset and azimuth. This is done by subdividing the domain of integration during Kirchhoff summation. Migrated cubes obtained by subdividing the domain of integration use only a subset of the data, and thus can be referred to as partial, or prestack, images. The term *prestack images* can be confusing, because it may refer to images

Figure 24. Migrated inline section of the SEG-EAGE salt data set, with narrow migration apertures in both the inline and crossline directions (α_x = 30° and α_y = 5°).

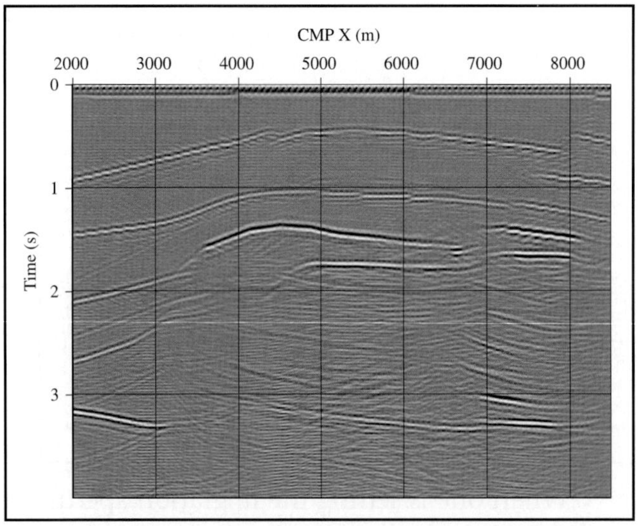

Figure 25. Migrated inline section of the SEG-EAGE salt data set, with a wide migration aperture in the inline direction and a narrow aperture in the crossline direction (α_x = 70° and α_y = 30°).

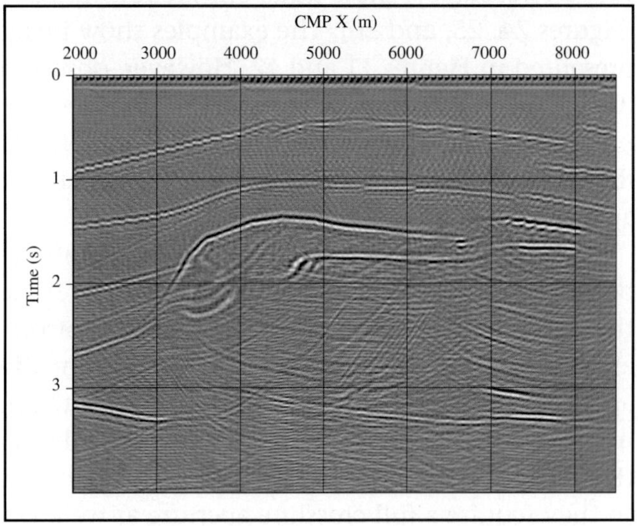

Figure 26. Migrated inline section of the SEG-EAGE salt data set, with wide migration apertures in both the inline and crossline direction (α_x = 70° and α_y = 60°).

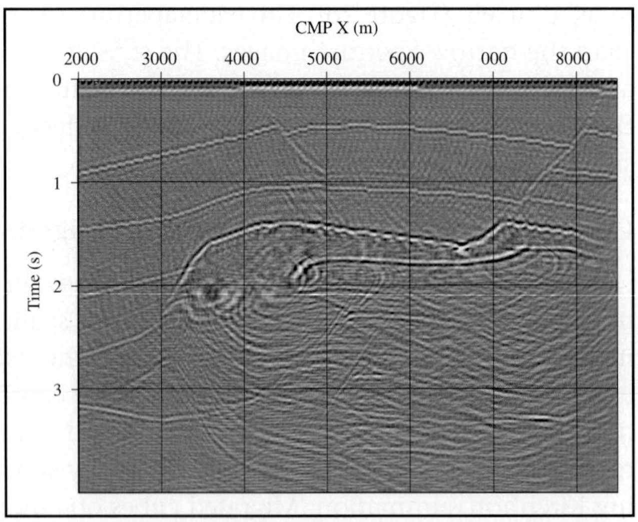

obtained by prestack migration; therefore, in the following discussion, I will use the term *prestack partial images*. The whole image is a hypercube (usually five-dimensional) made by the ensemble of all partial images. When the partial images are created according to the data offset, the two additional dimensions are either the absolute offset and azimuth (h, θ_h), or the inline and crossline offsets (x_h, y_h). When the partial images are created according to the reflection angles at the reflection point, the two additional dimensions are the reflection-opening angle (γ) and the reflection-azimuth angle (ϕ).

For velocity analysis and petrophysical analysis, measuring the variations between the partial images at a fixed image point is very useful. The subsets of the whole image with fixed surface location thus often are displayed to perform this kind of analysis. These subsets often are called common-image gathers (CIGs) or common-reflection-point gathers (CRP). In the following discussion, these subsets are referred to as CIGs. If the partial images are functions of the data offset, I will refer to the corresponding CIGs as offset-domain common-image gathers (ODCIGs). If the partial images are functions of the reflection angles, I will refer to the corresponding CIGs as angle-domain common-image gathers (ADCIGs).

ODCIGs are the easiest type of CIGs to compute by Kirchhoff migration. This choice is appropriate for time-migration problems, but in the presence of strong lateral variations in velocity, ODCIGs can be affected by artifacts that diminish their utility (Stolk and Symes, 2004). In such cases, ADCIGs tend to behave better (Xu et al., 2001; Brandsberg-Dahl et al., 2003).

Offset-domain common-image gathers

Offset-domain partial images are straightforward to produce by use of Kirchhoff migration. The migration integral is performed on separate subsets of the data, and the partial images are kept separate. The four-dimensional integral of equation 2.1 becomes the following two-dimensional integral over midpoints:

$$I(\xi,\mathbf{h}) = \int_{\Omega_{\xi},\mathbf{h}} W_{\mathbf{h}}(\xi,\mathbf{m},\mathbf{h}) D\left[t = t_D(\xi,\mathbf{m},\mathbf{h}),\mathbf{m},\mathbf{h}\right] d\mathbf{m}. \tag{2.14}$$

The whole integration domain Ω_ξ (as used in equation 2.1) is divided then into several subdomains $\Omega_{\xi,\mathbf{h}}$, where the vector offset $\mathbf{h}$ can be represented either in Cartesian coordinates (x_h, y_h) or in polar coordinates (h, θ_h). Integration is performed independently on each subdomain, and the results are analyzed separately for velocity and amplitude analysis. The extent of the subdomains $\Omega_{\xi,\mathbf{h}}$ depends on the acquisition geometry. They can be partially overlapping in the offset plane, or completely separate.

A graphical interpretation of the differences between the integral in equation 2.14 and the integral in equation 2.1 can be gained by considering the summation surfaces shown in Figure 5. For the integral in equation 2.1, the data are integrated simultaneously over both summation surfaces shown in Figure 5 (and all the other surfaces corresponding to the offsets within the range of the data acquisition). In contrast, for the integral in equation 2.14, the data are integrated over only one summation surface at a time. This graphical interpretation is more obvious if we consider the summation curves in 2D instead of the summation surfaces in 3D. In 2D, the summation curves for all offsets form a pyramidal surface that is often called Cheops pyramid (Claerbout, 1985). Figure 27 shows the common-offset summation curves superimposed onto a Cheops pyramid. Evaluation of the integral in equation 2.1 is equivalent to integrating the data on the whole pyramid, whereas evaluation of the integral in equation 2.14 is equivalent to integrating the data along each of the parallel lines superimposed on the pyramid.

Examples of ODCIGs extracted from prestack partial images of the SEG-EAGE salt data set are shown in Figures 28 and 29. Figure 28 shows a window from an inline section of the stacked image. Figure 29 shows a set of ODCIGs; the offset axis is the absolute half-offset $|\mathbf{h}|$. All ODCIGs shown in Figure 29 were taken at the same surface location (CMP X = 7.02 km along the section shown in Figure 28), but to illustrate the sensitivity of ODCIGs to velocity error, they were obtained by using different velocity functions. The third panel was obtained by using the correct velocity above the salt; thus the reflections are aligned along the offset axis. A migration velocity lower than the true one was used to produce panels (a) and (b). In those panels, the reflections smile upward. The misalignment is more pronounced in panel (a) than in panel (b), because the velocity error is larger. A higher migration velocity than the true one was used to produce panel (d), and the reflections frown downward.

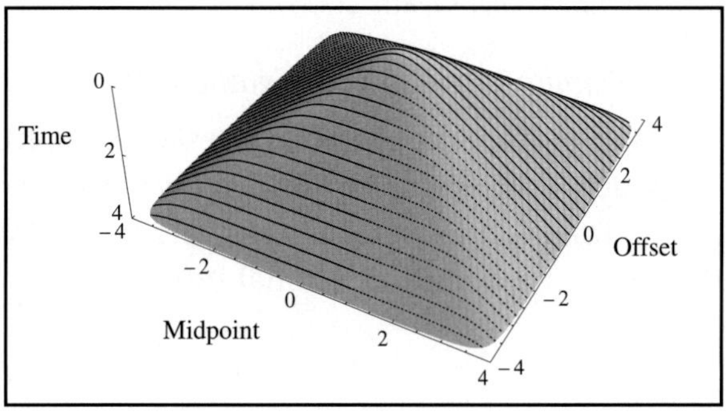

Figure 27. Common-offset integration curves superimposed onto the Cheops pyramid. (Courtesy of Sergey Fomel.)

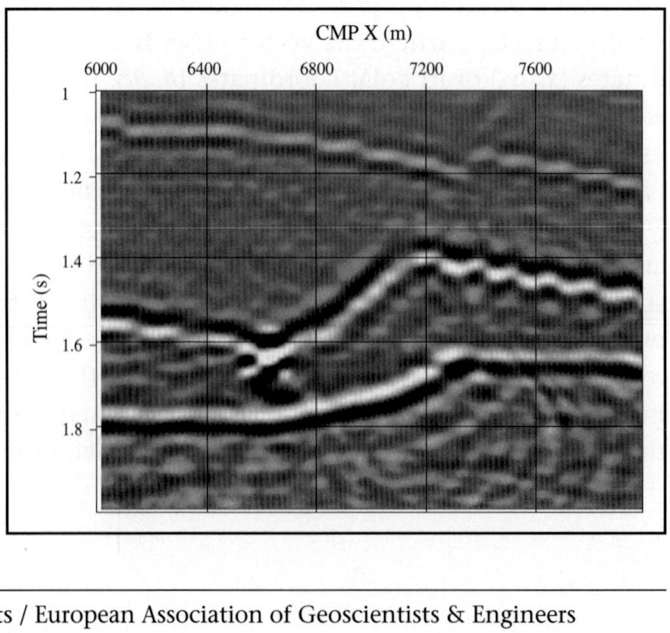

Figure 28. Migrated inline section from the SEG-EAGE salt data set, in the area where a major fault cuts through the salt body.

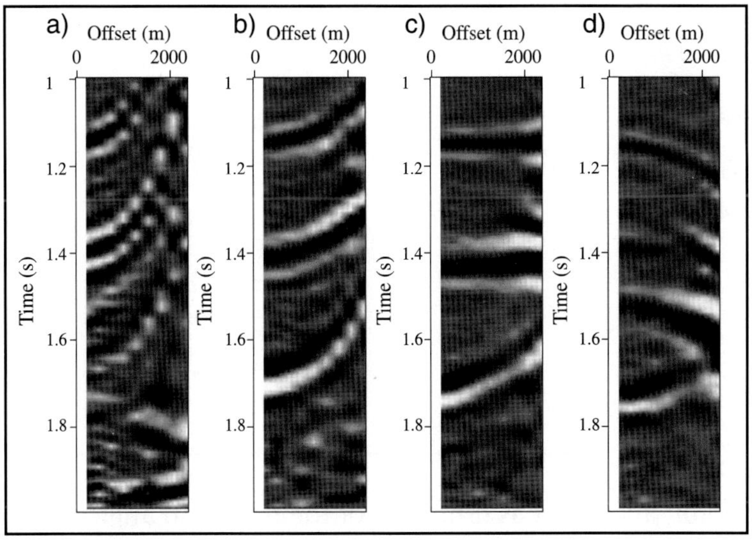

Figure 29. Four ODCIGs taken at the same surface location (CMP X = 7.02 km along the section shown in Figure 28). The migration velocity (v) increases from the panel on the left to the panel on the right as a constant fraction of the true velocity (v_o): (a) $v = 0.8\ v_o$, (b) $v = 0.92\ v_o$, (c) $v = v_o$, and (d) $v = 1.08\ v_o$.

Angle-domain common-image gathers

Angle-domain Kirchhoff imaging is based on reflector-centered parameterization of the migration summation surfaces, whereas conventional offset-domain imaging is based on surface-centered parameterization — which is a function of the data's midpoint and offset. The summation surfaces are parameterized as functions of the angles at the reflection point. Because the geometric complexity in 3D may hamper our understanding of the basic concepts, ADCIGs in 2D will be analyzed first. In Chapter 3, the section titled "Angle-domain common-image gathers by wavefield continuation" generalizes the 2D analysis to 3D ADCIGs obtained by wavefield-continuation methods. Kirchhoff-based 3D ADCIGs are described well by Brandsberg-Dahl et al. (2003).

The schematic in Figure 30 defines the angles relevant to 2D ADCIGs — the reflection opening angle γ and the geologic dip angle α_x (along the inline direction). The opening angle and the dip angle are related to the propagation angles of the source and receiver rays (β_s, β_r) by the following simple relationships:

$$\gamma = \frac{\beta_r - \beta_s}{2}, \text{ and } \alpha_x = \frac{\beta_r + \beta_s}{2} . \tag{2.15}$$

To perform angle-domain Kirchhoff imaging in 2D, the migration integral in equation 6.1 can be rewritten as integration over α_x at constant γ, as

$$I(\xi,\gamma) = \int_{\Omega_{\xi,\gamma}} W_\gamma(\xi,\alpha_x,\gamma) D\left[t = t_D(\xi,\alpha_x,\gamma), x_m = x_m(\xi,\alpha_x,\gamma), x_h = x_h(\xi,\alpha_x,\gamma)\right] d\alpha_x , \tag{2.16}$$

where $\Omega_{\xi,\gamma}$ are the integration subdomains defined as functions of the aperture angle γ.

The summation curves depend on the velocity model and are defined parametrically by the functions t_D (ξ, α_x, γ), x_m (ξ, α_x, γ), x_h (ξ, α_x, γ). In the general case, they must be computed numerically by ray tracing (see section titled "Migration in heterogeneous media"). As was true in the offset-domain imaging case, when velocity is constant, the

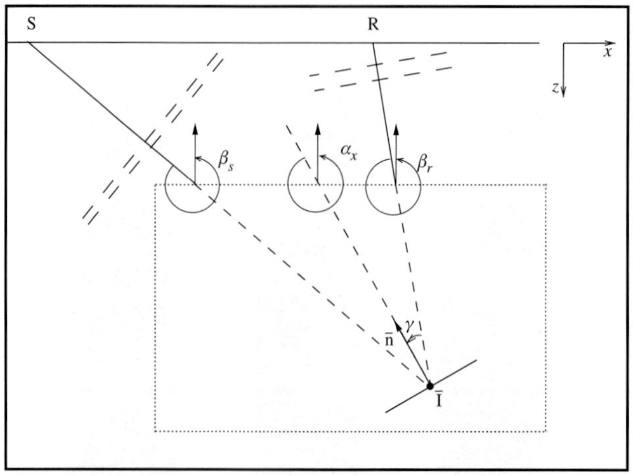

Figure 30. A schematic of the geometry of an angle-domain CIG in 2D. The angles can be either of the angles formed by the propagation direction of the rays. The arrows indicate positive angles; β_s, β_r, and α_x are negative, and γ is positive. This sign convention is consistent with upward-propagating rays.

expressions for traveltime, midpoint, and offset as functions of the reflection angle and the dip angle can be derived analytically by trigonometry (Fomel and Prucha, 1999).

In this case, the integration path for the migration integral in equation 2.16 is defined parametrically as

$$t_D = \frac{2z_\xi}{V}\frac{\cos\alpha_x\cos\gamma}{\cos^2\alpha_x - \sin^2\gamma}, \tag{2.17}$$

$$x_h = z_\xi\frac{\sin\gamma\cos\gamma}{\cos^2\alpha_x - \sin^2\gamma}, \text{ and} \tag{2.18}$$

$$x_m = x_\xi + z_\xi\frac{\sin\alpha_x\cos\alpha_x}{\cos^2\alpha_x - \sin^2\gamma}. \tag{2.19}$$

For the denominator in equations 2.17 through 2.19 not to vanish, the admissible range of α_x and γ is defined by the condition:

$$\cos^2\alpha_x > \sin^2\gamma \text{ or } |\alpha_x| + |\gamma| < \frac{\pi}{2}. \tag{2.20}$$

This analytical condition corresponds to the geometric condition that both the source ray and the receiver ray emerge at the surface.

Figure 31 shows common-reflection-angle curves superimposed onto the same Cheops pyramid that is shown in Figure 27. Comparison of Figure 31 with Figure 27 illustrates the similarities and differences between common-offset migration and common-angle migration. First, both families of curves cover the pyramid. Therefore, the images obtained by stacking the CIGs over offset or reflection angle are equivalent, if the integration weights properly take into account the Jacobian of the transformation defined by equations 2.17 through 2.19. Second, the zero-offset curve is the same as the normal-incidence curve. Third, the curves for narrow reflection angles are close to the curves for short offsets, but the curves for wider reflection angles differ substantially from the wide-offset curves.

Figure 31 also shows that the common-angle summation curves for one point diffractor at depth may span a wide range of offsets as a function of the traveltime. That

Figure 31. Common-reflection-angle integration curves superimposed onto the Cheops pyramid. (Courtesy of Sergey Fomel.)

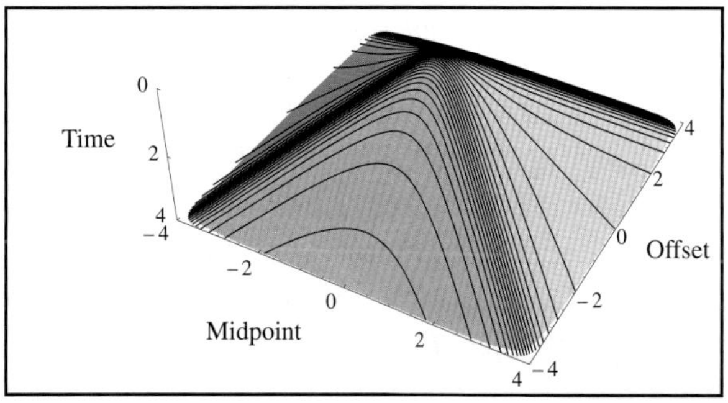

characteristic of the summation curves creates a challenge for efficient implementation of common-angle migration algorithms. In common-offset migration, input data are accessed according to a predictable and local pattern (each common-offset cube is accessed sequentially). In contrast, in common-angle migration, input data are accessed according to a more global pattern, which is also strongly model dependent for depth migration. For large 3D data sets, this quasi-random data access may make it difficult to avoid frequent data reading from disk devices.

References

Alkhalifah, T. and I. Tsvankin, 1995, Velocity analysis for transversely isotropic media: Geophysics, **60**, 1550–1566.

Aminzadeh, F., N. Burkhard, J. Long, T. Kunz, and P. Duclos, 1996, Three dimensional SEG/EAGE models — An update: The Leading Edge, **2**, 131–134.

Bear, L., T. Dickens, and P. Traynin, 2003, Incorporating nonseismic information for improved imaging with anisotropic PSDM: 73rd Ann. Internat. Meeting, Society of Exploration Geophysicists, Expanded Abstracts, 949–952.

Bleistein, N., 1984, Mathematical methods for wave phenomena: Academic Press, 341 p.

Bleistein, N., and S. H. Gray, 2001, From the Hagedoorn imaging technique to Kirchhoff migration and inversion: Geophysical Prospecting, **49**, no. 6, 629–643.

Brandsberg-Dahl, S., M. V. de Hoop, and B. Ursin, 2003, Focusing in dip and AVA compensation on scattering-angle/azimuth common image gathers: Geophysics, **68**, 232–254.

Červeny and Pšenčík, 1983, Gaussian beam and paraxial ray approximation in three-dimensional inhomogeneous media: Bulletin of the Seismological Society of America, **70**, 47–77.

Fomel, S., and M. Prucha, 1999, Angle-gather time migration: Stanford Exploration Project, Report 100, 141–151, http://sep.stanford.edu/research/reports.

Claerbout, J. F., 1985, Imaging the earth's interior: Blackwell Scientific Publications.

Hubral, P., 1977, Time migration — Some ray theoretical aspects: Geophysical Prospecting, **25**, no. 4, 738–745.

Krebs, J., L. Bear, and J. Liu, 2003, Integrated velocity model estimation for accurate imaging: 73rd Ann. Internat. Meeting, Society of Exploration Geophysicists, Expanded Abstracts, 2084–2087.

Popovici, A. M., and J. Sethian, 2002, 3-D imaging using higher-order fast marching traveltimes: Geophysics, **67**, 604–609.

Sava, P., and S. Fomel, 2001, 3-D traveltime computation using Huygens wavefront tracing: Geophysics, **66**, 883–889.

Schneider, W. A., 1978, Integral formulation for migration in two dimensions and three dimensions: Geophysics, **43**, 49–76.

Stolk, C. C., and W. W. Symes, 2004, Kinematic artifacts in prestack depth migration: Geophysics, **69**, 562–575.

Thomsen, L., 1986, Weak elastic anisotropy: Geophysics, **51**, 1954–1966.

———, 2002, Understanding seismic anisotropy in exploration and exploitation: Society of Exploration Geophysicists.

Tsvankin, I., 2001, Seismic signatures and analysis of reflection data in anisotropic media: Elsevier Science, 454 p.

van Trier, J., and W. W. Symes, 1991, Upwind finite-difference calculation of traveltimes: Geophysics, **56**, 812-821.

Vidale, J. E., 1990, Finite-difference calculation of traveltimes in three dimensions: Geophysics, **55**, 521–526.

Xu, S., H. Chauris, G. Lambaré, and M. S. Noble, 2001, Common-angle migration: A strategy for imaging complex media: Geophysics, **66**, 1877–1894.

Chapter 3 Principles of Migration Using Wavefield-continuation Methods

Introduction

For depth-migration problems, wavefield-continuation migration methods can yield better images than Kirchhoff methods do. Wavefield-continuation methods provide accurate solutions of the wave equation over the whole range of seismic frequencies, whereas Kirchhoff methods are based on a high-frequency approximation of the wave equation. Furthermore, wavefield-continuation methods naturally handle multipathing of the reflected energy induced by complex velocity functions. In contrast, when multipathing occurs, Kirchhoff methods require summation of the data over complex multi-valued surfaces. That process can be cumbersome and error-prone.

The following example illustrates a case in which severe multipathing of the wavefield makes wavefield-continuation methods advisable. Figure 1 shows a crossline section through the SEG-EAGE salt velocity model. Figure 2 compares the images obtained by 3D prestack migration with (a) a wavefield-continuation method and (b) a Kirchhoff method using the SEG-EAGE C3-NA data set. The location of this section coincides with that of the section shown in Figure 1. Improvements achieved by wavefield-continuation migration in the subsalt image (Figure 2a) are evident. Image (a) shows several reflectors that are not visible in image (b). Furthermore, several imaging artifacts that degrade image (b) are not in image (a). The wavefield-continuation image was computed by applying a computational algorithm called common-azimuth migration (Chapter 5), the cost of which is comparable to that of using Kirchhoff migration for full-volume images.

The superior image quality achieved by wavefield-continuation migration can be understood by analyzing results of wavefield modeling in the vertical section that corresponds to the images shown in Figure 2. Figure 3 shows the superposition of two snapshots of the wavefield, taken at times t = 0 and t = 1 s, when the source is below the salt body. Figure 4 shows the wavefield recorded at the surface. Notice the complex multipathing of the wavefield and the multibranching of the Green's function. The line superimposed on the wavefield represents the time-delay function computed using a finite-difference solution of the Eikonal equation. In this case, the Eikonal solution is a poor approximation of the true Green's function.

The previous example demonstrates that migration methods based on numerical propagation of the recorded wavefield can overcome some of the limitations of Kirchhoff migration methods. However, their computational cost, particularly for prestack imaging, can be substantially higher than that of Kirchhoff methods. Wavefield-continuation methods are the most efficient methods to use with regular and complete computational grids. For example, full-volume imaging of zero-offset data usually is made more efficient by use of wavefield methods than by use of Kirchhoff methods.

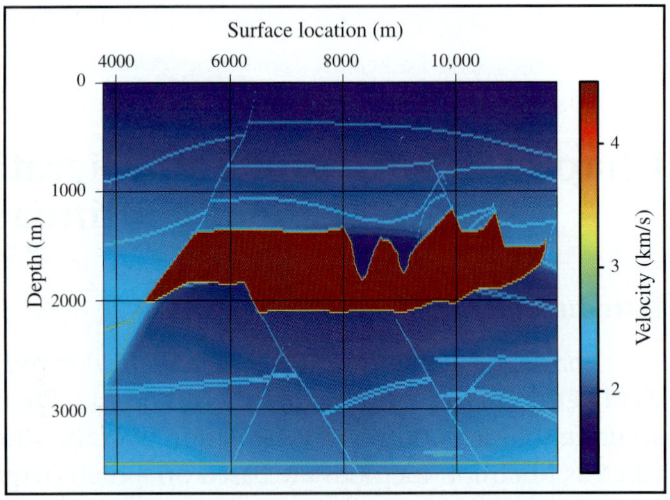

Figure 1. Crossline section of the interval velocity model for the SEG-EAGE salt data set.

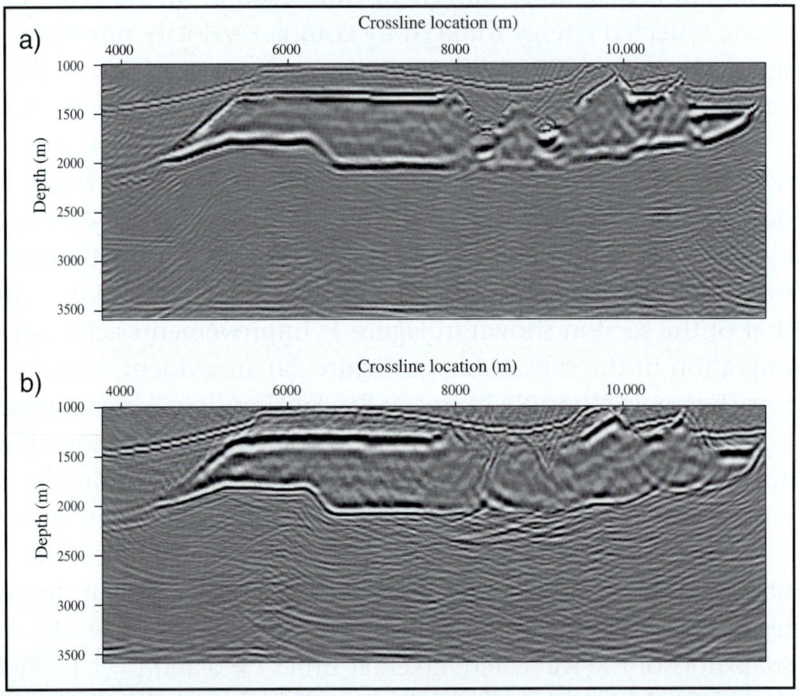

Figure 2. Sections of 3D prestack migration results of a synthetic data set with CMP X = 7500. (a) Wavefield-continuation migration and (b) Kirchhoff migration. The location of this section coincides with the section shown in Figure 1.

Figure 3. Snapshots of the wavefield at t = 0 s and t = 1 s, when the source is located below a salt body with a rugose top.

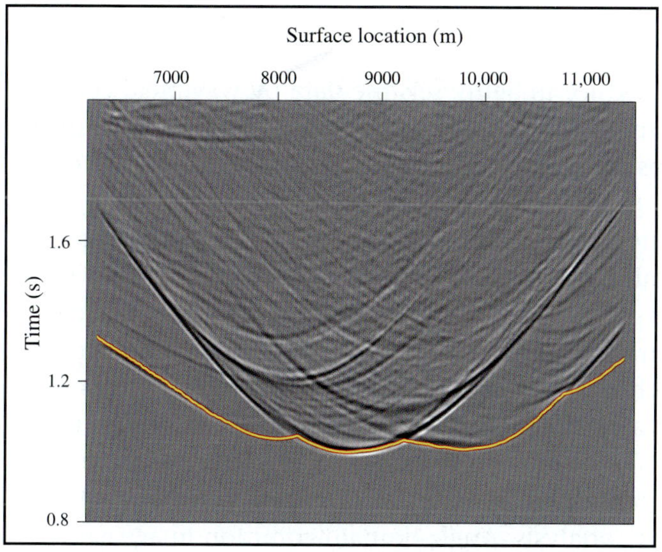

Figure 4. Wavefield recorded at the surface, corresponding to the wave modeling shown in Figure 3. The trajectory superimposed onto the wavefield represents the time-delay function computed by numerically solving the Eikonal equation. The Eikonal solution is a poor representation of the full-wavefield Green's function.

In contrast, for prestack migration, gaps in the recorded data must be filled with zero traces, because prestack recording geometries are far from being complete and regular (Chapter 1). Depending on the acquisition geometry and the migration method in such cases, many zero traces may be needed. The resulting increase in the size of the computational domain may make wavefield-continuation methods very expensive. Chapter 5 discusses several strategies to overcome this computational hurdle and presents a family of methods to perform wavefield-continuation migration of 3D marine data at an expense of computation competitive with that of Kirchhoff methods.

Another important computational advantage of Kirchhoff methods is their relative efficiency in imaging only a subset of the whole image cube (i.e., in target-oriented migration). For target-oriented migration, the computational cost of Kirchhoff migration is reduced in proportion to the reduction in image size.

In contrast, the computational cost of target-oriented migration by wavefield-continuation methods seldom is substantially lower than the cost of full-volume migration. The possibility of performing target-oriented migration is particularly advantageous in a migration-velocity-analysis (MVA) procedure (Chapter 7), in which migration is applied iteratively to improve the accuracy of the velocity model. For velocity-updating purposes, often the prestack images can be sampled coarsely along the horizontal axes, or only a layer can be analyzed at each iteration (layer stripping), or both.

Kirchhoff migration has the conceptual advantage of being easier to visualize and understand. That characteristic is the main reason for introducing migration with Kirchhoff methods (Chapter 2). Ray tracing (which is the basis of Kirchhoff methods) provides an intuitive representation of wave propagation. Furthermore, Kirchhoff-migration operators are multidimensional convolutional operators. Graphically analyzing the summation kernels of these convolutional operators gives us direct insight into their main properties and into the relation between the recorded data and the migrated image. In contrast, for wavefield-continuation methods, the relation between image and data is indirect, because these methods are based on two consecutive but distinct steps: (1) numerical propagation of the recorded wavefield (and possibly of the source function) and (2) formation of the image by application of an imaging condition to the propagated wavefields.

Many methods have been proposed for imaging seismic data by wavefield continuation. Each of them has the two main components of wavefield propagation image formation, but they differ in how those steps are carried out. The methods can be classified according to (1) the dimension along which the wavefields are continued (e.g., depth or time), (2) the spatial dimension of the computational domain (source-receiver or shot gathers), and (3) the numerical methods used to propagate the wavefields. In this chapter, items (1) and (2) are discussed, but not item (3). Application of an accurate and efficient numerical algorithm to propagate the wavefields is crucial for accurate imaging; however it is a fairly technical subject and it is not covered in this book. Chapter 5 in Biondi (2006a) provides a comprehensive discussion of the methods used for downward continuation.

As was discussed in Chapter 2, computation of prestack images is crucial to the linking between migration and velocity analysis. Angle-domain common image gathers (ADCIG) are the most natural choice for wavefield-continuation migration methods. Computation of ADCIGs by wavefield-continuation migration is based on a decomposition of the wavefield, either before imaging (de Bruin et al., 1990; Prucha et al., 1999) or after imaging (Biondi and Shan, 2002; Rickett and Sava, 2002; Sava and Fomel, 2004). ADCIGs computed before imaging are parameterized as a function of the inline and crossline offset ray parameters (p_{x_h}, p_{y_h}), whereas ADCIGs computed after imaging are parameterized directly in reflection angles (γ, ϕ).

In this chapter I will analyze first the basic idea for imaging prestack seismic data by propagating wavefields in the time domain. The family of migration methods based on time-domain propagation commonly is known as reverse-time migration. Then, the fundamental concepts underlying migration methods based on downward continuation of wavefields will be presented. The principles of wavefield-continuation migration are independent of the specific migration method. Indeed, when appropriate conditions are fulfilled, apparently different wavefield-continuation methods actually produce the same prestack image cube (Biondi, 2003).

Reverse-time Migration

Reverse-time migration (Baysal et al., 1983; Whitmore, 1983) probably is the most intuitive wavefield-continuation migration method, although it seldom is applied in practice, because it also is the most expensive computationally. The most straightforward way to introduce reverse-time migration is by defining the migration of data collected from one single experiment, that is, from a shot profile. Thus, that kind of migration is called shot-gather migration (Jacobs, 1982; Etgen, 1986). In shot-gather migration, each profile is migrated independently and then the image cubes obtained for all shots are summed.

Two wavefields are propagated independently in shot-profile migration: (1) the receiver wavefield is propagated starting from the recorded data, and (2) the source wavefield is propagated starting from an assumed source wavelet. Both the source wavefield and the recorded wavefield are continued along the time axis. The source wavefield is propagated forward in time, whereas the recorded wavefield is propagated backward in time. (This characteristic justifies the name *reverse-time* for this migration method.) The image is formed by crosscorrelating the two wavefields and evaluating the correlation at zero time (Claerbout, 1971).

Because of the linearity of the wave equation with respect to the source function, shot-profile migration can be generalized to image collections of traces (gathers) that are the result of combining several shot gathers. In this case the recorded data and the source function are combined to generate generalized-source gathers and generalized source functions. Generalized source functions can have a wide horizontal extent and a long duration in time. Chapter 5 presents several migration methods that exploit this possibility to improve the computational efficiency and accuracy of wavefield migration.

If $P^s(t, z, x, y; \mathbf{s}_i)$ and $P^g(t, z, x, y; \mathbf{s}_i)$ are, respectively, the source wavefield and the recorded wavefield for the ith source function $\mathbf{s}_i$, as functions of time t and subsurface location (z, x, y), the image is formed by applying the following imaging condition:

$$I(z_\xi, x_\xi, y_\xi) = \sum_i \sum_t P^g(t, z = z_\xi, x = x_\xi, y = y_\xi; \mathbf{s}_i) P^s(t, z = z_\xi, x = x_\xi, y = y_\xi; \mathbf{s}_i). \quad (3.1)$$

This expression distinguishes between the image-space coordinates (z_ξ, x_ξ, y_ξ) and the physical-space coordinates (z, x, y), although in practice, the two spaces often coincide. To simplify the notation in the following discussion, the two coordinate systems are distinguished only when it is necessary to avoid ambiguities.

Application of imaging condition 3.1 produces an image that has the correct kinematics (structural image) but not necessarily the correct amplitudes. In this respect, it is similar to the Kirchhoff migration formula in equation 2.2, in which the summation weights are ignored.

The following simple example illustrates the process of reverse-time migration and provides a sense of how the imaging condition in equation 3.1 works. Figure 5a shows the velocity model used to generate the shot profile shown in Figure 5b. To avoid artifacts during the migration, the direct arrival was removed from the shot profile. As expected, the reflection from the dipping event is asymmetric around zero offset, and

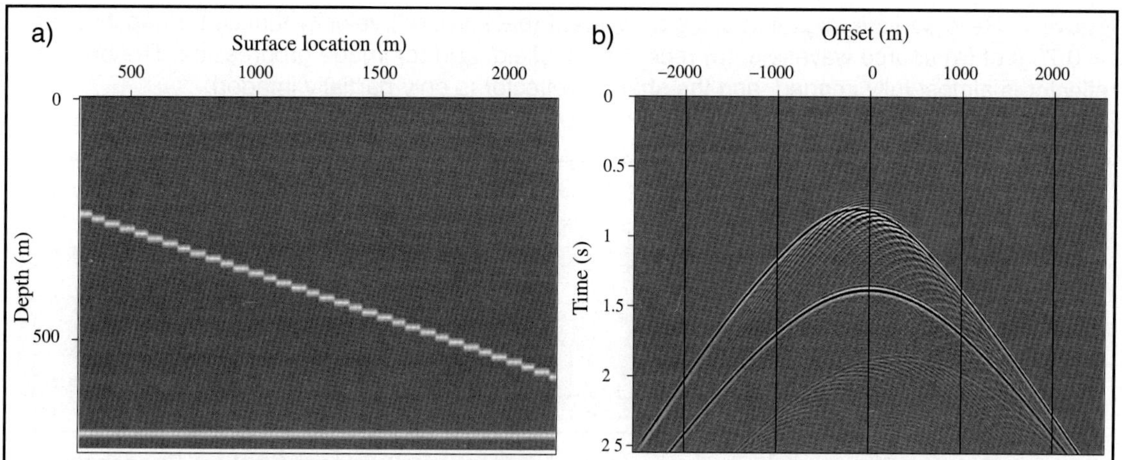

Figure 5. (a) Velocity model and (b) shot profile used for illustrating the reverse-time migration process.

the definition of the dipping reflector as a staircase creates several diffractions along that reflection.

Figures 6, 7 and 8 show three snapshots of the reverse-time migration process. The snapshots are shown in reverse-time order, to be consistent with the order in which the actual computations are performed on the data. Snapshots shown in Figure 6 were taken at a later time (t = 1.20 s), that is, at the beginning of the migration process. Snapshots in Figure 7 were taken in the middle of the migration process (t = 0.75 s), and snapshots in Figure 8 were taken at an early time (t = 0.30 s), after the reflectors were fully imaged. Panels in the middle (panels b) show the receiver wavefield [$P^g(z, x)$] as it is propagated backward, using the recorded shot profile (Figure 5b) as the boundary condition on the

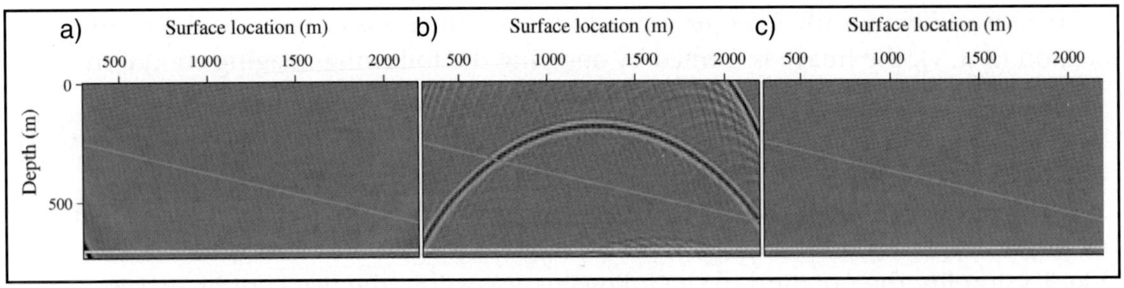

Figure 6. Reverse-time migration using a constant (background) velocity function. Snapshots at t = 1.20 s of (a) source wavefield, (b) receiver wavefield, and (c) image progression. Neither reflector has been imaged.

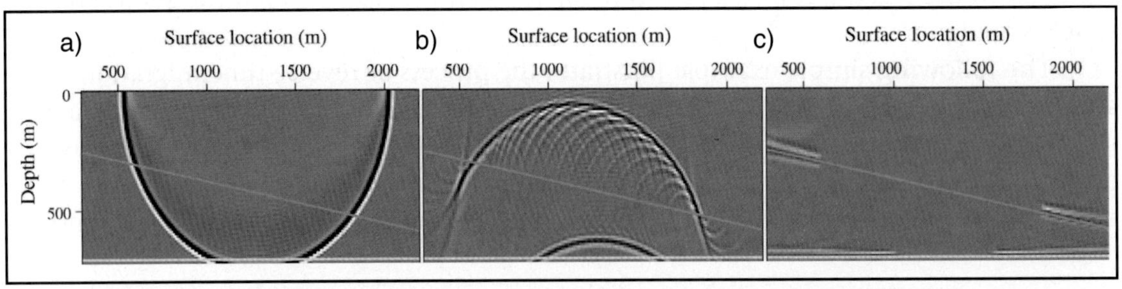

Figure 7. Reverse-time migration using a constant (background) velocity function. Snapshots at t = 0.75 s of (a) source wavefield, (b) receiver wavefield, and (c) image progression. The bottom reflector is almost fully imaged, and the shallow reflector is only partially imaged.

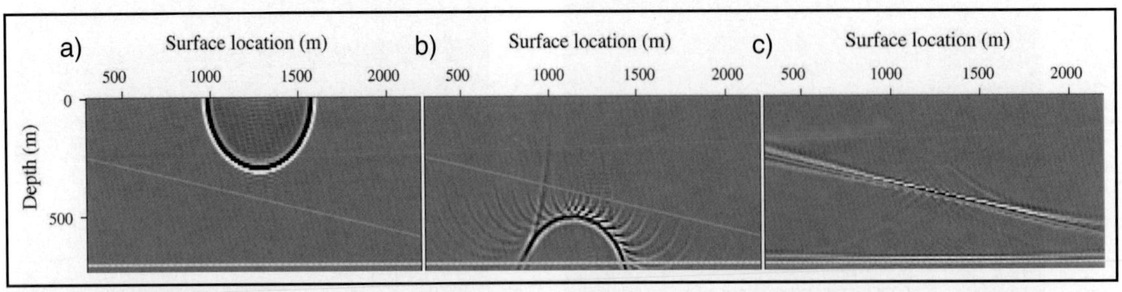

Figure 8. Reverse-time migration using a constant (background) velocity function. Snapshots at t = 0.30 s of (a) source wavefield, (b) receiver wavefield, and (c) image progression. Both reflectors are fully imaged.

top of the model. Panels on the left (panels a) show the source wavefield [$P^s(z, x)$] as it propagates through the model. Panels on the right (panels c) show the image as it progresses from being empty when the source wavefield does not interfere with the receiver wavefield (Figure 6c) to being fully formed after the two wavefields have passed each other (Figure 8c).

The migration velocity function used to generate Figures 6, 7 and 8 is the background velocity field, which is constant in this case; that is, it does not include the reflectors (which are represented by white lines in Figure 5a). Including the reflectors (or any velocity discontinuity) in the velocity model would generate internal reflections during propagation of both wavefields, and thus would cause artifacts in the image. Such a problem is illustrated in Figures 9, 10 and 11, which were generated using the velocity field shown in Figure 5a, including the reflectors. Snapshots shown in Figure 9 were taken at a later time (t = 1.120 s), that is, at the beginning of the migration process. Snapshots in Figure 10 were taken in the middle of the migration process (t = 0.75 s), and snapshots in Figure 11 were taken at an early time (t = 0.30 s), after the reflectors were fully imaged.

When the source wavefields (panels a) and the receiver wavefields (panels b) shown in Figures 9 to 11 are compared with corresponding wavefields in Figures 6 to 8, it is evident that adding the interfaces to the velocity model created internal reflections. When the two wavefields are crosscorrelated, the presence of these internal reflections creates the strong artifacts in the image that are evident in the right panels (panels c) of Figures 9 to 11.

A simple way to eliminate the internal reflections would be to smooth the velocity model. However, such smoothing might adversely affect the kinematics of the propagating wavefields when the actual velocity contrast is strong (e.g., in salt bodies). Baysal et al. (1984) proposed a method to reduce the internal reflections by matching the acoustic impedance at the velocity discontinuities. Because their method eliminates normal reflections, it is effective for zero-offset migration, but it is not effective for prestack migration. Even when computational cost is not an obstacle to widespread adoption of reverse-time migration, the problem related to internal reflections will need to be solved. Cunha (1992) and Karrenbach (1991) present some ideas on how to address this problem, in principle.

The imaging condition in equation 3.1 creates a stacked image cube. In many ways, it is equivalent to the result of Kirchhoff migration when the images for all the data offsets are stacked together. However, for both velocity estimation and amplitude-versus-angle (AVA) analysis, it often is useful to create a prestack image cube. The imaging condition in equation 3.1 can be generalized by introducing the subsurface half offsets $x_{\xi h}$ and $y_{\xi h}$. Then the image is obtained by correlating the source wavefield and the receiver wavefield shifted horizontally with respect to each other, as follows:

$$I\left(z_\xi, x_\xi, y_\xi, x_{\xi_h}, y_{\xi_h}\right) = \sum_i \sum_t P^g\left(t, z_\xi, x_\xi + x_{\xi_h}, y_\xi + y_{\xi_h}; \mathbf{s}_i\right) P^s\left(t, z_\xi, x_\xi - x_{\xi_h}, y_\xi - y_{\xi_h}; \mathbf{s}_i\right). \quad (3.2)$$

Obviously, the imaging condition in equation 3.1 is a special case of the imaging condition in equation 3.2, where $x_{\xi h}$ and $y_{\xi h}$ are zero.

Figures 12 and 13 show results of reverse-time migration on a field data set acquired in the North Sea, using equations 3.1 and equation 3.2 to perform the imaging step. Figure 12 shows the conventional image computed using equation 3.1, whereas Figure 13 shows the application of the generalized imaging condition introduced by equation 3.2. The left panel in Figure 13 shows the "zero offset" section extracted from the whole prestack image, and it is obviously equal to the result of conventional imaging shown in Figure 12. The right panel in Figure 13 shows the prestack image as functions of depth

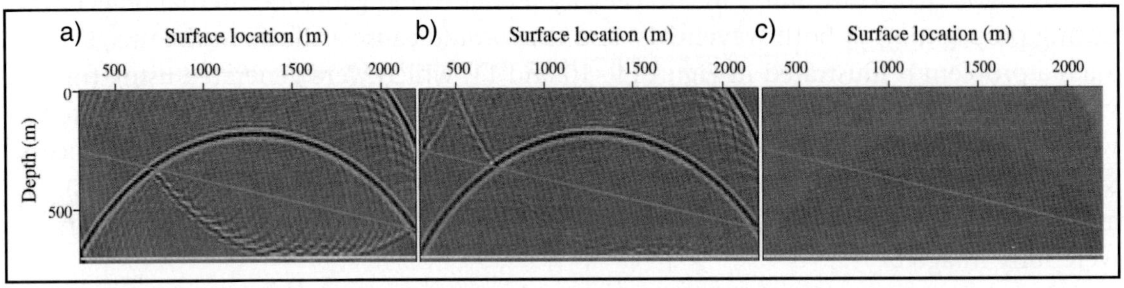

Figure 9. Reverse-time migration using the true velocity function (including reflectors). Snapshots at t = 1.20 s of (a) source wavefield, (b) receiver wavefield, and (c) image progression. In the source wavefield (a), notice the reflection from the flat interface. The crosscorrelation between this reflection and the backward-propagated receiver wavefield (b) creates a faint shadow in the image (c).

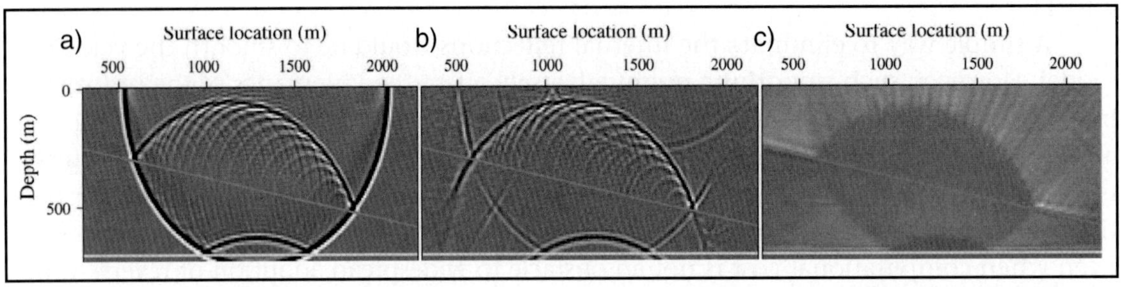

Figure 10. Reverse-time migration with the true velocity function (including reflectors). Snapshots at t = 0.75 s of (a) source wavefield, (b) receiver wavefield, and (c) image progression. The reflection from the dipping interface in the source wavefield (a) creates additional artifacts in the image (c).

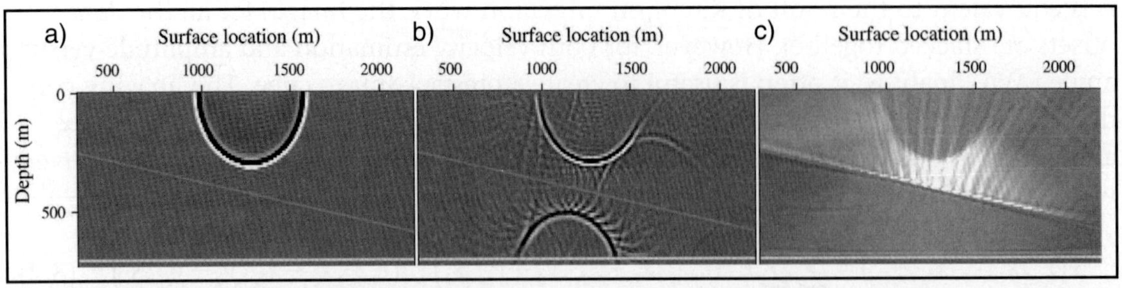

Figure 11. Reverse-time migration with the true velocity function (including reflectors). Snapshots at t = 0.30 s of (a) source wavefield, (b) receiver wavefield, and (c) image progression. In the receiver wavefield (b), constructive interference of the reflections from the two interfaces creates a circular wavefront. This wavefront correlates constructively with the direct-source wavefield (a) and causes additional artifacts in the image (c).

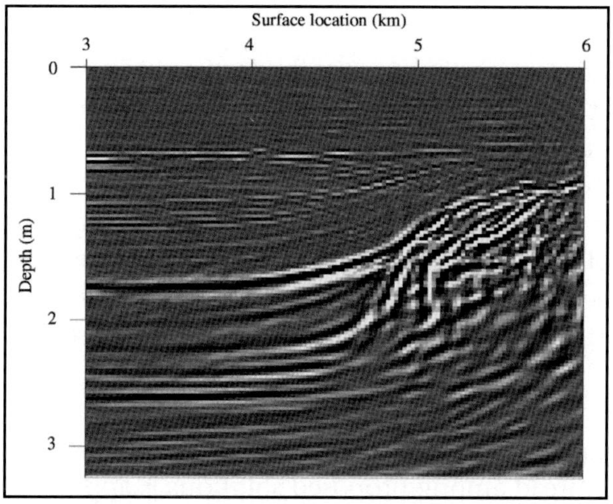

Figure 12. Image obtained by shot-profile reverse-time migration of a field data set acquired in the North Sea.

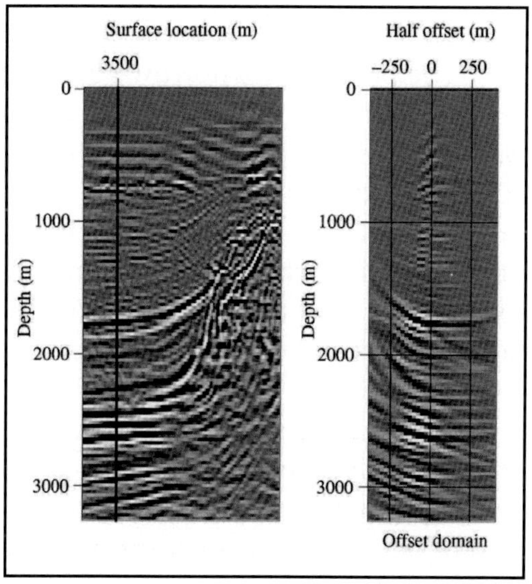

Figure 13. Prestack image obtained by shot-profile reverse-time migration of the same field data set used to obtain the image shown in Figure 12. The panel on the right shows the image as a function of the subsurface offset $x_{\xi h}$, at the fixed surface location indicated by the vertical line superimposed onto the left panel.

and subsurface offset $x_{\xi h}$, at fixed surface location x_{ξ}. The vertical line superimposed on the left panel shows the location of the right panel. Notice that in the right panel the events are focused toward zero offset, but they are not perfectly so. This partial misfocusing is caused by errors in the migration velocity and incompleteness of the recording geometry; that is by the asymmetric nature (in terms of data offsets) of conventional streamer geometry used to acquire the data (see Chapter 2).

The generalization of the imaging condition is important and it allows us to extract both velocity and AVA information from the migrated cube. Physical interpretation of the subsurface offsets is not immediate for the reverse-time migration case. However, it is immediate for the case of source-receiver migration presented in the upcoming section titled "Source-receiver migration." The insight one gains into the physical meaning of the subsurface offset — which one gains by analyzing source-receiver migration — can be applied to reverse-time migration because of the equivalence between shot-profile migration and source-receiver migration (Biondi, 2003).

The prestack image cube defined as a function of the subsurface offsets seldom is useful in itself. In this chapter, the section "Angle-domain common-image gathers by wavefield continuation" presents methods to transform the prestack image obtained by the imaging condition in equation 3.2 into another prestack-image-cube function of the aperture angle and azimuth at the reflection point.

Downward-continuation Migration

The principles of downward-continuation migration are similar to those of reverse-time migration. The main difference is that with downward-continuation migration, wavefields are propagated along the depth axis instead of along the time axis. That characteristic is the origin of both the advantages and limitations of downward-continuation migration methods. The main limitation is that by propagating the wavefield along only one direction of the depth axis, one cannot properly image overturned events — events for which at least one of the legs (either the source leg or the receiver leg) propagates both upward and downward. At least in principle, such events can be imaged by reverse-time migration.

In some situations, overturned events provide extremely useful information, so several authors recently have addressed the inability of downward-continuation migration to image overturned events. The common thread among proposed solutions is application of one-way extrapolators on a coordinate system that differs from the conventional Cartesian coordinate system aligned with the earth's depth axis. Such a coordinate system could be a tilted coordinate system (Shan and Biondi, 2004) or a coordinate system warped to conform with the raypaths (Sava and Fomel, 2004).

Figures 14 and 15 illustrate the potential shortcomings of downward-continuation migration with respect to reverse-time migration. Figure 14 shows the results of downward-continuation migration (in this case source-receiver migration) applied to the North Sea data set used to produce the image shown in Figure 12. The downward-continuation image obviously misses some of the steeply dipping reflectors and vertical reflectors that are well imaged by reverse-time migration. Figure 15 shows the same images as in Figures 12 and 14 — Figure 12 in the left panel and Figure 14 in the right panel. The ray fans superimposed onto the images represent propagation paths of reflections from a steeply dipping interface. This reflector is well imaged in the right panel, but is strongly attenuated in the left panel. This deficiency in the downward-continuation image is explained by color-coding of the ray. Colored rays correspond to wavefield components that are propagated properly by the migration methods, whereas black rays correspond to overturned events that are lost during downward continuation. Because

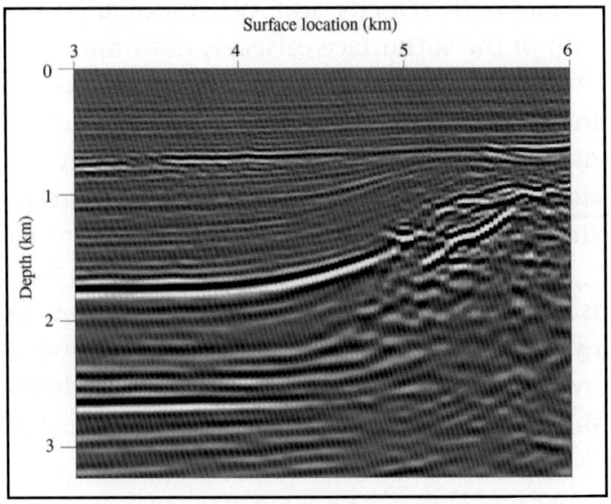

Figure 14. Image obtained by downward continuation of the field data set used to obtain the image shown in Figure 12. In this image the vertical event is missing and the steeply dipping reflectors are strongly attenuated because downward-continuation migration cannot image overturned events.

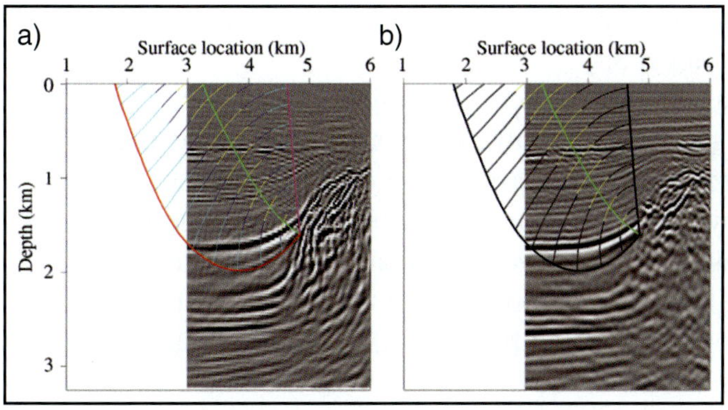

Figure 15. Images shown in Figure 12 (panel a) and 14 (panel b). Rays superimposed on the images show the propagation path of a steeply dipping event. Colored rays correspond to wavefield components that are propagated properly by the migration method, whereas black rays correspond to the overturned events that are lost during downward continuation.

either the source path or the receiver path overturns, downward-continuation migration fails to image the events with a reflection angle outside a narrow range around normal incidence.

On the other hand, propagating the wavefields along the depth axis has two meaningful advantages: (1) computations can be carried out in the temporal-frequency domain with substantial reduction of both computational and memory requirements, and (2) discontinuities in the velocity model almost never cause image artifacts such as those discussed in the previous section. Furthermore, downward continuation leads to definition of another useful family of migration methods — source-receiver migration. Source-receiver migration is based on the concept of survey sinking and thus is dependent on downward continuation of the recorded wavefield. Source-receiver migrations have characteristics different from those of shot-profile migrations, and they have several advantages for marine data.

Downward-continuation methods are based on a recursive solution of the one-way wave equation, a detailed derivation of which is beyond the scope of this book. Readers interested in more background material are encouraged to read Claerbout (1985). The basic continuation step used to compute the wavefield at depth $z + \Delta z$ using the wavefield at depth z can be expressed in the frequency-wavenumber domain as

$$P_{z+\Delta z}(\omega,k_x,k_y) = P_z(\omega,k_x,k_y)e^{ik_z\Delta z} \tag{3.3}$$

where ω is the temporal frequency, k_x and k_y are the horizontal wavenumbers, and k_z is the vertical wavenumber.

The vertical wavenumber can be expressed by the following dispersion relation, which often is called the single-square-root (SSR) equation:

$$k_z = \mathrm{SSR}(\omega,k_x,k_y) = -\sqrt{\frac{\omega^2}{v(z,x,y)^2} - \left(k_x^2 + k_y^2\right)}, \tag{3.4}$$

where $v\ (z, x, y)$ is the propagation velocity. The minus sign in front of the square root means that we want to downward continue upward-propagating waves, in accordance with the sign convention defined by Claerbout (1985). In equation 3.4, an apparent conflict exists between use of the horizontal wavenumbers k_x and k_y and expression of the propagation velocity v as a function of the horizontal coordinates x and y. Reconciling that conflict is the main challenge of numerical solution of the downward-continuation equation. Chapter 5 in Biondi (2006a) presents a detailed overview of several methods that have been proposed to solve equation 3.4.

Chapter 2 discusses the advantages of introducing an anisotropic velocity function in the migration operator when the subsurface is sufficiently anisotropic that the data cannot be focused properly by using isotropic migration. Wavefield-continuation anisotropic migration can be performed by modifying the dispersion relation in equation 3.4 in order to propagate the wavefield with different velocities, according to the propagation angle. Many expressions can be used to introduce anisotropy in the dispersion relation. The following dispersion relation, introduced by Alkhalifah (1998), is simple and accurate for wide propagation angles:

$$k_z = \mathrm{SSR}\left(\omega,k_x,k_y\right) = -\frac{\omega}{V_V(z,x,y)}\sqrt{\frac{\omega^2 - V_H^2(z,x,y)\left[k_x^2+k_y^2\right]}{\omega^2+\left[V_N^2(z,x,y)-V_H^2(z,x,y)\right]\left[k_x^2+k_y^2\right]}}, \qquad (3.5)$$

where $V_V(z,x,y)$, $V_H(z,x,y)$, and $V_N(z,x,y)$, are the interval values of the three velocities that characterize a VTI medium. More complex dispersion relations can be applied, as the complexity of the problem requires (e.g. TTI medium). As discussed in Chapter 2, the main challenge of anisotropic imaging is estimation of these interval velocity functions, rather than performance of the migration itself.

Shot-gather migration by downward continuation

Shot-gather migration by downward continuation is performed in a manner similar to shot-gather migration by time-domain propagation. The receiver wavefield at depth P_z^g is computed by downward-continuing the recorded data $P_{z=0}^g$ as follows:

$$P_z^g = P_{z=0}^g e^{ik_z z}, \qquad (3.6)$$

where the short-hand notation $e^{ik_z z}$ is used to denote the downward continuation operator that propagates a wavefield from the surface all the way to depth z. This operator is given by the cascade of all the $e^{ik_z \Delta z}$ for all the depth steps Δz necessary to go from the surface to z.

The source wavefield at depth P_z^s is computed by downward-continuing the assumed source wavelet $P_{z=0}^s$:

$$P_z^s = P_{z=0}^s e^{-ik_z z}. \qquad (3.7)$$

The negative sign in front of the exponential function in equation 3.7 is necessary, because the source wavefield propagates downward instead of upward, as the receiver wavefield does.

The image can be formed by crosscorrelating the two wavefields along the time axis (using multiplication by the complex conjugate in the frequency domain) and evaluating the correlation at zero time (summation over frequencies):

$$I\left(z_\xi,x_\xi,y_\xi,x_{\xi_h},y_{\xi_h}\right) = \sum_i \sum_\omega P_z^g\left(\omega,x_\xi+x_{\xi_h},y_\xi+y_{\xi_h};\mathbf{s}_i\right)\overline{P_z^s\left(\omega,x_\xi-x_{\xi_h},y_\xi-y_{\xi_h};\mathbf{s}_i\right)}. \qquad (3.8)$$

Source-receiver migration

Source-receiver migration is based on the concept of survey sinking. After each depth-propagation step, the propagated wavefield is equivalent to data that would have been recorded if all sources and receivers were placed at the new depth level. This task is accomplished by downward-continuing all the source and receiver gathers at each depth step (Schultz and Sherwood, 1980). Therefore, downward continuation is performed by applying the double-square-root (DSR) equation,

$$k_z = \mathrm{DSR}\left(\omega, \mathbf{k}_s, \mathbf{k}_g\right) = k_{z_s} + k_{z_g} = -\sqrt{\frac{\omega^2}{v(\mathbf{s},z)^2} - \mathbf{k}_s^2} - \sqrt{\frac{\omega^2}{v(\mathbf{g},z)^2} - \mathbf{k}_g^2}, \tag{3.9}$$

where $\mathbf{k_s}$ and $\mathbf{k_g}$ are the wavenumbers along the source and receiver axes, respectively, and where $v(\mathbf{s}, z)$ and $v(\mathbf{g}, z)$ are the propagation velocities at the source and receiver locations, respectively. The first square root in the DSR equation downward-continues the source wavefield, whereas the second square root in the DSR equation downward-continues the receiver wavefield.

Often, it is convenient to express the double-square-root equation as a function of the midpoint and offset coordinates, in place of the source and receiver coordinates. In that case, equation 3.9 becomes

$$k_z = \mathrm{DSR}\left(\omega, \mathbf{k}_m, \mathbf{k}_h\right) = -\sqrt{\frac{\omega^2}{v(\mathbf{s},z)^2} - \frac{1}{4}\left(\mathbf{k}_m - \mathbf{k}_h\right)\cdot\left(\mathbf{k}_m - \mathbf{k}_h\right)} - \sqrt{\frac{\omega^2}{v(\mathbf{g},z)^2} - \frac{1}{4}\left(\mathbf{k}_m + \mathbf{k}_h\right)\cdot\left(\mathbf{k}_m + \mathbf{k}_h\right)}, \tag{3.10}$$

where $\mathbf{k_m}$ and $\mathbf{k_h}$ are the midpoint wavenumber and the offset wavenumber, respectively.

Notice that the structure of the DSR equation is similar to that of equation 2.3 for the summation surfaces of prestack migration. However, the velocities in the DSR are interval velocities, not the average velocities of equation 2.3.

Applying the DSR operator sinks the whole survey $P_z\,(\omega, \mathbf{k_m}, \mathbf{k_h})$ together at increasing depth levels; that is,

$$P_{z+\Delta z}\left(\omega, \mathbf{k}_m, \mathbf{k}_h\right) = P_z\left(\omega, \mathbf{k}_m, \mathbf{k}_h\right) e^{ik_{z_s}\Delta z} e^{ik_{z_g}\Delta z}. \tag{3.11}$$

The image is extracted from the wavefield by applying an imaging condition at each depth level (Claerbout 1985) — that is, by evaluating the wavefield at zero time (by summation over frequencies) and at zero offset, as follows:

$$I\left(z_\xi, x_\xi, y_\xi\right) = \sum_\omega P_z\left(\omega, x_m = x_\xi, y_m = y_\xi, x_h = 0, y_h = 0\right). \tag{3.12}$$

The intuitive explanation for applying the imaging condition in equation 3.12 is that a reflector located at the depth level of the source and receivers would generate a reflection recorded at zero offset and zero time. Another intuitive justification of this imaging condition is that the wavefield recorded at the surface was generated by the reflectors exploding at zero time (Claerbout, 1985).

Notice that in the source-receiver imaging condition 3.12, the midpoint coordinates (x_m, y_m) are equated with the image coordinates (x_ξ, y_ξ). After imaging, the data midpoint coordinates (x_m, y_m) of the wavefield coincide with the coordinates of the image (x_ξ, y_ξ). Furthermore, as was discussed in the section titled "Reverse-time migration" at the beginning of this chapter, coordinates of the image often coincide with the physical-space coordinates. That is, often one has $(x = x_\xi = x_m)$ and $(y = y_\xi = y_m)$. As was indicated in the section on reverse-time migration, I will make an explicit distinction here between these sets of coordinates only when the distinction is necessary to avoid ambiguities.

As was true in the case of shot-profile migration, the imaging condition for source-receiver migration can be generalized to image the subsurface offsets also, as follows:

$$I\left(z_\xi, x_\xi, y_\xi, x_{\xi_h}, y_{\xi_h}\right) = \sum_\omega P_z\left(\omega, x_m = x_\xi, y_m = y_\xi, x_h = x_{\xi_h}, y_h = y_{\xi_h}\right). \tag{3.13}$$

With source-receiver migration, the subsurface offsets (x_{ξ_h} and y_{ξ_h}) coincide with data offsets of the downward-continued survey. This interpretation of the subsurface offsets in the image as the data offsets of a survey conducted using survey sinking is important. It allows direct physical interpretation of the prestack images obtained by use of the imaging condition in equation 3.13. Furthermore, because of the equivalence between prestack images obtained by shot-profile migration and by source-receiver migration (Biondi, 2003), it leads to the physical interpretation of prestack images obtained using the imaging conditions in equations 3.2 and 3.8.

When one uses source-receiver migration, the image is available immediately as a function of the subsurface offsets. In contrast, obtaining the prestack-image cube by shot-profile migration requires the additional wavefield correlations defined in equation 3.8. In comparison to shot-profile migration, source-receiver migration has the cost advantage of allowing one to generate the prestack image cube without computing additional correlations between wavefields. In 3D migration, such a cost saving can be substantial, because the offset space is a plane.

- Zero-offset migration

The equation for zero-offset downward continuation can be derived from the prestack equation 3.10 by setting the offset wavenumbers k_{x_h} and k_{y_h} to zero, and $\mathbf{m} = \mathbf{s} = \mathbf{g}$ (i.e., by setting the offset to zero). With these substitutions, the DSR equation simplifies to an SSR equation as follows:

$$k_z = \mathrm{SSR}\left(\omega, \mathbf{k_m}\right) = -\sqrt{\frac{4\omega^2}{v(\mathbf{m},z)^2} - \mathbf{k_m} \cdot \mathbf{k_m}}. \tag{3.14}$$

This equation is similar to SSR equation 3.4, with the important difference that now the migration velocity is one-half the true interval velocity.

When zero-offset data are migrated, the imaging condition of equation 3.12 then obviously simplifies to

$$I\left(z_\xi, x_\xi, y_\xi\right) = \sum_\omega P_z\left(\omega, x_m = x_\xi, y_m = y_\xi\right). \tag{3.15}$$

The reader should be aware that setting the offset wavenumbers to zero in equation 3.10 is not equivalent to evaluating the wavefield at zero offset. The justification for this apparently incorrect substitution is based on a stationary-phase approximation of the DSR, which is similar to the stationary-phase approximation that will be used to derive common-azimuth migration in Chapter 5.

Setting the offset wavenumbers to zero is equivalent to assuming that for all zero-offset events, the source raypath coincides with the receiver raypath. Therefore, strictly speaking, migration of zero-offset data by use of the SSR operator is a common-angle migration at normal incidence instead of a zero-offset migration (see Chapter 2, the section titled "Angle-domain common-image gathers" for an introduction to common-angle migration). Whereas it is true that all normal-incidence reflections are recorded at zero offset, not necessarily all the events recorded at zero-offset are normal-incidence reflections. Situations occur in which lateral velocity anomalies, multiple reflections, or both, cause our assumption about source and receiver raypaths to break down (Claerbout, 1985).

Figures 16 and 17 show an example of reflections that break the assumption that the upgoing path coincides with the downgoing path for all zero-offset events. Figure 16 shows the raypath corresponding to an event reflected by both walls of a deep valley etched into the upper part of a salt body. This particular kind of multiple reflection often is called prismatic reflection (Broto and Lailly, 2001). Prismatic reflections potentially carry useful information on the geometry of the salt body and on the velocity in the sediments. However, such information seldom is used, because of the difficulties in imaging prismatic reflections — although they commonly are recorded when the top-of-salt sur-

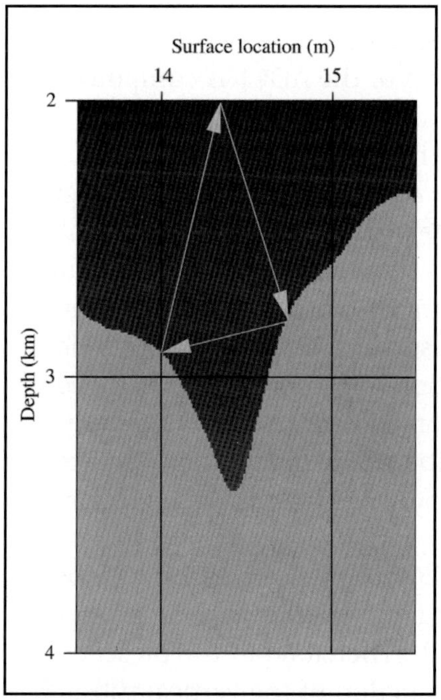

Figure 16. Sketch of a raypath corresponding to a prismatic reflection recorded at zero offset at the surface. These compose a particular type of multiple reflection, and they are recorded frequently over salt bodies that have rugged tops.

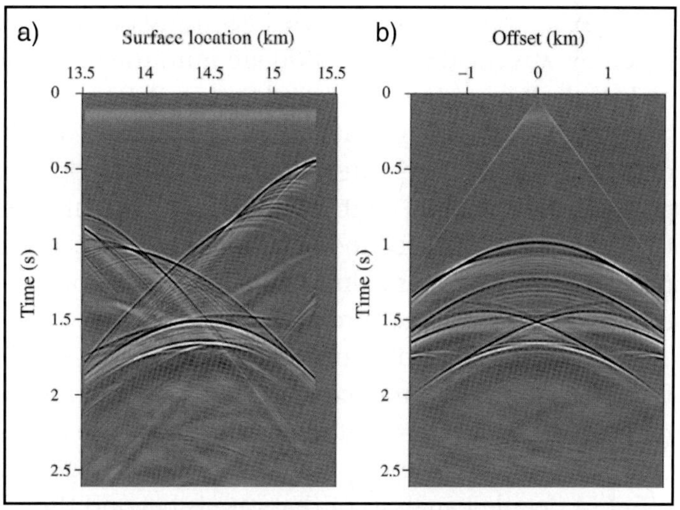

Figure 17. (a) Zero-offset section and (b) common-midpoint gather extracted from a synthetic survey modeled by assuming the velocity function shown in Figure 12. The common-midpoint gather (b) is symmetric around zero offset because of reciprocity, but the prismatic reflections cross the zero-offset line with a nonzero time dip.

face is rugged. Figure 17 shows the zero-offset section (a) and common-midpoint gather (b) extracted from a synthetic survey modeled assuming the velocity function shown in Figure 16. Because of reciprocity, the common-midpoint gather (Figure 17b) is symmetric around zero offset. However, the prismatic events cross the zero-offset line with a nonzero time dip, that is, with $\mathbf{k_h} \neq 0$. Therefore, the prismatic reflections would not be imaged properly if equation 3.14 were used to image the zero-offset section (Figure 17a).

Angle-domain Common-image Gathers by Wavefield Continuation

Computing surface-offset CIGs using wavefield-continuation methods is challenging, because the information on the surface offset is lost during wave propagation. Fortunately, the information on the propagation directions of both the incident and reflected waves is available from the wavefield at depth, and thus ADCIGs can be computed easily using wavefield-continuation migration.

This section presents two methods for computing ADCIGs in conjunction with wavefield-continuation migration. Both methods rely on decomposition of the wavefield into plane-wave components. Such decomposition is performed by slant stacks (Schultz and Claerbout, 1978) along the subsurface offset axes (see the definition of subsurface offset axes above, in this chapter). The two methods differ primarily in whether the plane-wave decomposition is performed before or after imaging. The method presented first is based on angle decomposition before imaging (de Bruin et al., 1990; Prucha et al., 1999), because it has the most direct physical interpretation. Physical interpretation of the ADCIGs computed after imaging (Biondi and Shan, 2002; Rickett and Sava, 2002; Sava and Fomel, 2004) follows directly from interpretation of the ADCIGs computed before imaging.

In the presence of multipathing and complex velocity, wavefield-continuation methods produce better images and consequently better ADCIGs than Kirchhoff migrations do, as was discussed at the beginning of this chapter. Furthermore, ADCIGs computed by wavefield continuation are immune to artifacts that can affect CIGs computed by Kirchhoff migration (Stolk and Symes, 2004). Those artifacts are generated by imaging subsets of the whole data set independently, which usually is done during Kirchhoff migration. In contrast, when we use wavefield-continuation migration methods, we compute ADCIGs using all contributions from the entire recorded data set simultaneously. Unfortunately, the need to include contributions from the whole data set may add complexity to data handling during computations for some wavefield-continuation migration methods (e.g., shot-profile migration).

Where geologic dips are steep, computation of ADCIGs using the horizontal subsurface offset does not produce useful gathers; the horizontal-offset CIGs degenerate, and their focusing at zero offset blurs (Biondi and Shan, 2002). Therefore, in the presence of complex structure, we may need to compute CIGs as a function of the vertical subsurface offset. A detailed analysis of that case is beyond the scope of this chapter, but the interested reader can find it in Biondi and Symes (2004).

ADCIG before imaging: ADCIG (p_{x_h}, p_{y_h})

A 3D prestack wavefield can be decomposed according to the offset ray parameters (p_{x_h}, p_{y_h}) by applying slant stacks along the offset axes. Such slant stacks produce plane-wave components called offset plane waves. If the velocity function is invariant horizontally, each offset plane wave can be propagated accurately and imaged independently. Each prestack partial image produced by this procedure corresponds to one offset ray parameter; thus ADCIG functions of the offset ray parameters [ADCIG(p_{x_h}, p_{y_h})] can be extracted from the whole prestack image cube. The offset ray parameters measured at the surface are functions of the surface take-off angles of the source and receiver rays and obviously are linked with the reflection opening angle γ and the reflection azimuth ϕ.

Ottolini and Claerbout (1984) presented a procedure for obtaining angle-dependent images based on the slant stack of 2D CMP gathers at the surface and on the propagation of offset plane waves. Mosher et al. (1997) generalized the 2D procedure to 3D marine data and derived a more accurate downward-continuation method for the offset plane waves. However, there is a limit to the accuracy of a migration based on an offset plane-wave decomposition at the surface. When the velocity function varies laterally, the source and receiver wavefields are refracted differently, according to the local velocities. Consequently, the independent propagation of each offset plane wave is inaccurate when strong lateral variations in velocity occur. In such situations, the idea of decomposing the wavefield into offset plane waves is still useful, but it must be generalized. The prestack wavefield must be slant-stacked recurrently at each depth level instead of only once at the surface.

Once we have introduced the concept of decomposing the prestack wavefield at depth, derivation of the procedure for computing ADCIGs before imaging using source-receiver downward continuation is straightforward (Prucha et al., 1999). As we saw above in this chapter, when source-receiver downward-continuation is used, the whole prestack wavefield is readily available at every depth level by downward-continuing the recorded data using the DSR. Then the image is obtained by extracting the values of the wavefield at zero time. The usual migration process can be represented schematically as

$$P(\omega,\mathbf{m},\mathbf{h};z=0) \overset{\text{DSR}}{\Rightarrow} P(\omega,\mathbf{m},\mathbf{h};z), \tag{3.16}$$

$$P(\omega,\mathbf{m},\mathbf{h};z) \overset{\text{Imaging}}{\Rightarrow} P(t=0,\mathbf{m},\mathbf{h};z). \tag{3.17}$$

To compute ADCIGs, we need to introduce a slant-stack transformation between downward continuation (step 3.16) and imaging (step 3.17). The results of slant stacks are functions of the offset ray parameter $\mathbf{p_h}$ and the intercept time τ. The image then is obtained by extracting the values of the wavefield at zero intercept time, i.e., at $\tau = 0$. The migration process used to produce ADCIGs can be represented schematically as

$$P(\omega,\mathbf{m},\mathbf{h};z=0) \overset{\text{DSR}}{\Rightarrow} P(\omega,\mathbf{m},\mathbf{h};z), \tag{3.18}$$

$$P(\omega,\mathbf{m},\mathbf{h};z) \overset{\text{Slant stack}}{\Rightarrow} P(\tau,\mathbf{m},\mathbf{p_h};z), \tag{3.19}$$

$$P(\tau,\mathbf{m},\mathbf{p_h};z) \overset{\text{Imaging}}{\Rightarrow} P(\tau=0,\mathbf{m},\mathbf{p_h};z). \tag{3.20}$$

Figure 18 illustrates the subsequent stages of the process represented in equations 3.18 through 3.20. The synthetic data set used to generate Figure 18 is the same one used in

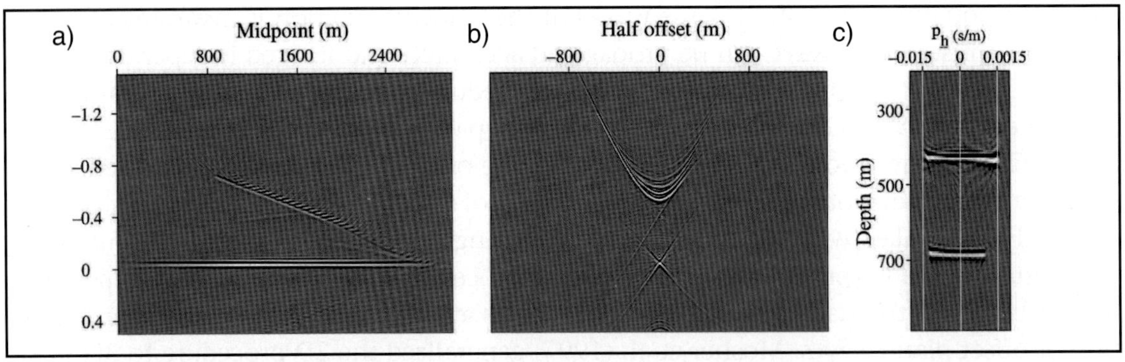

Figure 18. Illustration of the method for computing ADCIGs before imaging. Shown are orthogonal slices of the prestack wavefield after downward continuation with the correct velocity to the depth of 700 m, the depth of the flat reflector: (a) zero-offset section, (b) common-midpoint gather at 1410 m, and (c) the complete (i.e., for all depths) ADCIG at 1410 m.

this chapter to illustrate the principles of reverse-time migration. Data were generated assuming two reflectors (one dipping and one flat) immersed in a constant-velocity medium.

The downward-continuation process (step 3.16 or step 3.18) focuses the wavefield toward zero offset. If the continuation velocity is correct, a migrated image can be obtained by extracting the value of the wavefield at zero offset and zero time. Figure 18a shows the zero-offset section of the downward-continued wavefield at the depth of the flat reflectors, z = 700 m. As expected, the flat reflector is imaged well at zero time. Figure 18b shows the wavefield at the same depth as in Figure 18a, but now it is a function of offset and is at fixed midpoint (x_m = 1410 m). The flat reflector is focused well at zero offset and zero time, whereas the energy corresponding to the dipping reflector already has been defocused and therefore appears as a time-reversed hyperbola at negative times. The wavefield shown in Figure 18b is decomposed into plane waves by performing slant stacks for each midpoint location (step 3.19), and the results of these slant stacks are imaged at zero time (step 3.20). The steps for downward continuation, slant stacks, and imaging are repeated for all depth levels. Figure 18c shows the ADCIG at x_m = 1410 m for all depths. Both reflectors are imaged as flat events in the ADCIGs, because at each depth level, slant stacks transform band-limited impulses located at zero offset into flat events (such as the one shown in Figure 18b).

Figure 19 shows the effects of velocity errors on ADCIGs. It shows the same panels that Figure 18 shows (i.e., wavefield at zero-offset [a], wavefield at fixed midpoint [b], and ADCIG [c]), but with a migration velocity 10% higher than the correct one. At the correct depth of the flat reflector (z = 700 m), the energy for both reflectors has started defocusing, and it forms time-reversed hyperbolas. Most of the energy for the flat reflector has not reached the zero-time line yet, thereby causing the flat reflector to be imaged more deeply than the correct depth. In the ADCIG, both events frown downward, indicating that migration velocity is too high. Thus, the residual moveout caused by velocity errors in ADCIGs is qualitatively similar to the moveout observed in ODCIGs generated by Kirchhoff migration (refer to Figure 2-29, Chapter 2). If the velocity function is too low, the reflections will smile upward; if the velocity function is too high, the reflections will frown downward. The section in Chapter 7 on residual moveout (RMO) analysis

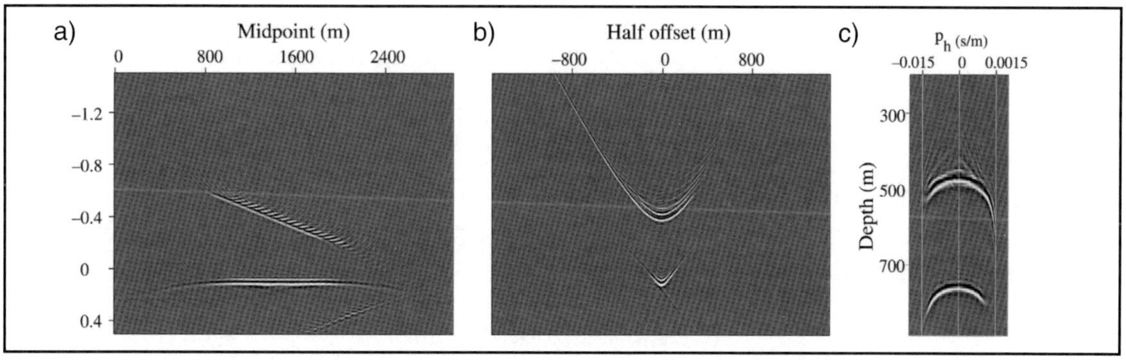

Figure 19. Illustration of the sensitivity of ADCIGs to velocity errors. Shown are orthogonal slices of the prestack wavefield after downward continuation with a velocity 10% more than the correct one, to the depth of 700 m: (a) zero-offset section, (b) common-midpoint gather at 1410 m, and (c) the complete (i.e., for all depths) ADCIG at 1410 m.

presents a quantitative analysis of residual moveout in ADCIGs. There, our purpose is to use the information contained in the ADCIGs to estimate migration velocity, on the basis of the theory developed by Biondi and Symes (2004).

Computing ADCIGs before imaging is more difficult when one is using shot-profile migration than when using source-receiver migration, because the prestack wavefield at depth is not easily available as a function of the subsurface offset. In theory, the prestack wavefield could be computed by appropriate correlations (or deconvolutions) of the source and receiver wavefields (de Bruin et al., 1990). That procedure generates ADCIGs equivalent to those obtained by source-receiver migration (Biondi, 2003). In practice, computing ADCIGs before imaging for shot-profile migration is computationally demanding, because it requires either decomposing the wavefield for each shot independently and accumulating the prestack images for all shots (which is computationally inefficient) or accumulating the contributions of each shot to the whole prestack wavefield before decomposing it (which is costly in both storage and input/output operations). Therefore, ADCIGs rarely are computed before imaging, when one is using shot-profile migration. In an upcoming section, we will see how computing ADCIGs after imaging addresses this problem.

- Geometric interpretation of 2D ADCIG (p_{xh})

Regardless of whether they have been computed using source-receiver migration or shot-profile migration, ADCIGs computed before imaging have the same geometric interpretation. Strictly speaking, they are functions of the offset ray parameters (p_{x_h}, p_{y_h}) and not of the reflection opening angle γ and the reflection azimuth ϕ. However, in 2D, the inline offset ray parameter p_{x_h} easily can be related to the reflection opening angle γ and the geologic dip angle α_x (as defined in Figure 2-30, Chapter 2) by a simple trigonometric relationship.

The offset ray parameter can be expressed as a function of the source and receiver ray parameters (p_{x_s}, p_{x_g}) as follows:

$$p_{x_h} = p_{x_g} - p_{x_s}. \tag{3.21}$$

The source and receiver ray parameters (p_{x_s}, p_{x_g}) are functions of the source and receiver ray propagation angles (β_s, β_r) and of the velocity at the reflector $v(z, x)$, as

$$p_{x_s} = \frac{\sin\beta_s}{v(z,x)} \quad \text{and} \quad p_{x_g} = \frac{\sin\beta_g}{v(z,x)}. \tag{3.22}$$

By substituting the relationships in equations 3.22 into equation 3.21 and then using the angular relationship expressed in equation 6.2 and applying fundamental trigonometric identities, we obtain the desired relationship

$$p_{x_h} = \frac{\sin\beta_r - \sin\beta_s}{v(z,x)} = \frac{\sin\left(\frac{\beta_r - \beta_s}{2}\right)\cos\left(\frac{\beta_r + \beta_s}{2}\right)}{v(z,x)} = \frac{2\sin\gamma\cos\alpha_x}{v(z,x)}, \tag{3.23}$$

which links the offset ray parameter p_{x_h} directly to the reflection opening angle γ. Mapping of the p_{x_h} axis into the γ axis depends on both the local velocity and the geologic dip. When spatial variations in velocity are significant and the geologic structure is complex, unraveling this dependence might be a challenge because of the well-known difficulties inherent in reliably measuring rapidly changing local dips.

Figure 20 shows an example of a 2D ADCIG and compares it with the ODCIGs extracted at the same surface-location. The ODCIG was obtained by an amplitude-preserving Kirchhoff migration (Ecker et al., 1996). Both the amplitudes and kinematics (including a slight upward smile) of the image function of the offset ray parameter (Figure 20a) are in qualitative agreement with the image function of offset (Figure 20b). However, the appearance of the two gathers differs in important ways. The range of well-illuminated offset ray parameters p_{x_h} decreases with depth (two-way traveltime), because a fixed maximum recording offset at the surface illuminates a decreasing range of reflection opening angles at depth. In addition, notice the amplitude discontinuity clearly visible in the ODCIG at about 1.8-km offset. This discontinuity was caused by a change in receiver spacing in the recording cable. In the ADCIG, this amplitude discontinuity has been smoothed over a Fresnel zone.

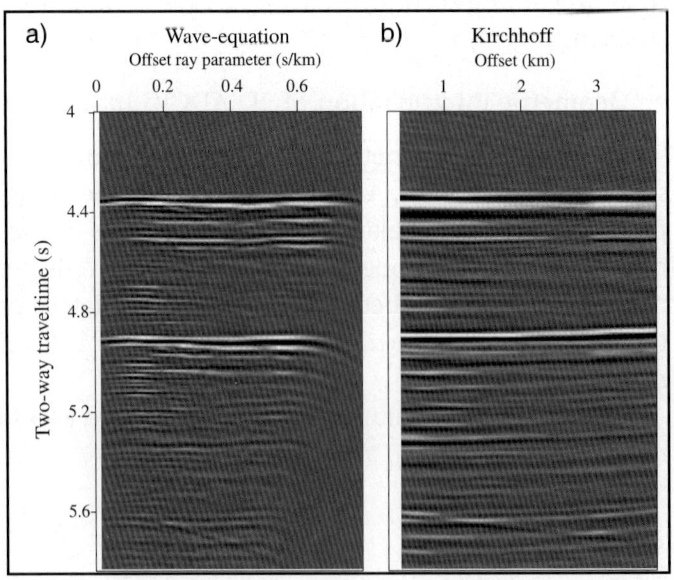

Figure 20. (a) ADCIG obtained by wavefield-continuation migration. (b) ODCIG gather obtained by Kirchhoff migration. Diagram (b) is from Ecker et al. (1996).

- Geometric interpretation of 3D ADCIG (p_{x_h}, p_{y_h})

To generalize the 2D relationship between reflection angle and offset ray parameter I refer to the schematic shown in Figure 21, which is the 3D generalization of the 2D schematic shown in Figure 2-30, in Chapter 2. The relationships in equation 3.23 also are valid in 3D when the reflection azimuth ϕ is equal to zero and the cross-line dip α_y. However, in the general 3D case we need to add another relationship linking the crossline offset ray parameter p_{y_h} to p_{x_h} and the angles γ and ϕ at the reflection point. The reflection azimuth angle ϕ is defined as the angle formed by the inline axis x with the line defined by the intersection of the coplanar plane with a constant depth plane. After rotation by ϕ, the horizontal coordinates x and y become x' and y'. Once rotated by ϕ, the coplanar plane is tilted with respect to the vertical by the crossline dip angle $\alpha_{y'}$. Notice that the rotated angles and coordinate are denoted with a prime.

In 3D, the 2D relationship between the inline offset ray parameter and the aperture angle (equation 3.23) becomes:

$$p_{x'_h} = \frac{2\sin\gamma\cos\alpha_{x'}}{v(z',x',y')}. \tag{3.24}$$

In addition, the crossline offset ray parameter is related to the reflection angles as follows:

$$p_{y'_h} = -\frac{2\sin\alpha_{y'}\sin\gamma\sin\alpha_{x'}}{v(z',x',y')}. \tag{3.25}$$

These two relationships can be used to map the image from the offset ray parameter domain (p_{x_h}, p_{y_h}) into the angle domain (γ, ϕ). More detailed derivation of these 3D relationships can be found in Chapter 6 of Biondi (2006a).

As in the 2D case, mapping the ray parameters into angles depends on both the local velocity and the geologic dips. For some applications, this dependence does not cause difficulties. For example, in migration velocity analysis (Clapp and Biondi, 2000), there is no need to map from ray parameters to angles, because ray parameters are adequate initial conditions for ray tracing. For other applications, the reflection angles are needed, and thus mapping must be performed. When such mapping is difficult, it might be convenient to use the ADCIGs computed after imaging, which are presented in the

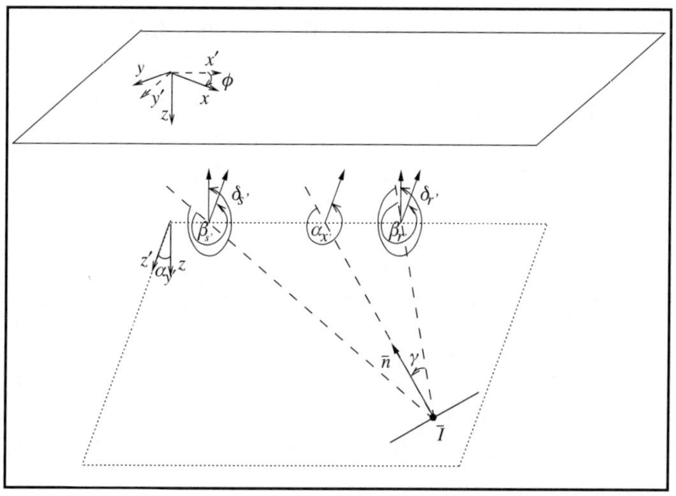

Figure 21. A schematic of the geometry of an angle-domain CIG gather in 3D. The geometry is analogous to the 2D case illustrated in Figure 30, in Chapter 2. In contrast with the schematic of that figure, the plane of coplanarity is not vertical but is tilted by $\alpha_{y'}$ and rotated by ϕ, with respect to the horizontal coordinates. Notice also that the angles δ_s' and δ_r' formed by the rays with respect to the true vertical axis, are different from the angles β_s' and β_r' formed by the rays with respect to the tilted vertical axis.

next section. That kind of ADCIG circumvents the need to estimate local dips by transforming the image directly into angles.

- ADCIG after imaging: ADCIG (γ, ϕ)

Computation of ADCIGs before imaging has a straightforward physical interpretation; thus it was the first methodology to be developed for both shot-profile migration (de Bruin et al., 1990) and source-receiver migration (Prucha et al., 1999). However, as was discussed in the previous section, computation of ADCIGs before imaging with shot-profile migration can be demanding, either computationally or in terms of input/output. Sava and Fomel (2004) recognized that for source-receiver migration, plane-wave decomposition can be performed after applying the imaging condition as well as before. Rickett and Sava (2002) applied this insight to compute ADCIGs after shot-profile downward continuation and imaging, with the generalized imaging condition expressed in equation 3.8. Their method is based theoretically on the equivalence of shot-profile and source-receiver migrations. Therefore, the ADCIGs computed after imaging with source-receiver migration and with shot-profile migration are equivalent. Biondi and Shan (2002) have extended their method further to computation of ADCIGs by reverse-time migration after imaging with the generalized imaging condition expressed in equation 3.2. That last extension enables us to compute ADCIGs for overturned waves. Recently, Biondi (2006b) extended the computation of ADCIGs to the increasingly important case of anisotropic migration.

Computation of the ADCIGs after imaging is based on a plane-wave decomposition by slant stacks along the offset axes, exactly as was the case in the computation before imaging, which we discussed above. The only difference between the two methods (computing before versus after imaging) is that dips along the offset axes are affected by the transformation from time to depth that is implicit in the imaging step. Therefore, the offset-dip parameters are linked to the reflection opening angle γ and azimuth ϕ differently from the previous case. In the next two sections, I will derive and discuss the analytical relationships between reflection angles and offset dips after imaging. The simpler 2D case will be discussed (Sava and Fomel, 2003), then the general 3D case will be addressed (Biondi et al., 2003; Biondi and Tisserant, 2004).

- ADCIG after imaging in 2D: ADCIG (γ)

Application of the imaging condition transforms a wavefield propagating in time into an image cube that is a function of depth. Transformation from time to depth depends on the local dips in the wavefield and on the local propagation velocity. In the frequency-wavenumber domain, this transformation is represented by the DSR operator, which can be expressed in 2D as a function of the angles β_s and β_r, as follows:

$$k_z = -\frac{\omega}{v(z,x)}\left(\cos\beta_r + \cos\beta_s\right). \tag{3.26}$$

Recalling the relationship between the inline offset wavenumber k_{x_h} and the inline offset ray parameter p_{x_h},

$$k_{x_h} = p_{x_h}\omega , \tag{3.27}$$

and substituting both equation 3.26 and equation 3.23 into equation 3.27, we obtain the relationship

$$k_{x_h} = -p_{x_h} \frac{k_z v(z,x)}{\cos\beta_r + \cos\beta_s} = -\frac{2\sin\gamma\cos\alpha_x}{v(z,x)} \frac{k_z v(z,x)}{2\cos\gamma\cos\alpha_x} = -k_z \tan\gamma. \tag{3.28}$$

This relationship directly links the dips in the depth-offset domain (k_{x_h} / k_z) to the aperture angle ($\tan\gamma$). Notice that both the local velocity $v(z, x)$ and the inline dip angle α_x have been canceled from equation 3.28, so that the offset-to-angle transformation can be performed accurately in the wavenumber domain. Further, because it is independent from the geologic dips, the transformation can be performed on each ODCIG independently. However, it is important to notice that absence of the local velocity $v(z, x)$ from the expression used to compute ADCIGs after imaging does not make the result independent from the local migration velocity. This dependence is indirect, through the vertical wavenumber k_z; in other words, the estimate of γ depends on the apparent vertical wavelength of the imaged reflectors. Therefore, advantages derived from the absence of $v(z, x)$ in equation 3.28 are purely computational. Sensitivity of the estimate of γ on the accuracy of the local velocity $v(z, x)$ is the same, regardless of whether γ is estimated indirectly through p_{x_h} by using equation 3.23 or directly by using equation 3.28. This concept is illustrated well by the following example.

Figures 22 and 23 illustrate computation of 2D ADCIGs after imaging with a synthetic example. They are analogous to Figures 18 and 19, which illustrate computation of 2D ADCIGs before imaging. Figure 22a shows the ODCIG after migration with the correct velocity. The energy is focused well at zero offset for both the dipping and the flat reflectors. Slant stacks transform the impulses at zero offset into flat events in the angle domain (Figure 22b).

When the migration velocity is 10% higher than the correct velocity, energy is defocused in the ODCIGs and the events have a hyperbolic moveout (Figure 23a); apexes of the hyperbolas are deeper than depths of the true reflectors. Slant stacks transform the hyperbolas into frowns (Figure 23b), similar to those shown in Figure 23c. However, there is a subtle but important difference between the two cases. The p_{x_h} range does not change between the ADCIG shown in Figure 22c and the one shown in Figure 23c, whereas the γ range increases when the velocity is too high (Figure 23b). Such an increase results from increase in the apparent vertical wavelength in the image, which correspondingly causes a decrease of k_z in equation 3.28. Or from a different viewpoint,

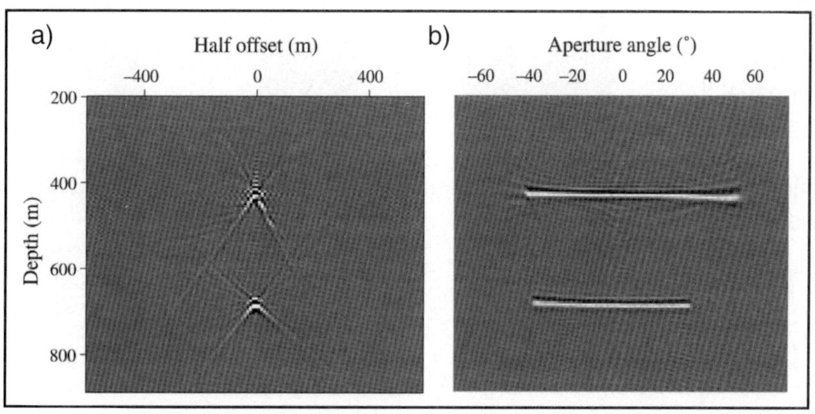

Figure 22. (a) ODCIG and (b) ADCIG (γ), after migration with the correct velocity. These CIGs were taken from the same data and at the same surface location as the ADCIG shown in Figure 18c.

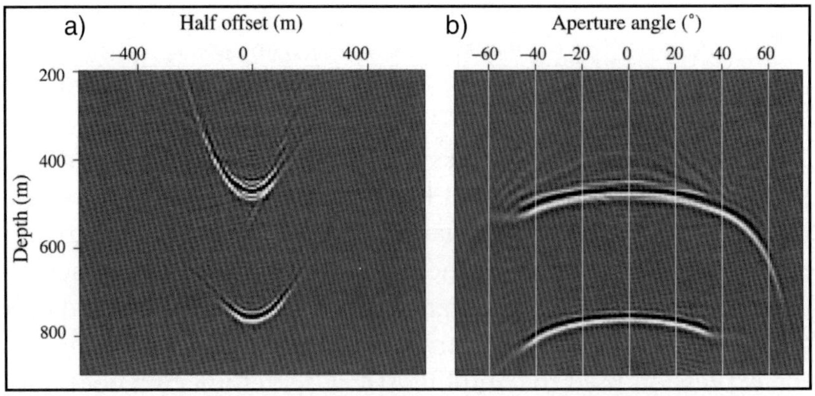

Figure 23. (a) ODCIG and (b) ADCIG (γ) after migration with a velocity of 10% higher than the correct one. Notice that the range of γ increases from Figure 22b to Figure 23b.

if we were to use equation 3.23 to map p_{x_h} into γ, the mapping would be affected by both the increase in $v(z, x)$ and the increase in apparent geologic dip α_x — and by the corresponding decrease in cos α_x.

Figure 24 shows application of the offset-to-angle transformation defined in equation 3.28 to the field-data prestack image shown in Figure 13. The right panel in Figure 24 shows the ADCIG obtained by applying equation 3.28 to the ODCIG displayed in the middle panel. The vertical line superimposed onto the image on the left indicates the location of these CIGs. Figure 25 shows the effects of migration velocity on ADCIGs. The left panels show the migrated sections obtained with two different velocity functions. The top-left panel is the same as the image shown in Figure 24, whereas the bottom-left panel is the image obtained by use of a slower velocity (original velocity scaled by 0.95). The corresponding ADCIGs are shown in the middle and right-hand panels. The ADCIG obtained with the original velocity (middle) is frowning downward, indicating that the migration velocity is too fast around the ADCIG location. Indeed, the ADCIG obtained with the slower velocity (right) is substantially flatter than the original-velocity ADCIG.

The examples above illustrate qualitatively the link between kinematics in the ADCIGs and migration velocity. Analysis of the prestack images in the angle-domain provides critical information about correctness of the migration velocity function and, if needed, about how to improve it. Chapter 7 presents detailed methods for extracting quantitative information on migration-velocity errors from migrated ADCIGs, and the chapter explains how to use that information to update the migration-velocity function.

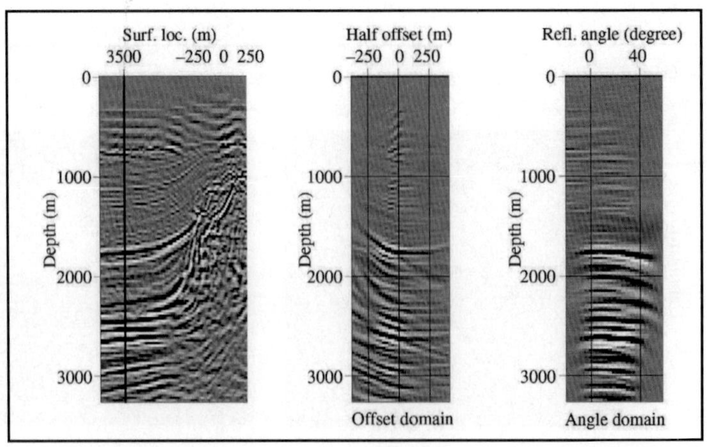

Figure 24. Transformation into the angle domain of the ODCIG obtained from a field data set and shown in Figure 13. The right panel shows the application of equation 3.28 to the ODCIG shown in the middle panel. The left and middle panels are the same as the ones shown in Figure 13.

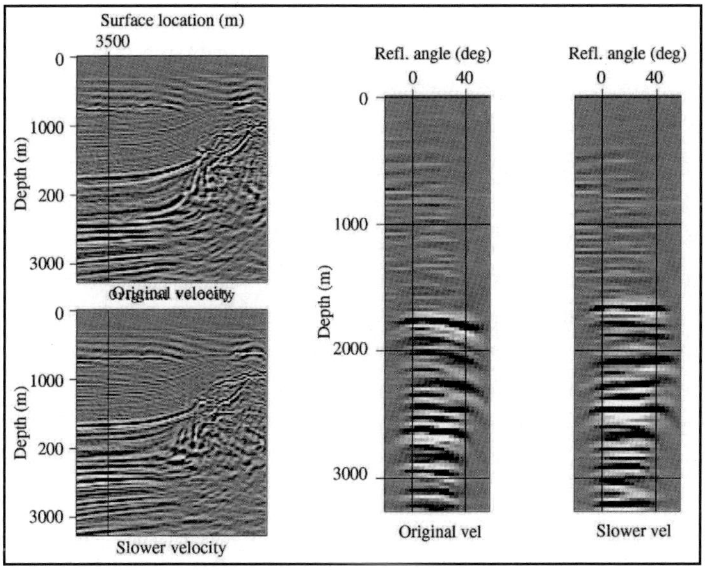

Figure 25. Illustration of the effects of migration velocity on ADCIGs. The left panels show the migrated sections obtained with two different velocity functions. The top-left panel shows the same image as the one shown in Figure 24, whereas the bottom-left panel shows the image obtained using a slower velocity. The middle and right-hand panels show the corresponding ADCIGs. The ADCIG obtained with the slower velocity (right) is flatter than the ADCIG obtained with the original velocity (middle), which is frowning downward.

ADCIG after imaging in 3D: ADCIG (γ, φ)

The after-imaging 2D offset-to-angle transformation can be also generalized to 3D by using the coplanarity condition and the geometric model shown in Figure 21. The resulting relationships (Biondi, 2006a) are the following:

$$k_{x'_h} = -k_{z'} \tan\gamma = -\sqrt{k_z^2 + k_{y'}^2}\,\tan\gamma, \tag{3.29}$$

and

$$k_{y'_h} = -k_{y'} \tan\gamma \, \tan\alpha_{x'} = -\frac{k_{y'}k_{x'_h}k_{x'}}{k_{z'}^2} = -\frac{k_{y'}k_{x'_h}k_{x'}}{k_z^2 + k_{y'}^2}. \tag{3.30}$$

Combination of equations 3.29 and 3.30 enables us to compute 3D ADCIGs after imaging. By use of these two relationships, we can map the prestack image from the offset domain (x_h, y_h) to the angle domain (γ, ϕ). Notice that the 3D expression (equation 3.29) for the aperture angle γ is not independent of geologic dips, as was its 2D equivalent (equation 3.28). Therefore, in the presence of significant crossline dips, it is important to use the correct 3D expression in place of the approximate 2D one. The reflection azimuth ϕ is included implicitly in these equations, because with the exception of k_z, all the wavenumbers are dependent on ϕ. Therefore, during computation, the wavenumbers must be transformed properly by rotating the horizontal axes by the azimuth angle ϕ.

In contrast with the transformation described by equations 3.23 through 3.25, the transformation described by equations 3.29 and 3.30 is independent of the local velocity v (z, x, y). Therefore, estimation of the geologic dips locally is not needed — only estimation globally. That estimate can be performed accurately and efficiently in the wavenumber domain.

- Examples of 3D ADCIG (γ, φ)

In 3D imaging, ADCIGs are defined in a five-dimensional space, so to gain an intuitive understanding of their behavior can be challenging. Shown in this section are 2D slices of the 5D space generated by computing 3D ADCIGs from the migrated cube obtained from a simple synthetic data set. The data set contains five planes, dipping at 0°, 15°, 30°, 45°, and 60° toward increasing x and y. The azimuth of the planes is oriented at 45° with respect to direction of the acquisition. Velocity is $V(z) = (1500 + 500z)$ m/s, which roughly corresponds to gradients that are typical in the Gulf of Mexico. Acquisition geometry had a single azimuth oriented along the x axis, and the maximum source-receiver offset was 3000 m. Figure 26 shows geometries of the reflectors. The data were imaged with a full-azimuth source-receiver 3D prestack migration.

Because of the velocity gradient and the oblique azimuthal orientation of the reflectors, azimuths of the reflections are not equal to the azimuth of the acquisition ($\phi = 0$). The reflection azimuths are within the range of $0° \leq \phi \leq 15°$ and depend on the reflector dip and on the aperture angle γ. The steeper the reflector dip and the wider the aperture angle, the larger the azimuth rotation is at the reflection point.

All the following figures show slices of the ADCIGs at one fixed horizontal location with $x = y = 450$ m; that is, they show slices through the 3D image cube described as ADCIG(z, γ, ϕ). The most familiar of these slices display the image as a function of the depth (z) and the aperture angle (γ). Figure 27 shows two such ADCIGs, for two reflection azimuths — $\phi = 10°$ (Figure 27a) and $\phi = 15°$ (Figure 27b). Reflections from the deepest and steepest reflector ($z = 1400$ m) are well focused within the range delimited by these two azimuths. In contrast, reflections from the other two reflectors (dipping at 30° and 45°) are not well focused at these azimuths; thus they frown downward even if the migration velocity is correct.

Figure 28 shows a slice taken at the constant depth of ($z = 1430$ m); that depth corresponds to the deepest reflector. Thus, the reflection amplitudes are shown as functions of both the aperture angle (γ) and the reflection azimuth (ϕ). Because of the poor azimuthal resolution that occurs near normal incidence, the azimuthal range is wide for small γ; it narrows at about $\phi = 15°$ as γ increases.

The increase of azimuthal resolution with aperture angle is demonstrated clearly in Figure 29. The three panels in Figure 29 display the image as a function of depth (z) and reflection azimuth (ϕ) and at constant aperture angle. The aperture angles are (a)

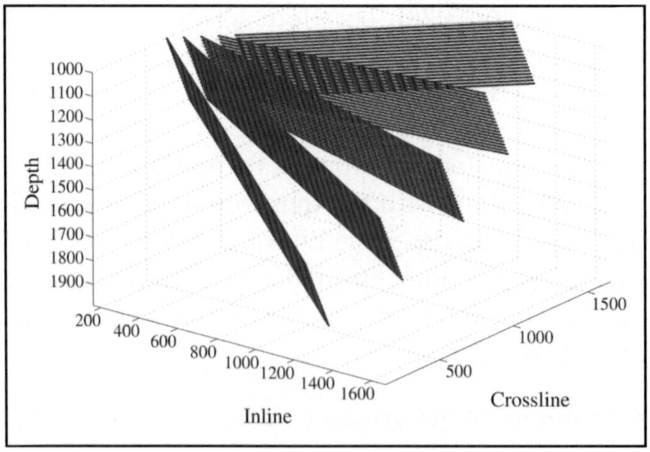

Figure 26. Geometries of reflectors for the synthetic data set used to illustrate 3D ADCIGs. The reflectors are slanted planes, dipping at 0°, 15°, 30°, 45°, and 60° toward increasing x and y. They are oriented with an azimuth of 45°, with respect to the inline direction.

$\gamma = 4°$, (b) $\gamma = 20°$, and (c) $\gamma = 30°$. Curvature of the reflectors as a function of the azimuth increases with increasing aperture angle, thereby indicating that the azimuthal resolution increases as the aperture angle widens. In other words, the common-azimuth data illuminate all the reflection azimuths for narrow aperture angles, but they illuminate only a narrow range of reflection azimuths at wide aperture angles.

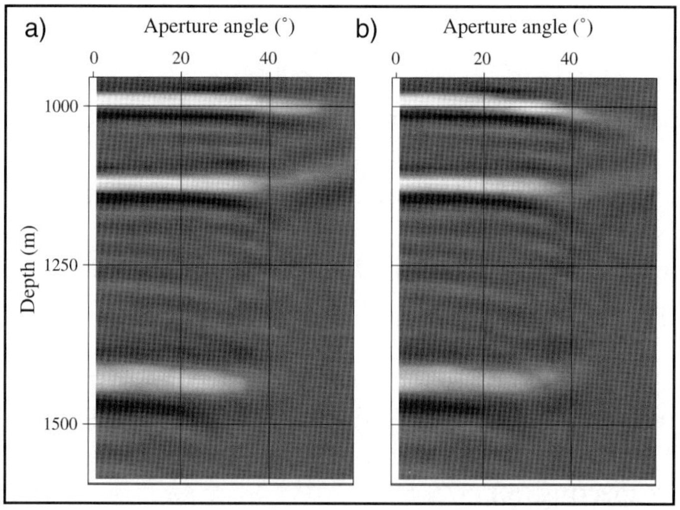

Figure 27. ADCIGs as functions of depth (z) and aperture angle (γ) for two reflection azimuths and at constant horizontal location (x = y = 450 m): (a) $\phi = 10°$ and (b) $\phi = 15°$.

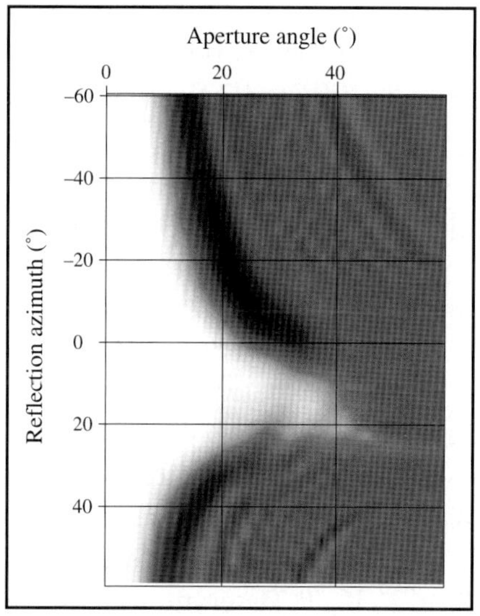

Figure 28. ADCIG as a function of aperture angle (γ) and reflection azimuth (ϕ) at constant depth (z = 1430 m) and horizontal location (x = y = 450 m).

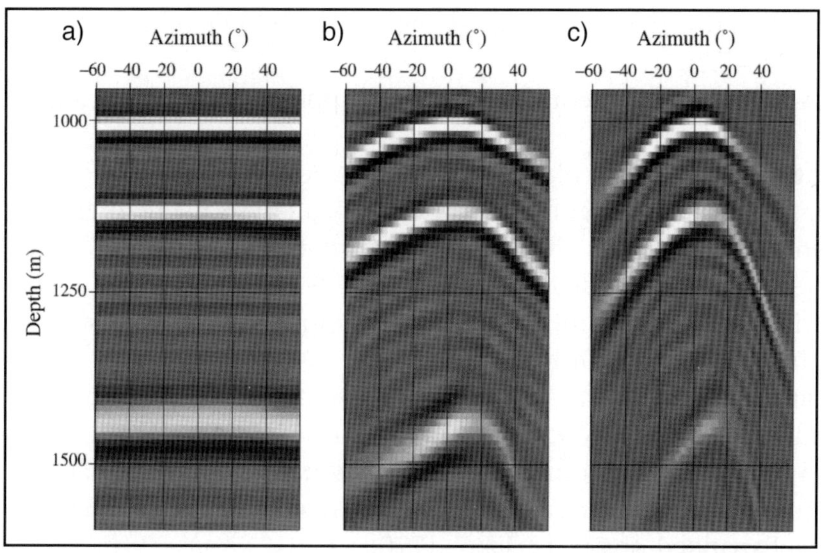

Figure 29. ADCIGs as functions of depth (z) and reflection azimuth (ϕ) for three aperture angles and at constant horizontal location (x = y = 450 m): (a) $\gamma = 4°$, (b) $\gamma = 20°$, and (c) $\gamma = 30°$.

References

Alkhalifah, T., 1998, Acoustic approximations for processing in transversely isotropic media: Geophysics, **63**, 623–631.

Baysal, E., D. D. Kosloff, and J. W. C. Sherwood, 1983, Reverse time migration: Geophysics, **48**, 1514–1524.

———, 1984, A two-way nonreflecting wave equation: Geophysics, **49**, 132–141.

Biondi, B., 2003, Equivalence of source-receiver migration and shot-profile migration: Geophysics, **68**, 1340–1347.

———, 2006a, 3D Seismic Imaging: Society of Exploration Geophysicists.

———, 2006b, Angle-domain common image gathers from anisotropic migration: Geophysics, *under revision*.

Biondi, B., and G. Palacharla, 1996, D prestack migration of common-azimuth data: Geophysics, **61**, 1822–1832.

Biondi, B., and G. Shan, 2002, Prestack imaging of overturned reflections by reverse time migration: 72nd Annual International Meeting, Society of Exploration Geophysicists, Expanded Abstracts, 1284–1287.

Biondi, B., and W. W. Symes, 2004, Angle-domain common-image gathers for migration velocity analysis by wavefield-continuation imaging: Geophysics, **69**, 1283–1296.

Biondi, B., and T. Tisserant, 2004, 3D angle-domain common-image gathers for migration velocity analysis: Geophysical Prospecting, **52**, 575–591.

Biondi, B., T. Tisserant, and W. W. Symes, 2003, Wavefield-continuation angle-domain common-image gathers for migration velocity analysis: 73rd Annual International Meeting, Society of Exploration Geophysicists, 2104–2107.

Brandsberg-Dahl, S., M. V. de Hoop, and B. Ursin, 2003, Focusing in dip and AVA compensation on scattering-angle/azimuth common image gathers: Geophysics, **68**, 232–254.

Broto, K., and P. Lailly, 2001, Towards the tomographic inversion of prismatic reflections: 71st Annual International Meeting, Society of Exploration Geophysicists, Expanded Abstracts, 726–729.

Claerbout, J. F., 1971, Toward a unified theory of reflector mapping: Geophysics, **36**, 467–481.

———, 1985, Imaging the earth's interior: Blackwell Scientific Publications.

Clapp, R., and B. Biondi, 2000, Tau domain migration velocity analysis using angle CRP gathers and geologic constraints: 70th Annual International Meeting, Society of Exploration Geophysicists, Expanded Abstracts, 926–929.

Cunha, C. A., 1992, Elastic modeling and migration in earth models: Ph.D. dissertation, Dept. of Geophysics, Stanford University, 126 p.

de Bruin, C., C. Wapenaar, and A. Berkhout, 1990, Angle-dependent reflectivity by means of prestack migration: Geophysics, **55**, 1223–1234.

Ecker, C., J. Dvorkin, and A. Nur, 1996, Sediments with gas hydrates: International structure from seismic AVO: 67th Annual International Meeting, Society of Exploration Geophysicists, Expanded Abstracts, 1767–1770.

Etgen, J., 1986, Prestack reverse time migration of shot profiles: Stanford Exploration Project, Report 50, 151–170, http://sep.stanford.edu/research/reports.

Jacobs, B., 1982, The prestack migration of profiles: Ph.D. thesis, Stanford University.

Jeannot, J. P., 1988, Full prestack versus shot record migration: Practical aspects: 58th Annual International Meeting, Society of Exploration Geophysicists, Expanded Abstracts, 966–968.

Karrenbach, M., 1991, Prestack reverse-time migration in anisotropic media: Stanford Exploration Project, Report 70, 132–144, http://sep.stanford. edu/research/reports.

Mosher, C. C., D. J. Foster, and S. Hassanzadeh, 1997, Common angle imaging with offset plane waves: 67th Annual International Meeting, Society of Exploration Geophysicists, Expanded Abstracts, 1379–1382.

Ottolini, R., and J. F. Claerbout, 1984, The migration of common-midpoint slant stacks: Geophysics, **49**, 237–249.

Prucha, M., B. Biondi, and W. W. Symes, 1999, Angle-domain common-image gathers by wave-equation migration: 69th Annual International Meeting, Society of Exploration Geophysicists, Expanded Abstracts, 824–827.

Rickett, J., and P. Sava, 2002, Offset and angle-domain common image-point gathers for shot-profile migration: Geophysics, **67**, 883–889.

Sava, P., and S. Fomel, 2004, Wavefield extrapolation in Riemannian coordinates: 74th Annual International Meeting, Society of Exploration Geophysicists, Expanded Abstracts, 1049–1052.

Schultz, P. S., and Claerbout, J. F., 1978, Velocity estimation and downward-continuation by wavefront synthesis: Geophysics, **43**, 691–714.

Schultz, P. S., and J. W. C. Sherwood, 1980, Depth migration before stack: Geophysics, **45**, 376–393.

Shan, G., and B. Biondi, 2004, Imaging overturned waves by plane-wave migration in tilted coordinates: 74th Annual International Meeting, Society of Exploration Geophysicists, Expanded Abstracts, 969–972.

Stolk, C. C., and W. W. Symes, 2004, Kinematic artifacts in prestack depth migration: Geophysics, **69**, 562–575.

Wapenaar, C. P. A., and A. J. Berkhout, 1987, Full prestack versus shot record migration: 69th Annual International Meeting, Society of Exploration Geophysicists, Expanded Abstracts, 761–764.

Whitmore, N. D., 1983, Iterative depth migration by backward time propagation: 53rd Annual International Meeting, Society of Exploration Geophysicists, Expanded Abstracts, 382–385.

Chapter 4 Approximations of Full Prestack Migration

Introduction

Ideally, every seismic data set should be imaged by use of 3D prestack migration. In practice, the computational cost of such migration limits its applications. Often, one can apply less expensive methods that yield satisfactory results in less time and with fewer resources.

However, computational complexity is not the only consideration. Prestack migration is very sensitive to the choice of velocity function and to irregular sampling of the data. This chapter introduces approximation methods for imaging prestack data — methods that are less expensive and often are more robust than full prestack migration. Therefore, approximate prestack-imaging operators can be applied successfully when full prestack migration might fail. Furthermore, prestack-imaging operators might be used to precondition (e.g., to regularize or to perform partial stacks of) the data before application of full prestack migration and before an accurate knowledge of the velocity model is gained.

The simplest of such partial-prestack-migration methods is based on the normal moveout (NMO) operator, which approximately transforms prestack data into equivalent zero-offset data. As one of the most basic seismic-processing operators, NMO is a simple stretching of the seismic traces. In the presence of dipping reflectors, NMO is not sufficiently accurate, and a more complex operator, dip moveout (DMO), becomes necessary to transform prestack data into equivalent zero-offset data. The DMO operator moves energy across the midpoint axes, partially focusing (migrating) the data. Therefore, it belongs to a family of partial-prestack-migration operators.

Another partial-prestack-migration operator that can be applied to reduce the computational cost of prestack migration is the azimuth-moveout (AMO) operator. AMO is more general than DMO, because AMO transforms prestack data into equivalent data with arbitrary offset and azimuth. In contrast, DMO is capable only of transforming nonzero-offset data into zero-offset data. Because of its flexibility, AMO can be applied to several 3D prestack-processing problems. This chapter discusses three applications of AMO: (1) partial stacking of prestack data, (2) transformation of marine data into common-azimuth data, and (3) data interpolation and/or extrapolation. Partial stacking is useful to reduce the size of prestack data and to regularize irregular acquisition geometries. Transformation of marine data into common-azimuth data enables us to apply efficient depth-imaging methods, such as those described in Chapter 5. Finally, prestack data interpolation and extrapolation improve the result of 3D surface-multiple prediction.

The last part of the chapter presents the concept of two-pass migration. Two-pass migrations achieve computational efficiency by decoupling the migration process along the two horizontal coordinate axes; they are typical examples of full-separation methods. Full-separation methods are exact only in ideal situations (e.g., in constant-velocity situations), but they may achieve satisfactory accuracy in several practically relevant cases.

Normal Moveout

In seismic processing, NMO is important, both as an approximate imaging operator and as a tool for estimating propagation velocity. NMO is a trace-to-trace transformation that remaps the time axis of data traces according to the average propagation velocity and the trace offset. For a trace with absolute offset h, the mapping is expressed analytically as

$$t_N = \sqrt{t_D^2 - \frac{4h^2}{V_N^2}}, \tag{4.1}$$

where t_N is the new time coordinate after the application of NMO, and V_N is the NMO or stacking velocity. Stacking velocity can be a function of the NMO time and of the midpoint coordinates and has the general functional form of $V_N\ (t_N, x_m, y_m)$.

To the first order, NMO transforms data collected at a finite offset to equivalent zero-offset data. The transformation to zero-offset data is only kinematic and approximate. Even in constant velocity, the NMO time t_N is exactly equivalent to zero-offset time only when reflectors have no dip component along the source-receiver azimuth of the trace.

Often, NMO is used in combination with stacking (NMO + stack) and zero-offset migration to approximate full prestack migration. This approximation procedure, which often is called poststack imaging, for a long time was the only way to image seismic data. Even today, when more-accurate imaging procedures are available to the explorationist, NMO + stack still is used for preliminary analysis of the vast majority of seismic data sets.

Poststack imaging is based on the assumption that the stacked cube obtained by stacking NMO-transformed traces is a good approximation of the true zero-offset cube. When this assumption is true, the subsurface image can be obtained by applying either zero-offset time migration or zero-offset depth migration. The large cost advantage of this procedure compared with full prestack migration derives from the data reduction achieved by NMO + stack. The stacked cube has only three dimensions (time and two midpoints), compared with the five dimensions of prestack data. More important, the number of data traces is reduced by a factor equivalent to the average data fold. Because the cost of performing NMO + stack is negligible compared with that of performing the subsequent zero-offset migration, the whole chain of processes is cheaper than full prestack migration by a factor proportional to the average data fold.

In addition to its lower computational cost, poststack imaging is attractive because it is a robust procedure both with respect to uncertainties in the velocity estimates and to irregularities in the data sampling. The NMO + stack sequence can be used as an approximate partial-prestack migration that focuses data along the offset axes. The only velocity information that NMO + stack needs for focusing data along offsets is the average velocity function V_N (t_N, x_m, y_m), which can be measured directly from the data.

The same stacking velocity function also can be used, possibly with some corrections, to focus the data along the midpoint axis by zero-offset time migration. Alternatively, if the interval velocity can be estimated with some confidence, the stacked cube can be depth-imaged with zero-offset depth migration. Zero-offset depth migration can be performed by either a Kirchhoff method (Chapter 2) or a wavefield-continuation method (Chapter 3). Because NMO is a trace-to-trace transformation, with no spreading of energy across midpoints, NMO + stack is immune to operator aliasing across midpoints.

In addition, as I will discuss in Chapter 8, the offset axis can be undersampled severely without compromising the quality of the stack. Irregular spatial coverage may cause uneven amplitudes in the stacked cube. In Chapter 9, therefore, I will discuss simple techniques to balance amplitudes by compensating for irregular coverage.

However attractive and robust it may be, NMO + stack cannot image all data sets with the required accuracy. When the velocity function is so complex that it causes nonhyperbolic moveouts across offsets, the simple hyperbolic NMO equation obviously is inappropriate to correct for the time delays of prestack traces.

The most common failure of the NMO + stack procedure is in the presence of dipping reflectors. To analyze the approximations that are made implicitly when the data are imaged by NMO + stack followed by zero-offset migration, I evaluate analytically the summation (and spreading) surface of the prestack imaging operator that is equivalent to the application of NMO + stack followed by zero-offset migration. I then compare the surface of this equivalent operator with the full prestack-migration summation surface, as derived in Chapter 2. The derivation starts by substituting the time after NMO t_N for t_D in the zero-offset migration equation 2.5. It yields the following summation surface in NMO time:

$$t_N = 2\frac{\sqrt{z_\xi^2 + \left|\overline{\mathbf{xy}}_\xi - \mathbf{m}\right|^2}}{V}. \tag{4.2}$$

The equivalent summation surface in the data space is derived immediately by eliminating t_N using the NMO equation 4.1 and by setting $V_N = V$:

$$t_D = 2\frac{\sqrt{z_\xi^2 + \left|\overline{\mathbf{xy}}_\xi - \mathbf{m}\right|^2 + h^2}}{V}. \tag{4.3}$$

Equation 4.3 defines a hyperboloid of rotation that is obviously different from the surface of full prestack migration defined by equation 2.4. Figure 1 shows both of these summation surfaces. The flattened outer surface is the full prestack-migration surface, and the inner, circularly symmetric surface is the one defined by equation 4.3. The two surfaces are tangent at the top and all along the curve oriented in the crossline direction y and passing through the apex. The interpretation of this obser-

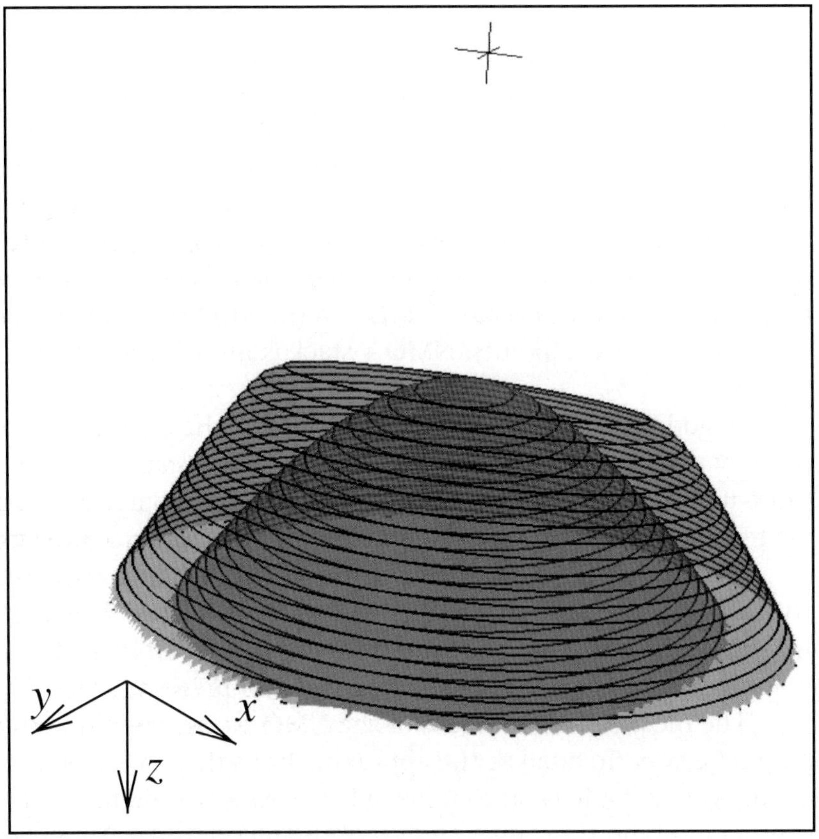

Figure 1. Summation surfaces for full constant-offset prestack migration and NMO + stack followed by zero-offset migration.

vation is that flat reflections and, generally, reflections that have no dip component along the inline direction are migrated perfectly by the NMO + stack followed by zero-offset migration procedure.

All other reflections in the data are migrated inaccurately when full prestack migration is approximated by its less costly alternative. The degree of inaccuracy varies and is greatest for steeply dipping shallow reflections.

To analyze the approximations introduced by NMO + stack followed by zero-offset migration, one also can compare the spreading surface of full prestack migration with the spreading surface of the poststack imaging procedure. The zero-offset migration hemisphere can be written in terms of NMO time as

$$\frac{4\left(x_\xi - x_m\right)^2}{t_N^2 V^2} + \frac{4\left(y_\xi - y_m\right)^2}{t_N^2 V^2} + \frac{4z_\xi^2}{t_N^2 V^2} = 1 . \tag{4.4}$$

Substituting NMO equation 4.1 into equation 4.4, I obtain the equivalent spreading surface

$$\frac{4\left(x_\xi - x_m\right)^2}{t_D^2 V^2 - 4h^2} + \frac{4\left(y_\xi - y_m\right)^2}{t_D^2 V^2 - 4h^2} + \frac{4z_\xi^2}{t_D^2 V^2 - 4h^2} = 1 . \tag{4.5}$$

This surface is spherical and obviously different from the ellipsoid of full prestack migration presented in equation 2.6. Figure 2 shows both the full prestack-migration ellipsoid (outer surface) and its spherical approximation (inner surface). Regarding the summation surfaces, the exact operator and the approximate operator are tangent along the curve passing through the bottom of the surfaces and oriented in the crossline direction. Therefore, only reflectors that are not dipping with respect to the inline direction are imaged perfectly by the approximation procedure. The larger the dip component is along the inline direction, the larger the error is.

Stacking velocity from dipping reflectors

The errors introduced by the NMO + stack procedure also can be analyzed and partially corrected by making the stacking velocity a function of the reflectors' dip. Consistent with the results of the previous section, the necessary correction to the NMO velocity is greatest when the trace azimuth is aligned with the structural-dip direction. The velocity correction is null when the trace azimuth is aligned with the structural-strike direction.

The analytic derivation of the stacking velocity as a function of the reflector dip and data azimuth, which is based on the schematic shown in Figure 3, was first presented by Levin (1971). Figure 3 shows the geometry of a reflection from a dipping plane. The source is at the origin of the axis and the receiver is on the inline axis x at a distance $2x_h$ from the source. The derivation is based on the fact that the length of the whole raypath, from the source to the reflection point and back to the receiver, is equal to the path from the source to the image receiver $\tilde{g}$.

The final result is interpreted better using Figure 4, which defines the angle η as the angle between the direction of the trace azimuth and the dip direction of the reflector. The angle α is defined as the angle between the vertical axis and the orthogonal to the reflector. Chapter 3 in Biondi (2006) presents a complete derivation of the following expression of the NMO velocity V_N:

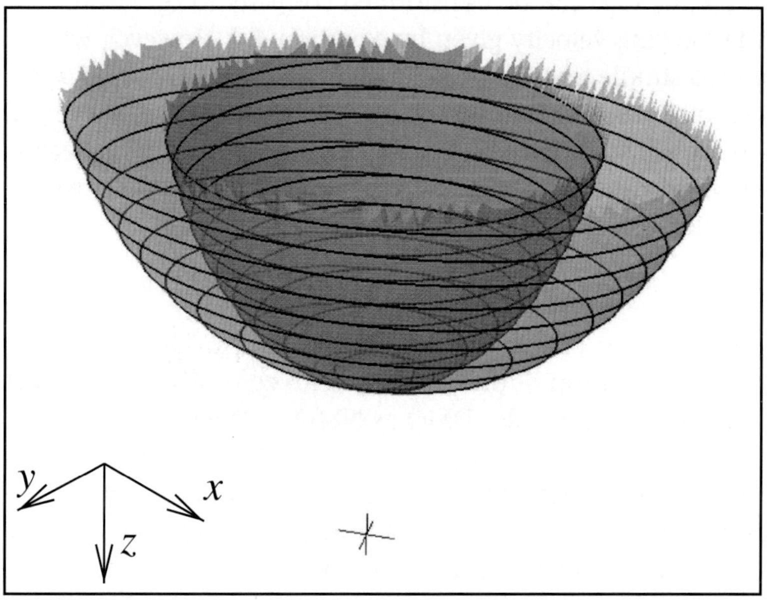

Figure 2. Spreading surfaces for full constant-offset prestack migration and NMO + stack followed by zero-offset migration.

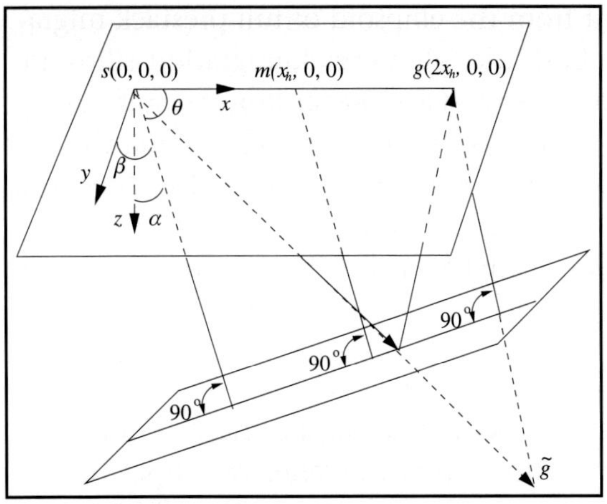

Figure 3. Geometry of a reflection from a dipping plane. The image receiver $\tilde{g}$ belongs to the perpendicular to the reflector passing through the actual receiver position.

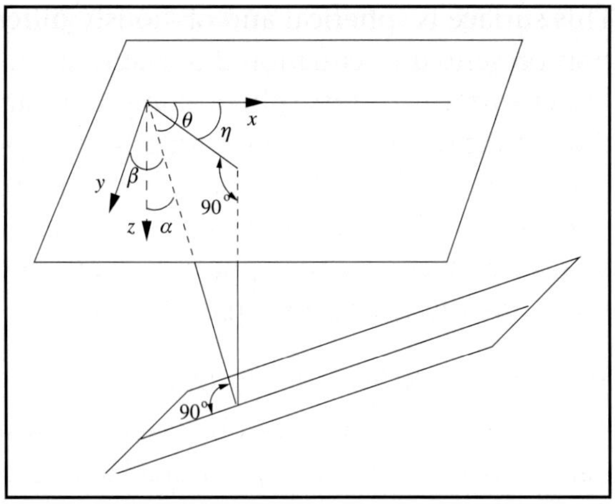

Figure 4. Definition of the angles η and α that enter the expression for the stacking velocity from a dipping reflector (equation 4.6).

$$V_N = \frac{V}{\sqrt{1-\sin^2\alpha\cos^2\eta}}. \tag{4.6}$$

Consistent with the analysis of the migration surfaces discussed in the section titled "Normal moveout," stacking velocity is equal to the medium velocity when (1) the reflector is flat (α = 0) or (2) the dip direction is perpendicular to the trace azimuth (η = ± 90°). In such cases, the poststack imaging procedure migrates the reflections exactly.

In contrast, when the dip direction is parallel to the trace azimuth (η = 0) and the reflector is vertical (α = ± 90°), the NMO velocity tends to infinity. In these cases, no NMO should be applied to transform the data to zero offset. This result explains why the summation and spreading surfaces for zero-offset and constant-offset migration shown in Chapter 2 (Figures 3 and 6) become tangent for vertical inline dips.

Prestack data without conflicting dips can be transformed correctly to zero-offset data by applying NMO with the stacking velocity given by equation 4.6. However, when the data contain conflicting dips, a simple trace-to-trace transformation such as NMO is insufficient, because it would require using different stacking-velocity functions at the same time on the same data. In such cases, a partial-prestack-migration operator is necessary to transform to zero offset all the dips in the data. The next section introduces the dip moveout (DMO) operator, which can be used to accomplish that task.

Dip Moveout

To overcome the limitations of the simple NMO operator, DMO introduces a dip-dependent correction into the transformation of prestack data into zero-offset data. I will refer to this improved transformation that includes DMO as NMO + DMO + stack, to distinguish it from the less-accurate NMO + stack.

Several DMO formulations exist. The original DMO (Deregowski and Rocca, 1981) was designed to be applied after NMO. More recently, Forel and Gardner (1988) presented an alternative formulation that can be applied before NMO. In either case, if the propagation velocity is assumed to be constant, the DMO operator is independent from the actual value of the velocity. Such DMO velocity independence is attractive because it means that NMO, which is considerably less expensive than DMO, is the only velocity-dependent component of the transformation to zero offset. The advantage of Forel and Gardner's formulation is that the inexpensive and velocity-dependent NMO operator is applied after the relatively expensive and velocity-independent DMO operator. Therefore, NMO velocity can be updated without reapplying DMO. On the other hand, applying a residual NMO after using Deregowski and Rocca's DMO introduces only small errors, if the residual moveout is sufficiently small.

Note that DMO is not a simple trace-to-trace transformation, as NMO is, but it performs a *partial prestack migration* of the data by spreading (or summing) energy across the midpoint axes. The constant-velocity DMO spreading surface is derived analytically by evaluating the impulse response of the chain of prestack migration followed by zero-offset modeling. Biondi (2006) presented a simple geometric derivation — derived by generalizing to 3D the derivation presented in Hale (1992) — that yields the following expression for the DMO impulse response:

$$\frac{t_{zo}^2}{t_N^2}+\frac{x_{zo}^2}{h^2}=1, \tag{4.7}$$

where t_{zo} is the zero-offset time (after DMO), t_N is the NMO time, h is the absolute offset of the impulse, and x_{zo} is the zero-offset (after DMO) inline midpoint. For a given offset h and NMO time t_N, equation 4.7 represents an ellipse defined in the zero-offset time-midpoint space. The sketch in Figure 5 shows both the DMO impulse response and the prestack-migration impulse response for the same impulse in the data, which is identified by the asterisk in the figure. The DMO ellipse is vertical and coplanar with the offset vector. The time of the bottom of the ellipse is equal to the NMO time, as expected, because flat reflectors do not require a DMO correction. The ellipse passes through the source and receiver locations at the surface. The shape of the ellipse is velocity independent. However, the effective aperture of the operator is velocity dependent, because the zero-offset midpoint x_{zo} is bound to satisfy the following inequality (Hale, 1991):

$$|x_{zo}|\le\frac{2h^2}{t_D V}\le h\cdot \tag{4.8}$$

Poststack imaging of the SEG-EAGE salt data set

To illustrate the capabilities and limitations of poststack imaging, I analyze the results obtained by processing a subset of the SEG-EAGE salt data set. I focus on the area around the crest of the salt body. Figure 6 shows an inline section of the interval-velocity function in this area, and Figure 7 shows the corresponding rms velocity. The reflections from the steeply dipping flanks present an insurmountable challenge for the simple NMO + stack procedure. Furthermore, the presence of a rugose salt top on the right side of the salt crest also presents problems for the NMO + DMO + stack pro-

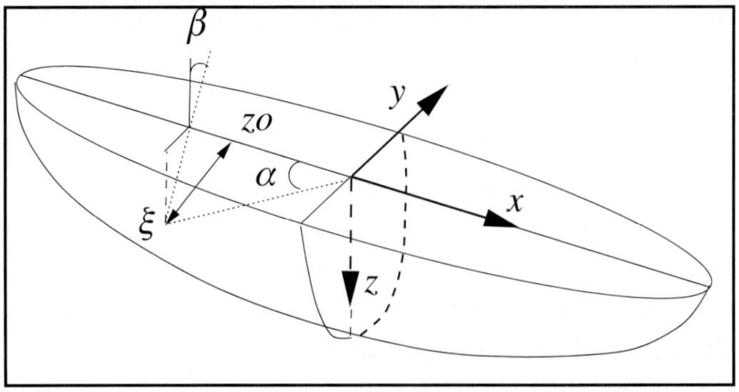

Figure 5. DMO impulse response and prestack-migration impulse response.

cedure. Some of the dipping events are recorded at later times than are the reflections from the top of the salt, as is evident from the stacked sections shown in Figures 8 and 9. Therefore, when we are determining the appropriate NMO velocity for correcting the events before DMO, a potential conflict exists between the NMO velocity required for the flattish events and the NMO velocity required for the dipping events. The dipping reflections recorded on the right of the crest might be undercorrected if the true rms velocity, which is displayed in Figure 7, is used. In practice, if our aim is to image only the reflectors above the salt body and the top of the salt itself, we can avoid this problem by ignoring the salt when we set the NMO velocity. In this example, I used the true rms velocity to illustrate the problems that the NMO + DMO + stack procedure encounters in complex situations.

Figure 8 shows the zero-offset cube obtained by NMO + stack. The reflections from the steeply dipping salt flanks that are oriented along the inline direction (CMP X) have been strongly attenuated, whereas the reflections oriented along the strike direction (CMP Y) have been stacked properly. That result is consistent with theoretical analysis of both the azimuthal dependency of stacking velocity from dipping reflectors (see the section titled "Stacking velocity from dipping reflectors"), and the errors introduced by NMO + stack (see the section titled "Normal moveout").

Figure 9 shows the zero-offset cube obtained by NMO + DMO + stack. The reflections from the shallow, steeply dipping salt flanks (above 1.5 s) have been stacked prop-

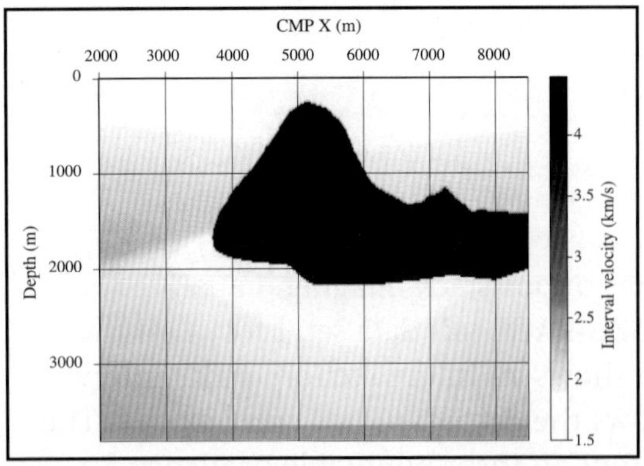

Figure 6. Inline section (CMP Y = 8 km) of the interval velocity function.

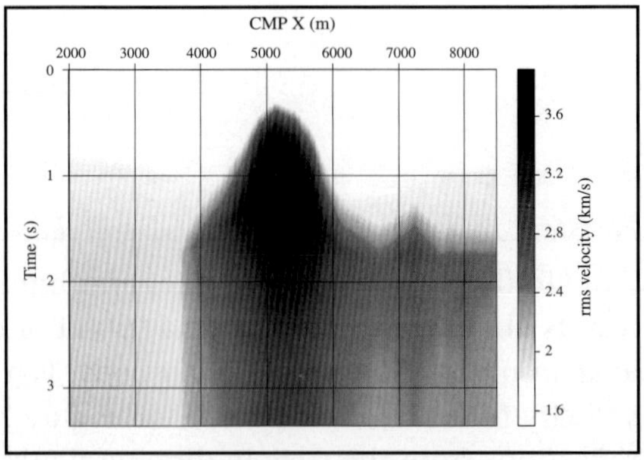

Figure 7. Inline section (CMP Y = 8 km) of the rms velocity function.

erly regardless of their azimuthal orientation. However, the reflections from the deeper portion of the right flank are not stacked properly because of the NMO-velocity conflict discussed above.

Limitations of the stacking procedure in the presence of complex geology have direct consequences on final migrated images. Figure 10 is an inline section (CMP Y = 8 km) obtained by full prestack time migration using the true rms velocity function. It is the best image that one can obtain by time imaging, so I use it as a reference.

Figure 11 shows the result of migrating by zero-offset time migration the data cube obtained by NMO + stack (shown in Figure 8). As one expects, the steeply dipping salt flanks are imaged poorly. The image of the steeply dipping salt flank on the left can be improved substantially by applying DMO before stacking. Figure 12 shows the result of migrating by zero-offset time migration the data cube obtained by NMO + DMO + stack (shown in Figure 9). However, whereas the left flank is imaged better in Figure 12 than in Figure 11, the lower part of the right flank and the small peak at CMP X = 7 km are imaged more poorly by the migration after application of DMO than by the migration without it.

In the next section, I will introduce another partial-prestack migration operator (azimuth moveout, or AMO), which allows us to overcome DMO limitations by enabling coherent stacking of only a subrange of offsets. Because final imaging requires prestack migration of the partial stacks, the use of AMO instead of DMO increases computational cost. However, it allows substantial savings compared with the computational cost requirements of full prestack time migration, which is needed to create the image shown in Figure 10.

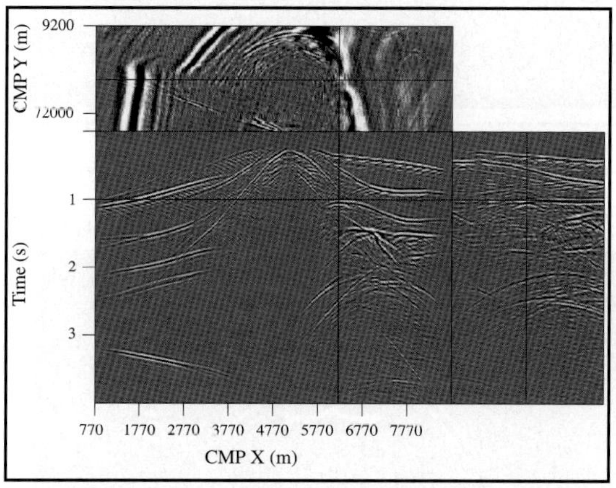

Figure 8. Zero-offset cube obtained by NMO + stack. The reflections from the steeply dipping salt flanks that are oriented along the inline direction (CMP X) have been strongly attenuated, whereas the reflections oriented along the strike direction (CMP Y) have been stacked properly.

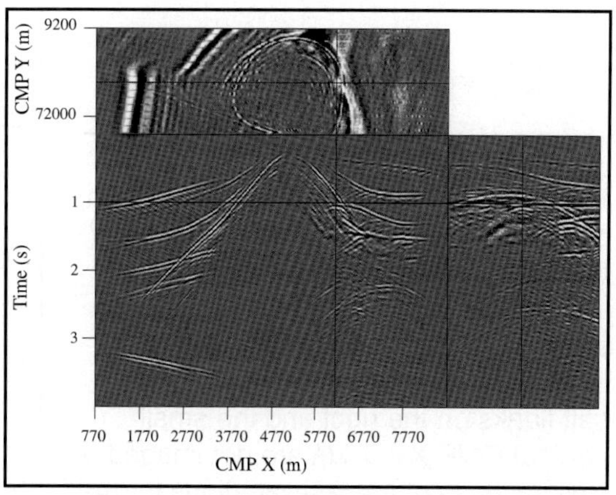

Figure 9. Zero-offset cube obtained by NMO + DMO + stack. The reflections from the shallow parts of the steeply dipping salt flanks have stacked properly, regardless of their azimuthal orientation. However, the deeper reflections on the right of the salt crest (below 1.5 s) are affected by the NMO-velocity conflict discussed in the text.

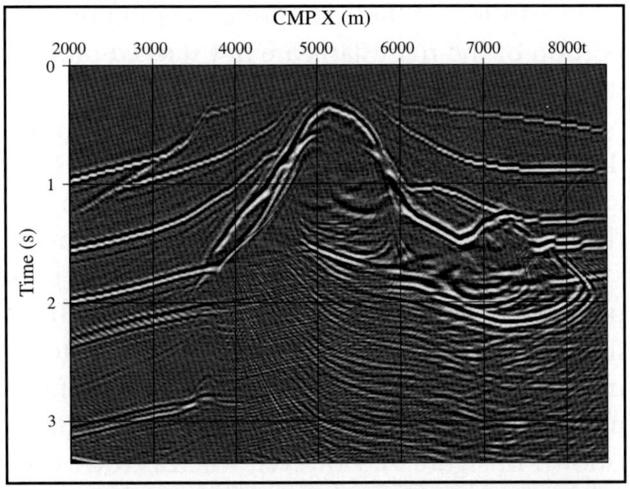

Figure 10. Inline section (CMP Y = 8 km) obtained by full prestack time migration.

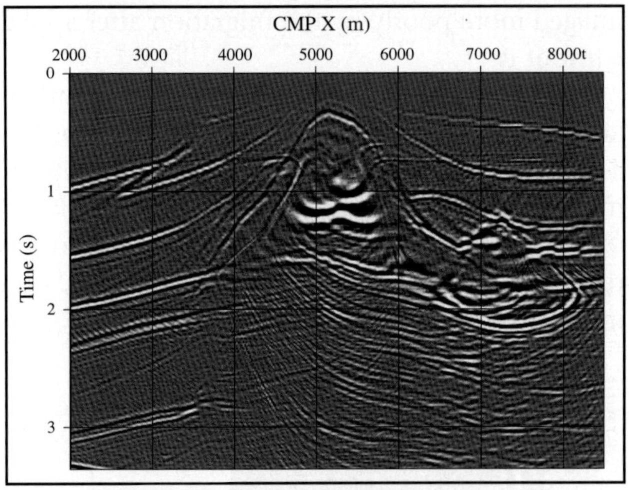

Figure 11. Inline section (CMP Y = 8 km) obtained by zero-offset time migration of the data cube obtained by NMO + stack (shown in Figure 8). The steeply dipping salt flanks are imaged poorly.

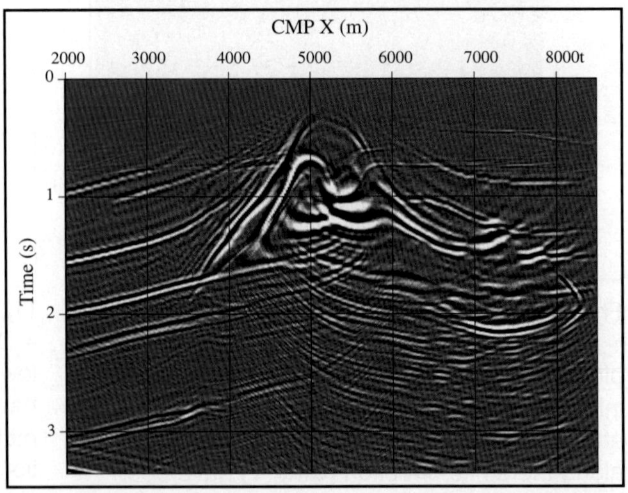

Figure 12. Inline section (CMP Y = 8 km) obtained by zero-offset time migration of the data cube obtained by NMO + DMO + stack (shown in Figure 9). The steeply dipping salt flank on the left is imaged better than in Figure 11. However, the salt flanks on the right and the small peak at CMP X = 7 km are not imaged well because of the fast variations in rms velocity associated with the salt body.

Azimuth Moveout

Partial-prestack-migration operators are useful tools for transforming prestack data and thereby reducing computational cost and increasing the accuracy of seismic imaging. The DMO operator introduced in the section titled "Dip moveout" transforms prestack data to equivalent zero-offset data and thus improves the quality of the stack. Now I introduce a more general partial prestack migration called azimuth moveout (AMO). Azimuth moveout is more general than DMO is, in the sense that AMO transforms prestack data into equivalent data with arbitrary offset and azimuth. In contrast, DMO can transform only nonzero-offset data into zero-offset data.

Azimuth moveout is a partial prestack migration that transforms 3D prestack data with a given offset and azimuth into equivalent data with a different offset and azimuth. The AMO operator is derived by collapsing into a single step the cascade of an imaging operator and its corresponding forward-modeling operator (Biondi et al., 1998). In principle, any 3D prestack imaging operator can be used for defining AMO. AMO has been derived both as a cascade of DMO and inverse DMO (Ronen, 1987; Canning and Gardner, 1992) and as the cascade of full 3D prestack constant-velocity migration and its inverse. Azimuth moveout is applied after NMO, thus accounting for velocity heterogeneities, at least to first order.

Azimuth moveout is not a single-trace-to-single-trace transformation; rather, it is a partial-prestack-migration operator that moves events across midpoints according to their dip. Its impulse response is a saddle in the time-midpoint domain. Figure 13 shows an example of the impulse response of the AMO operator.

To define the analytical expression of the AMO saddle, I analyze the schematic diagram of the AMO transformation in the midpoint plane, as displayed in Figure 14. An input trace with offset vector $\mathbf{h}_1$ and its midpoint at the origin is transformed into equivalent data with offset vector $\mathbf{h}_2$ and with its midpoint shifted by the vector $\Delta\mathbf{m} = (\Delta m_x, \Delta m_y) = \Delta m\,(\cos\Delta\phi, \sin\Delta\phi)$. Biondi (2006) presents the geometric derivation of the analytic expression for the kinematics of AMO transformation. The input time t_1 (after NMO) is related to the output time t_2 (before inverse NMO) as follows:

$$t_2 = t_1 \frac{h_1}{h_2} \sqrt{\frac{h_1^2 \sin^2(\theta_1 - \theta_2) - \Delta m^2 \sin^2(\theta_2 - \Delta\phi)}{h_2^2 \sin^2(\theta_1 - \theta_2) - \Delta m^2 \sin^2(\theta_1 - \Delta\phi)}}. \tag{4.9}$$

Equation 4.9 defines a skewed saddle on the $(\Delta m_x, \Delta m_y)$ plane; its shape and spatial extent are controlled by the input time t_1, by the absolute offsets h_1 and h_2, and by the azimuth rotation $\Delta\theta = \theta_1 - \theta_2$. Consistent with intuition, the spatial extent of the operator is largest when $\Delta\theta = 90°$, and it vanishes when both the azimuth rotation $\Delta\theta$ and the offset continuation $\Delta h = h_1 - h_2$ tend to zero. Furthermore, one can verify easily that $t_2 = t_1$ for the zero-dip components of the data; that is, the kinematics of zero-dip data after NMO do not depend on azimuth and offset.

The kinematic expression in equation 4.9 is velocity-independent, but the lateral aperture of the operator depends on velocity. An upper bound on the spatial extent of the AMO operator is defined by the region where the expression in equation 4.9 is valid. This region is delimited by the parallelogram with main diagonal $(\mathbf{h}_1 + \mathbf{h}_2)$ and minor

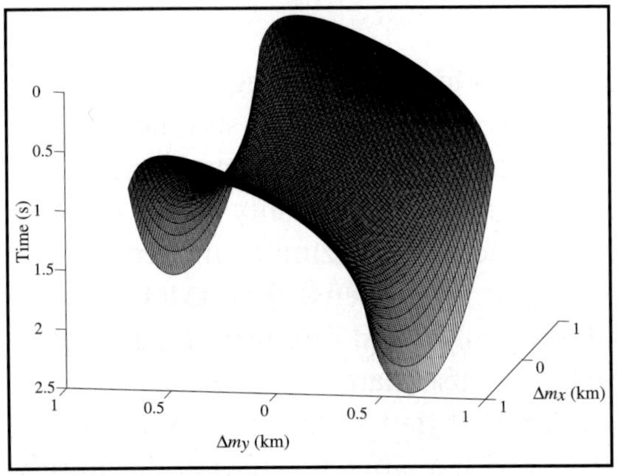

Figure 13. The full AMO impulse response ($V_{min} \simeq 0$) when t_1 = 1 s, h_1 = 2 km, h_2 = 1.8 km, θ_1 = 0°, and θ_2 = 30°.

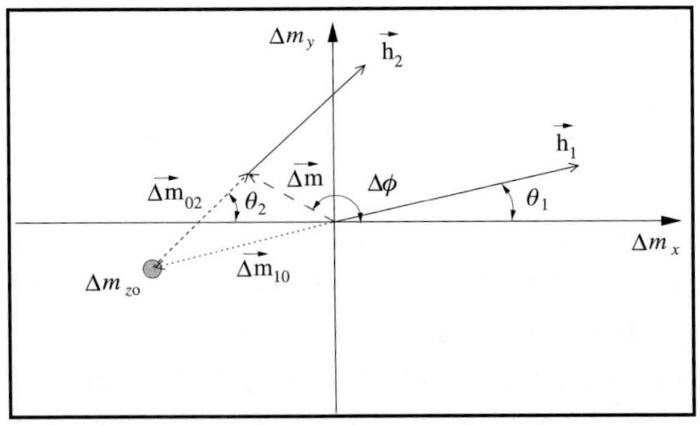

Figure 14. Schematic diagram of AMO transformation. An input trace with offset vector $\mathbf{h}_1$ and its midpoint at the origin is transformed into equivalent data with offset vector $\mathbf{h}_2$ and with its midpoint shifted by the difference vector $\Delta\mathbf{m}$.

diagonal ($\mathbf{h}_1 - \mathbf{h}_2$), as shown in Figure 15. The effective AMO aperture often is much narrower than the parallelogram and, for given $\mathbf{h}_1$ and $\mathbf{h}_2$, is a function of the minimum velocity V_{min} and of the input traveltime (Biondi et al., 1998).

Figure 16 shows the effective AMO impulse response when the velocity-dependent aperture limitation, corresponding to a realistic minimum velocity of 2 km/s, is applied to the impulse response shown in Figure 13. The surface shown in Figure 16 is significantly narrower than the whole impulse response shown in Figure 13. This velocity-dependent aperture limitation is important for efficient use of AMO, and it contributes to the cost-effectiveness of AMO compared with that of full 3D-prestack migration.

Notwithstanding the constant-velocity assumption underlying its derivation, AMO can be applied effectively to data recorded in areas where the velocity function is complex. The first-order effects of velocity variations are removed by NMO, which is applied before AMO. Azimuth moveout can transform data successfully to nearby offsets and azimuth when velocity variations are too strong for more global transformations, such as DMO, to transform data correctly. Because AMO is correct to the first order, its results are accurate if the amounts of azimuth rotation $\Delta\theta$ and offset continuation Δh are sufficiently small.

AMO application to coherent partial stacking

Azimuth moveout has several applications to 3D prestack processing. In this section, I discuss the use of AMO in conjunction with partial stacking to reduce the computational cost of 3D prestack depth imaging. As I did with our previous definitions, I will refer to the sequence of NMO followed by AMO and then by partial stacking as NMO + AMO + partial stack. By reducing the amount of data to be migrated, partial stacking reduces the cost of 3D prestack imaging, because the cost of migration is approximately proportional to the amount of data to be migrated (equation 2.12). However, for partial stacking to enhance reflections and suppress noise, reflections must be coherent across the traces to be stacked. Normal moveout increases the coherency of reflections over offset by a first-order correction of the traveltime. However, a simple

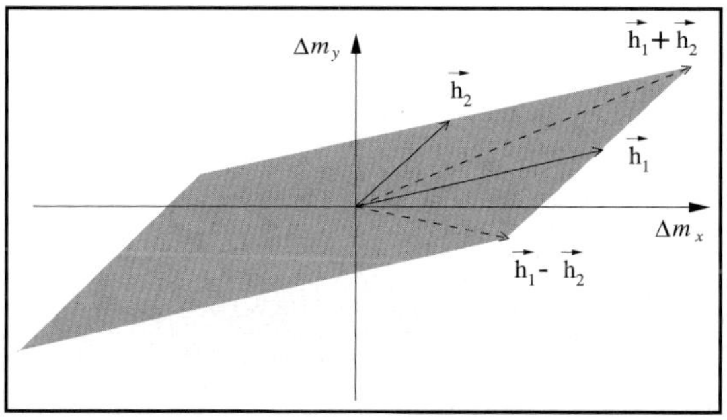

Figure 15. Maximum aperture of the AMO operator as a function of the input offset vector $\mathbf{h}_1$ and the output offset vector $\mathbf{h}_2$.

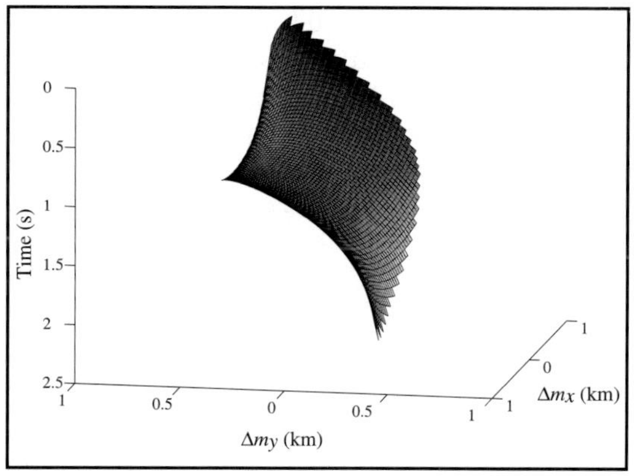

Figure 16. The effective AMO impulse response when V_{min} = 2 km/s, t_1 = 1 s, h_1 = 2 km, h_2 = 1.8 km, θ_1 = 0°, and θ_2 = 30°. This surface is a subset of the saddle shown in Figure 13.

trace-to-trace transformation such as NMO is insufficient when the reflections have conflicting dips or when diffractions occur. By correctly moving the dipping energy across midpoints, AMO ensures that all the dips in the data are preserved during partial stacking.

The following examples show AMO applied to the partial stacking of a data set recorded in the North Sea. To be imaged properly, the data require 3D prestack depth migration because of the existence of a shallow chalk layer with variable thickness and a deeper salt layer with a central upswelling (Hanson and Witney, 1995). The data-acquisition configuration was double source and triple streamer. The nominal common-midpoint spacing was 9.375 m in the inline direction and 25 m in the crossline direction. The cable length was 2.2 km, with maximum feathering of approximately 17°. Figure 17 shows the offset-azimuth distribution of a small subset of the data traces. As is typical for such acquisition geometry, the offset-azimuth distribution is characterized by six distinct trends that are most distinguishable at small offsets and that correspond to each possible source-streamer pair.

Two distinct partial-stacking methods have been applied to six different subsets of the data — NMO followed by partial stacking (NMO + partial stack) and NMO followed by AMO and partial stacking (NMO + AMO + partial stack). The subsets were determined according to the absolute value of the offset. Starting from zero, each offset range was 0.4 km wide. The boundaries between the offset ranges are shown as vertical bars in Figure 17. For each offset range, the output data were a regularly sampled cube, with nominal offset equal to the midpoint of the range; that is, 0.2, 0.6, 1.0, 1.4, 1.8, and 2.2 km. The data-reduction rate achieved by partial stacking is approximately seven. In general, the longer the offset and the larger the azimuthal rotation are, the more significant the effect of AMO is on the data. For the geometry of this data set, the most significant effects are visible for the longer offset ranges, starting from the 0.8- to 1.2-km range.

Figure 18 compares two time slices obtained with the two flows described above, for the 0.8- to 1.2-km-offset range. Figure 18a shows the slice obtained by simple NMO

Figure 17. Offset-azimuth distribution of the North Sea data set. The vertical bars show the boundaries among the offset ranges that were used for partial stacking.

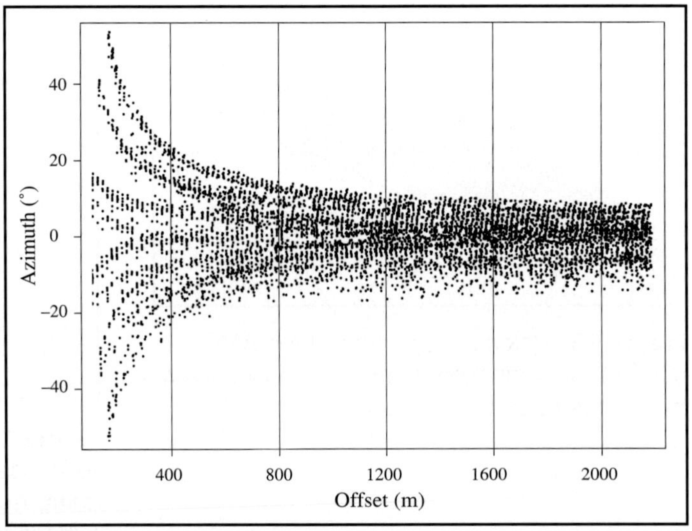

+ partial stack, whereas Figure 18b shows the results of NMO + AMO + partial stack. The difference plot (Figure 18c) clearly shows that most trends of high-frequency diffractions were not preserved correctly by the conventional process. The most evident differences tend to occur for reflections that are perpendicular to the inline direction. This observation is consistent with the fact that the conventional NMO-stacking process is most inaccurate for reflections with dips that are aligned with the nominal azimuth, and thus those reflections appear orthogonal to the inline direction in time slices.

Figure 19 shows windows of an inline section for the 1.2- to 1.6-km-offset range. This inline section is located at 20.940 km and is centered around reflections from the left side of the salt swell. The dipping-salt-flank reflection is preserved better by the application of AMO. Further, as was the case for the shallower section, some mildly dipping reflections appear to be cleaner after AMO. A possible explanation for this phenomenon is that the incoherent stacking of the diffractions and of other steeply dipping reflections contributes to the general level of background noise in the data obtained by simple NMO stacking.

AMO transformation to common-azimuth data

Another useful application of AMO to 3D-prestack imaging is the transformation of marine data into equivalent common-azimuth data. Common-azimuth data can be depth-imaged effectively with efficient algorithms, such as the common-azimuth migration described in Chapter 5. To transform marine data into common-azimuth data, AMO performs a simple azimuth rotation. In contrast, both an azimuth rotation and an offset continuation are required when one is reducing the data size by partial stacking, as described previously in the section on AMO application in coherent partial stacking. Therefore, the effects of including AMO in the common-azimuth transformation are less significant than the effects of including AMO in the partial-stacking process described previously. The width of the azimuth range in the data geometry is an important factor in determining whether the inclusion of AMO is required for accurate imaging. In general, the more streamers we use for the marine acquisition, the more impact AMO has on the accuracy of common-azimuth imaging. As for partial stacking, the standard alterna-

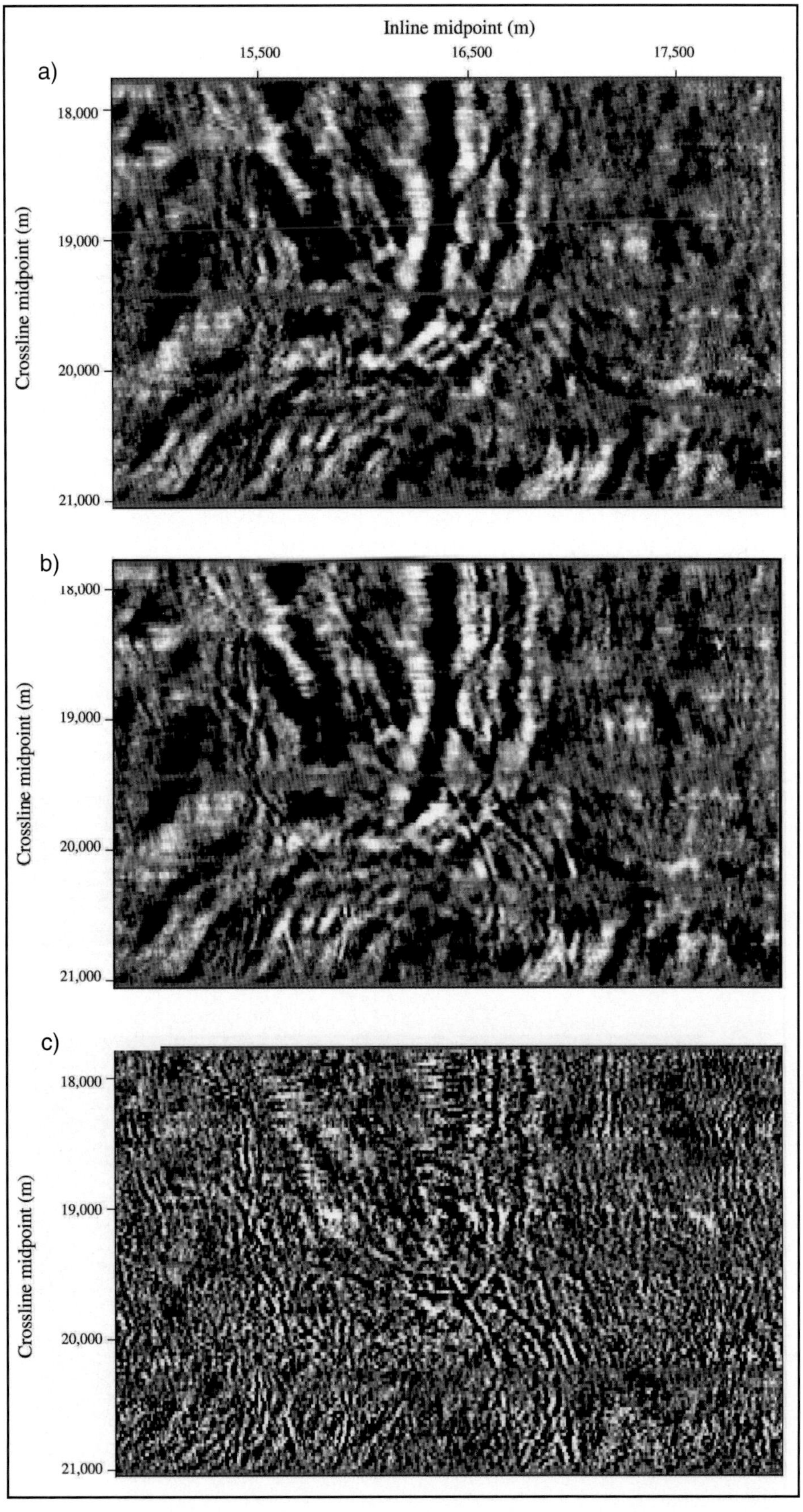

Figure 18. Time slices (1.068 s) for the 0.8- to 1.2-km-offset range, obtained by (a) NMO-stacking, (b) NMO-AMO-stacking, and (c) subtracting (a) from (b).

Figure 19. Inline sections (CMP Y = 20.940 km) for the 1.2- to 1.6-km-offset range, obtained by (a) NMO stacking; (b) NMO-AMO stacking; and (c) subtracting (a) from (b).

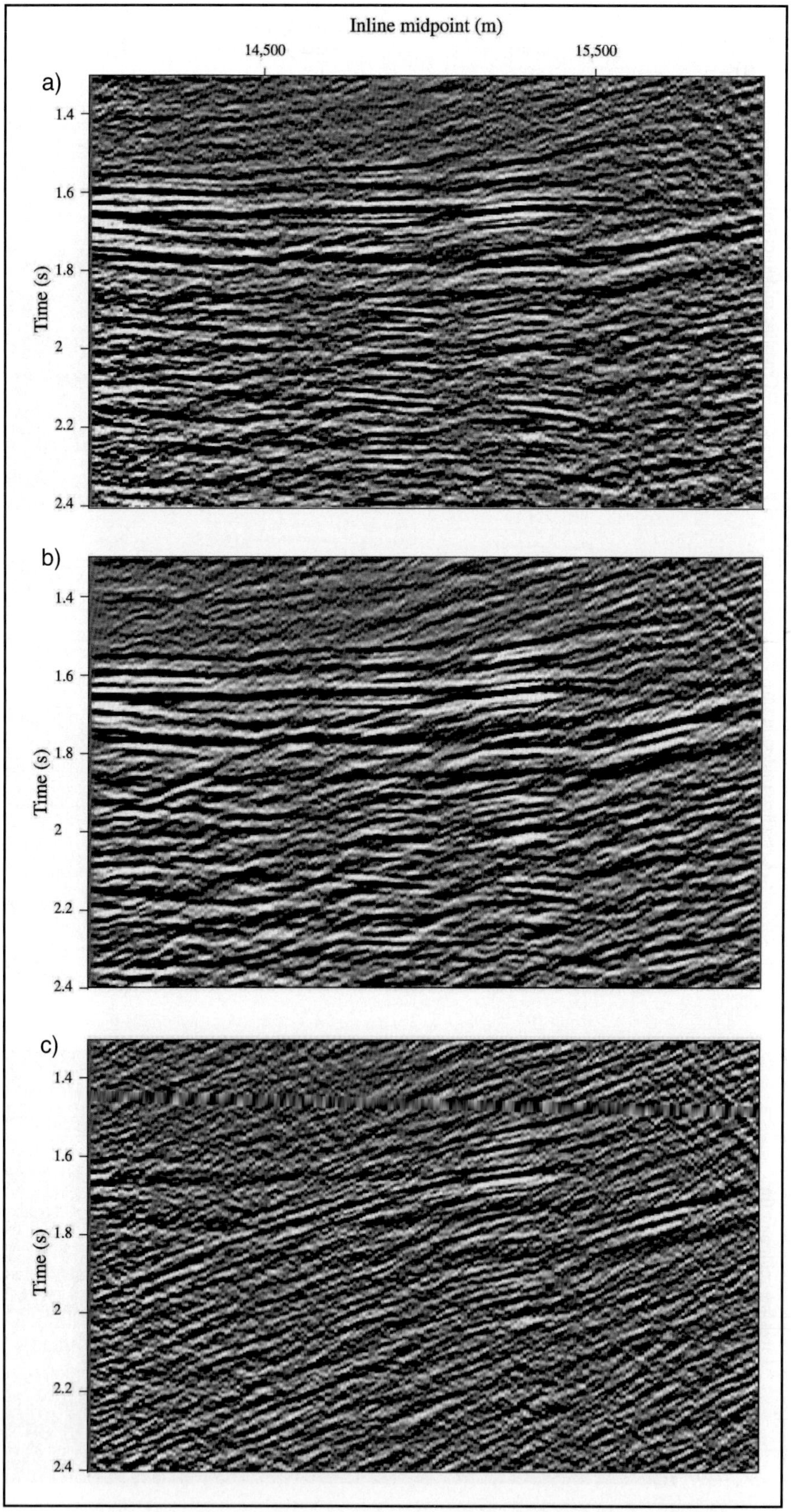

tive to AMO for transforming prestack data to common-azimuth data is application of residual NMO followed by midpoint interpolation.

Figures 20 and 21 show the effects of including AMO in the common-azimuth imaging procedure. The figures show sections extracted from image cubes obtained by the common-azimuth depth migration described in Chapter 5. The sections on the left were obtained with AMO in the processing sequence; the ones on the right resulted from transforming the data into common-azimuth data using residual NMO followed by midpoint interpolation. Figure 20 compares depth slices. The fault-plane reflections indicated by the arrows are imaged better in the results obtained with AMO. These fault-plane reflections correspond to the diffractions shown in Figure 18, which were preserved better when AMO was used to partially stack the data. Figure 21 compares inline sections. The image obtained with AMO better preserves the salt-flank reflection (arrow A) and the reflection from a dolomite rafting inside the salt layer (arrow B). In general, inclusion of AMO improves the accuracy of common-azimuth depth imaging, even for data sets with a relatively small azimuth range.

Partial prestack migration applied to 3D prediction of multiples

The presence of strong multiple reflections is a common challenge to seismic imaging in many regions of the world. In marine data, surface-related multiples often are the most harmful because of both their strength and their timing relative to the desired imaging targets. The potentially most accurate method for predicting, and thus for attenuating, multiple reflections in complex areas is the surface-related-multiple-elimination (SRME) method initially developed in 2D by Verschuur et al. (1992).

Unfortunately, application of full 3D SRME to data acquired with conventional marine-streamer acquisition geometry is hampered by the limited crossline-offset range afforded by the streamer array and by the sparseness of the shot grid usually employed in such surveys (see Chapter 1 for a discussion on marine-streamer acquisition). In theory, to predict multiple reflections accurately, SRME requires dense areal arrays of both sources and receivers. In practice, marine-data geometries do not fulfill these requirements and when only the data that are actually recorded are used as input to SRME (2.5D SRME), multiple reflections are only partially suppressed. The areas that are most challenging for 2.5D SRME are the ones in which the interfaces generating the most-energetic multiple reflections (e.g., the water bottom or top of a salt body) have strong crossline dip components. Figure 22 shows the water-bottom map of one such area. The crossline undulations indicated by the red arrows create "true" 3D multiple reflections that are reflected at the water surfaces outside of the receiver arrays formed by the recording streamers.

The AMO operator's ability to synthesize data that have a geometry that differs from the actual recording geometry has been exploited recently to augment the recorded data (by both extrapolation and interpolation) and thus to enable application of fully 3D SRME methods (Matson et al., 2004; Matson and Abma, 2005). Baumstein and Hadidi (2006) followed a similar strategy, but, for computational and data-management reasons, they directly used forward shot-profile DMO (Biondi and Ronen, 1987) followed by inverse shot-profile DMO (Baumstein, 2004). As an example of this increasingly important application of AMO and other partial-prestack-migration operators, I will discuss some of the results presented by Baumstein and Hadidi (2006).

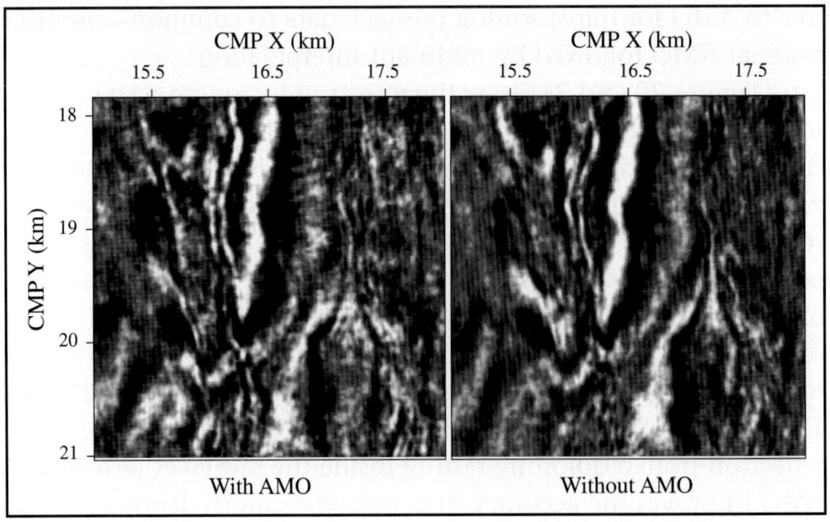

Figure 20. Depth slices extracted from two cubes imaged by common-azimuth migration, including (left) and excluding (right) AMO.

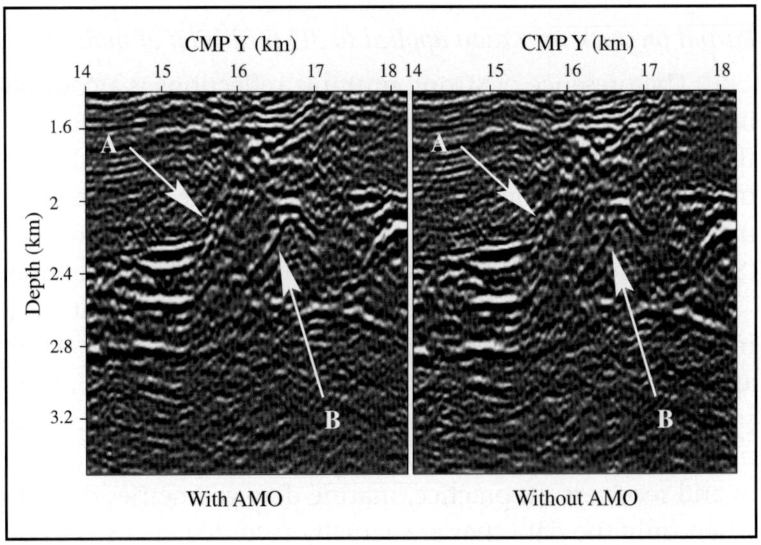

Figure 21. Inline sections extracted from two cubes imaged by common-azimuth migration, including (left) and excluding (right) AMO.

Figure 23 shows post-NMO field data recorded in the area above the water bottom displayed in Figure 22. The acquisition geometry employed eight streamers each 5 km long with a dual-source configuration. Figure 23 shows the data belonging to the same shot profile; the traces recorded by the far streamer are on the left panel, and the traces recorded by the near streamer are on the right panel. For comparison, Figure 24 shows the data for the same two streamers as the data shown in Figure 23, here reconstructed, by the DMO + inverse DMO procedure. The primaries indicated by the solid arrows have been well reconstructed by the procedure, whereas the multiples indicated by the dashed arrows are only partially reconstructed. The poor reconstruction of the multiples is expected because their NMO velocity is substantially lower than the NMO velocity of the primaries recorded at the same zero-offset time. Fortunately, SRME uses the primary events to predict the first-order multiples and uses the first-order multiples to predict the second-order multiples. Therefore, accurate reconstruction of the primaries ensures an accurate prediction of the first-order multiples.

Figure 25 shows a time slice (t_N = 3.5 s) of the reconstructed data with the recorded data overlaid on them. The recorded data are displayed with higher amplitudes and thus they are easily distinguishable from the reconstructed data. Notice that the imprint of the eight streamers is slanted because of cable feathering. The solid arrows point to parts of the data that are correctly reconstructed, whereas the dashed arrows point to areas

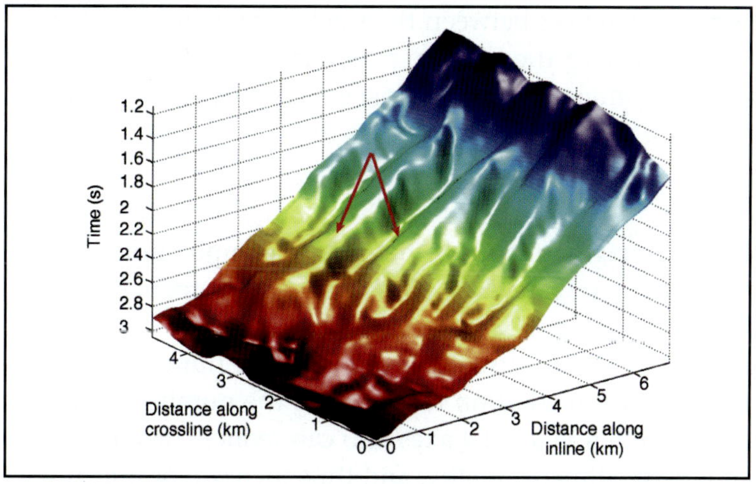

Figure 22. Water-bottom map of the area where the field data shown in Figure 23 were acquired. The primary cause of 3D multiples is the crossline undulation of the water bottom, which is indicated with red arrows. Figure is from Baumstein and Hadidi (2006).

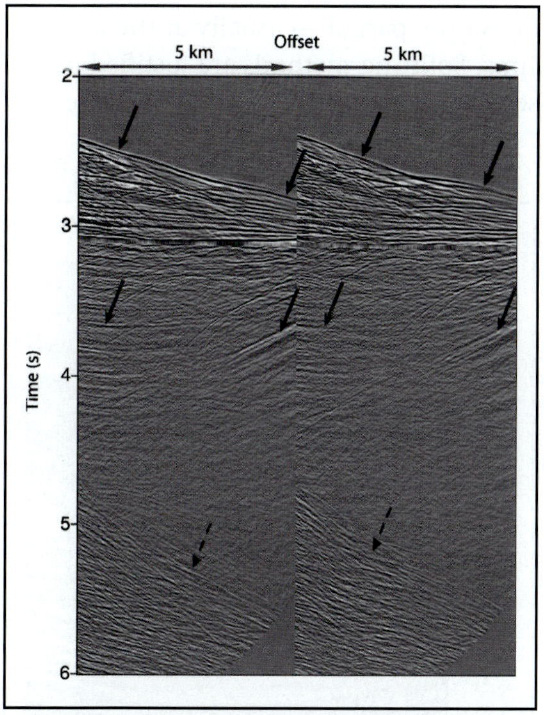

Figure 23. Field data (after NMO) acquired with an eight-streamer acquisition geometry. The far streamer is the left panel and the near streamer is the right panel. Figure is from Baumstein and Hadidi (2006).

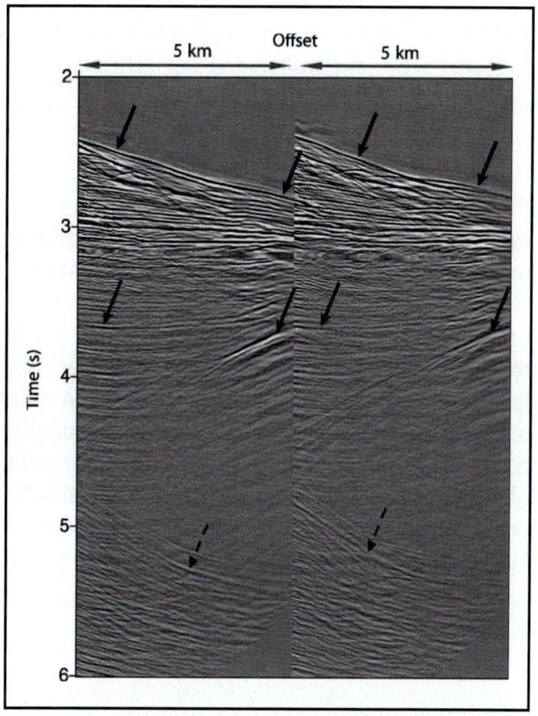

Figure 24. Reconstruction of the field data shown in Figure 23. The solid lines point to the primaries that are well reconstructed by the method, whereas the dashed arrows point to multiples that are poorly reconstructed. Figure is from Baumstein and Hadidi (2006).

where the recorded data and the reconstructed data are slightly shifted with respect to each other. This timing error is larger at far offset because NMO-velocity errors have larger effects on far-offset data.

Figure 26 compares the final results of applying 2.5D SRME with the results of 3D SRME using the reconstructed data. The top panels show the stacks after 2.5D SRME (a) and 3D SRME (c). The multiples have been better attenuated by 3D SRME than by 2.5D SRME, although some multiple-reflection energy is still present even in the stack after 3D SRME. The bottom panels show the differences between the stacked data without any SRME applied and the stacked data shown in the two top panels, after SRME. The difference section after 3D SRME (panel d) contains more multiple energy than does the section after simple 2.5D SRME (panel b).

Two-pass 3D Prestack Migration

The previous sections presented imaging methods that reduce computational cost by decreasing the number of offsets to be migrated. NMO + DMO + stack reduces the number of offsets to one (zero offset), whereas NMO + AMO + partial stack reduces the number of offsets to an arbitrary number that is usually lower than the number of offsets in the full-prestack data set. Instead of (or in addition to) reducing the number of offsets, separating the computation between different spatial axes also can reduce the cost of 3D imaging. This separation reduces the dimensionality, and thus the cost, of imaging operators. Full-separation methods completely decouple the computation along different spatial axes. In splitting methods, decoupling is only partial — usually at the level of a depth step in downward-continuation algorithms. Splitting methods are justified only on the grounds of computational savings. Full-separation methods also may increase the

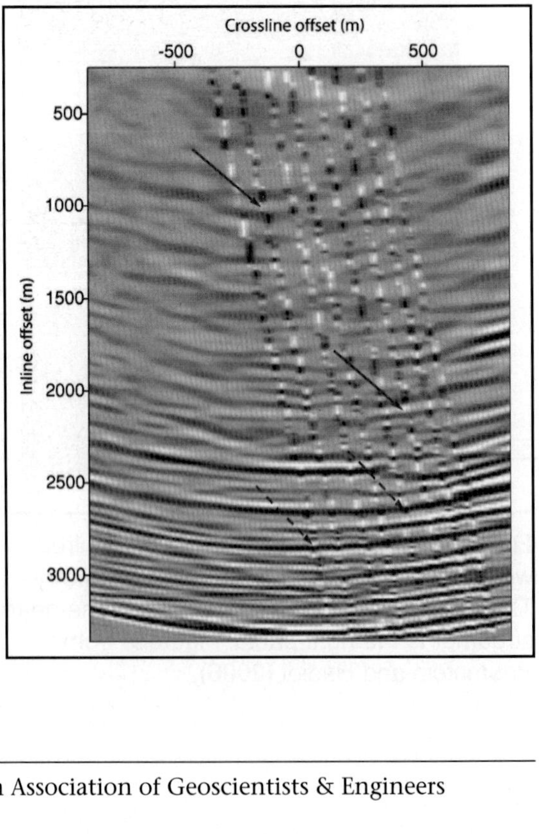

Figure 25. Time slice (at $t_N = 3.5$ s) of the reconstructed data with the field data overlaid on them. The imprint of the eight streamers is slanted because of cable feathering. The solid arrows point to parts of the data that are correctly reconstructed, whereas the dashed arrows point to areas where the recorded data and the reconstructed data are slightly shifted with respect to each other. These timing errors are larger at far offset because NMO errors have larger effects on far-offset data. Figure is from Baumstein and Hadidi (2006).

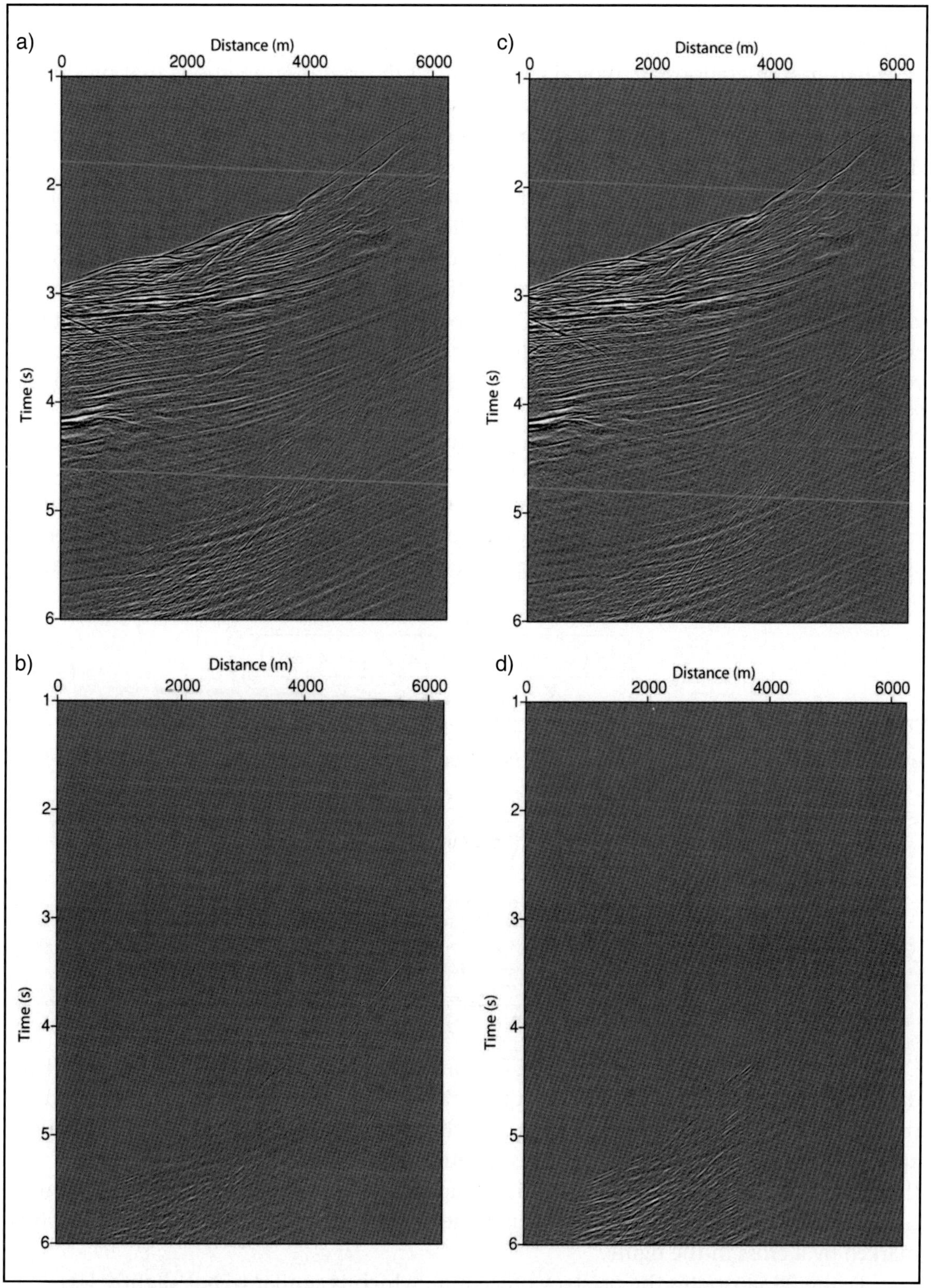

Figure 26. SRME results: (a) a section cut through the 2.5D SRME stack, (b) the difference between input data and the 2.5D SRME stack, (c) a section cut through the 3D SRME stack, and (d) the difference between input data and the 3D SRME stack. Figure is from Baumstein and Hadidi (2006).

robustness of the imaging procedure, because partially migrated results are more easily interpretable than the original data are.

Two-pass migrations are typical examples of full-separation methods that produce simplified intermediate results. The first pass reduces the data to an ensemble of equivalent parallel 2D lines that then can be migrated and interpreted separately. Two possible alternative strategies exist for two-pass 3D prestack migration. The simplest, and the one I discuss in this section, is to perform 2D prestack migration along the inline axis, followed by poststack migration along the crossline axis. Canning and Gardner (1993) and Devaux et al. (1996) proposed to invert the order of the inline and crossline migrations. That is, they proposed to perform first a modified prestack crossline migration and to follow that by standard inline migration. The advantage of this alternative is that the first crossline migration is fairly insensitive to migration velocity, and thus the problem of 3D velocity analysis is reduced to an ensemble of independent 2D velocity-analysis problems for parallel lines.

Constant-velocity prestack migration can be separated exactly into a cascade of prestack migration along the inline axis followed by poststack migration along the crossline axis. The analytical proof of full separation is immediate both for summation surfaces and for spreading surfaces. If, for simplicity, the data azimuth is assumed to be aligned with the inline axis x, it is straightforward to verify that the cascade of prestack migration in the inline direction

$$t_D = \frac{\sqrt{z_{\bar{x}}^2 + \left(x_\xi - x_m + x_h\right)^2}}{V} + \frac{\sqrt{z_{\bar{x}}^2 + \left(x_\xi - x_m - x_h\right)^2}}{V}, \tag{4.10}$$

with zero-offset migration in the crossline direction

$$z_{\bar{x}} = \sqrt{z_\xi^2 + \left(y_\xi - y_m\right)^2}, \tag{4.11}$$

yields the expression of full prestack migration:

$$t_D = \frac{\sqrt{z_\xi^2 + \left(x_\xi - x_m + x_h\right)^2 + \left(y_\xi - y_m\right)^2}}{V} + \frac{\sqrt{z_\xi^2 + \left(x_\xi - x_m - x_h\right)^2 + \left(y_\xi - y_m\right)^2}}{V}. \tag{4.12}$$

Figure 27 illustrates this procedure for the special case of zero-offset migration. The data first are summed along the parallel hyperbolas aligned along the inline direction (thinner lines in the figure). The results of each of these partial migrations are set at the apexes of the hyperbolas. The crossline migration hyperbola, which is drawn with a thicker line in the figure, passes through the apexes of all the inline hyperbolas. Summation along this crossline hyperbola focuses the data at the correct image point, which is marked by a cross in the figure.

Figure 28 illustrates the prestack procedure, which is similar to but slightly less obvious than the zero-offset one. In this case, results of the summations along the parallel flat-topped hyperbolas (thinner lines in the figure) are set at focal points above the

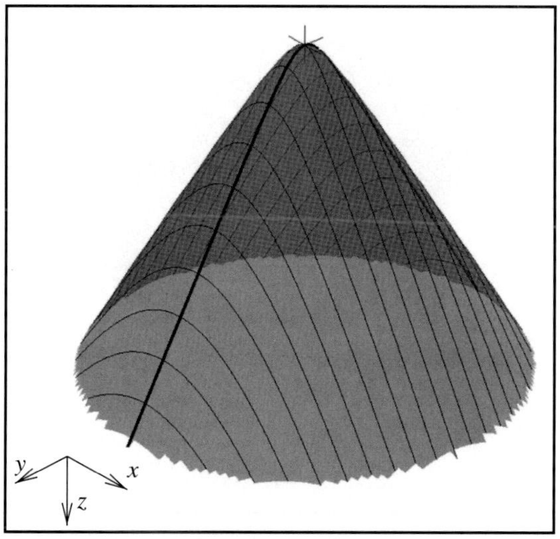

Figure 27. Full separation of the summation surface for zero-offset migration.

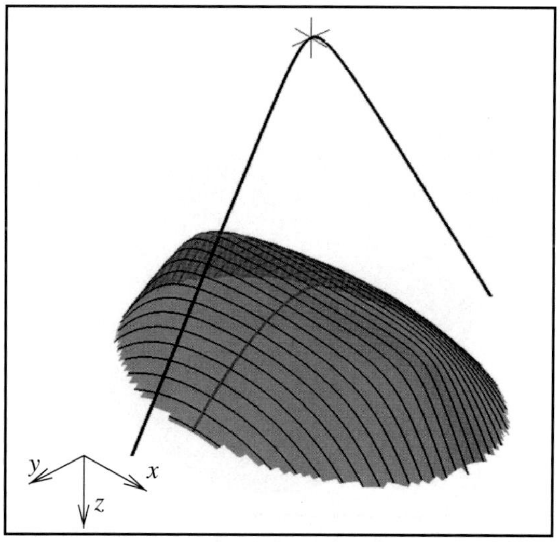

Figure 28. Full separation of the summation surface for constant-offset migration.

summation curves themselves. These focal points belong to the same crossline hyperbola as in the zero-offset case, and their energy is focused at the correct image location by use of crossline migration.

Derivation of the dual result — that is, proof that prestack migration by spreading also fully separates — is even simpler than the previous one. First we spread energy along the inline ellipse,

$$\frac{4\left(x_\xi - x_m\right)^2}{t_D^2 V^2} + \frac{4 z_{\bar{x}}^2}{t_D^2 V^2 - 4x_h^2} = 1 , \tag{4.13}$$

and then we spread energy along the crossline circles,

$$\frac{\left(y_\xi - y_m\right)^2}{z_{\bar{x}}^2} + \frac{z_\xi^2}{z_{\bar{x}}^2} = 1 , \tag{4.14}$$

to obtain the ellipsoid of full prestack migration

$$\frac{4\left(x_\xi - x_m\right)^2}{t_D^2 V^2} + \frac{4\left(y_\xi - y_m\right)^2}{t_D^2 V^2 - 4x_h^2} + \frac{4 z_\xi^2}{t_D^2 V^2 - 4x_h^2} = 1 . \tag{4.15}$$

Figure 29 illustrates the two-pass spreading of an impulse over the zero-offset hemisphere. First the impulse (a cross in the figure) is spread along the inline circle (thick line in the figure). Second, each point on the circle is spread along parallel circles (thin lines in the figure) oriented in the crossline direction.

The prestack case is similar to the poststack case. Figure 30 illustrates the two-pass spreading of an impulse over the prestack ellipsoid. First the impulse (a cross in the fig-

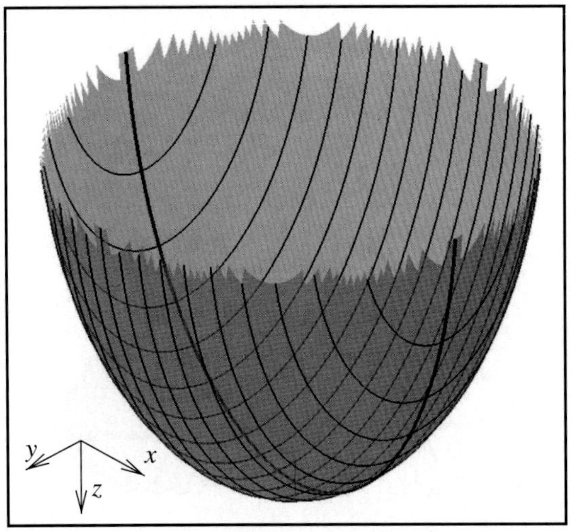

Figure 29. Full separation of the spreading surface for zero-offset migration.

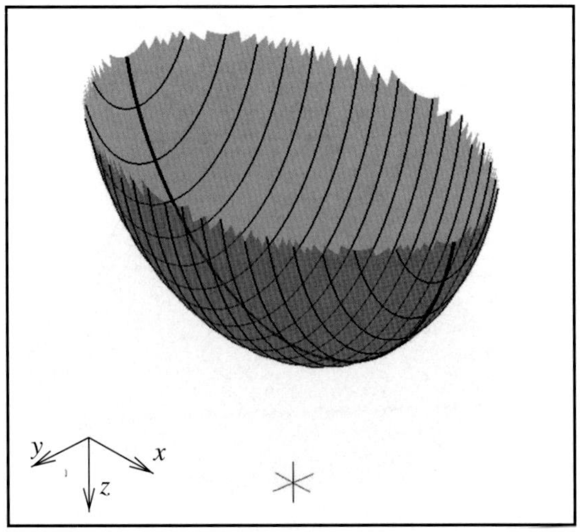

Figure 30. Full separation of the spreading surface for constant-offset migration.

ure) is spread along the inline ellipse (thick line in the figure). Second, each point on the ellipse is spread along parallel circles (thin lines in the figure) oriented in the crossline direction.

Applying the cascade of equation 4.13 with equation 4.14 or, equivalently, the cascade of equation 4.10 with equation 4.11, is less intensive computationally than applying the full-migration operators (respectively, equation 4.15 and equation 4.12), because a cascade of two summations over 2D curves replaces one summation over a 3D surface. Furthermore, the crossline migration is only a zero-offset migration.

Two-pass migrations are computationally convenient, not only because of the low flop count shown in the previous equation but also because they allow access to the data a single line at a time (inline or crossline). When the size of the whole data set is large compared with the computational resources available, this favorable data-access pattern can be as strong an incentive for the use of two-pass migrations as the low flop count is.

The considerable reduction in cost achieved by full separation justifies the popularity it enjoyed for zero-offset migration at the beginning of the 3D seismic era (Gibson et al., 1983). However, two-pass migrations are inaccurate when the velocity is not constant, even if it varies only with depth. The errors are larger at steeper dips and for strong velocity gradients (Dickinson, 1988). Further, the generalization of two-pass methods to true depth-migration problems is arduous. The next section presents a geometric interpretation of two-pass migration, which provides an intuitive insight into its limitations in accuracy. Nowadays, accurate and reasonably inexpensive alternatives to two-pass zero-offset migrations are available.

Geometric interpretation of two-pass migration

Two-pass prestack migration in a constant-velocity medium has a simple geometric interpretation, which is illustrated by Figure 31. Because the velocity is constant and no ray bending occurs, a reflection recorded at the surface with a source location **s** and receiver location **g** must propagate along one of the planes passing though the shot and

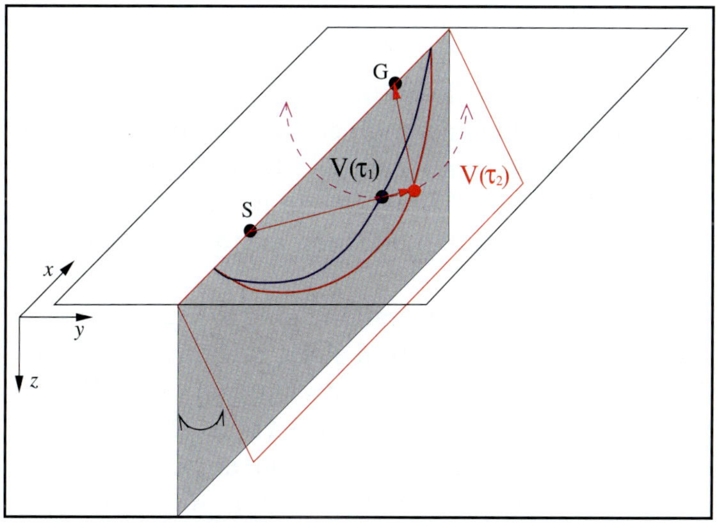

Figure 31. Geometric interpretation of two-pass migration.

receiver locations. Generally, this propagation plane is oblique. Migration should spread reflections along ellipses belonging to their respective propagation planes. The spreading can be achieved in two steps — first by spreading along ellipses lying on the vertical plane (i.e., inline prestack migration), and then by rotating the vertical ellipse until it is coplanar with the actual propagation plane (i.e., crossline zero-offset migration). When the velocity is constant, the two-step procedure is perfectly equivalent to a single-step procedure. The shape of the prestack ellipse is independent of the dip of the actual propagation plane. Thus, the arbitrary choice of the vertical plane for the first step is inconsequential.

However, when the velocity is not constant, the shape of the prestack ellipse depends on the dip of the actual propagation plane. Consequently, any single choice cannot be accurate for all the reflections. Even in the simplest case of time migration with velocity varying only as a function of depth, the V_{rms} of a reflection propagating along the vertical plane can be significantly different from the V_{rms} of a reflection propagating along a steep oblique plane. For example, if the velocity increases monotonically with depth, the rms velocity $V(\tau_1)$ measured along the vertical plane is always larger than the rms velocity $V(\tau_2)$ measured along the oblique plane shown in the figure. The steeper the propagation plane is, the larger the error is.

Such ambiguity in the shape of the first inline migration ellipse makes it difficult to generalize full-separation methods to depth migration. However, the basic idea that operators can be separated along the inline and the crossline directions is useful when applied to depth-migration methods that are based on downward continuation. For such methods, splitting is applied at the level of each downward-continuation step. Chapter 5 in Biondi (2006) discusses splitting methods for downward-continuation migration.

Two-pass migration of the SEG-EAGE salt data set

The analysis of two-pass migration presented in this chapter and the geometric interpretation of its inaccuracies are illustrated by the following example obtained by time-migrating the SEG-EAGE salt data set.

Figure 32 shows an inline section of full 3D prestack time migration of the first 1.8-km offsets of the salt data set. All the reflectors above the salt, including the dipping faults oriented at 45° to the acquisition direction, are imaged well.

Figure 33 shows the result of 2D prestack time migration along the inline direction, for the same crossline location as the section shown in Figure 32. The salt flank on the left of the section is imaged well because it is oriented along the inline direction. However, the dipping faults oriented at 45° to the acquisition direction are mispositioned grossly.

Figure 34 shows the result of 2D zero-offset time migration along the crossline direction of the cube obtained by 2D prestack time migration along the inline direction. In other words, Figure 34 shows the results of 3D two-pass prestack migration of the data set, according to the two-pass migration theory developed in this section. The dipping faults oriented at 45° to the acquisition direction are imaged well and are positioned very close to the correct location. However, careful comparison of the section shown in Figure 34 (approximate) with the section shown in Figure 32 (exact) shows that the dipping fault is slightly overmigrated and thus mispositioned. The overmigration is caused by the effects of the increase in V_{rms} with depth, as predicted by the geometric interpretation above.

Figure 35 shows the time slice cut at 1.3 s through the image obtained by full 3D prestack migration, corresponding to the migration result shown in Figure 32. Figure 36 shows the same time slice as in Figure 35 but corresponding to the 2D inline migration shown in Figure 33. Figure 37 shows the same time slice as in Figures 35 and 36 but corresponding to the two-pass 3D migration results. Notice the gross mispositioning of the dipping fault in the 2D inline migration results and the slight overmigration in the two-pass 3D migration results.

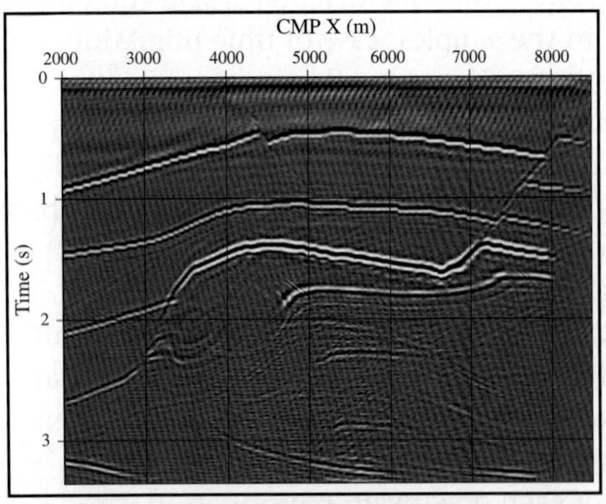

Figure 32. Full 3D prestack time migration of the salt data set. All the reflectors above the salt, including the dipping faults oriented at 45° to the acquisition direction, are imaged well.

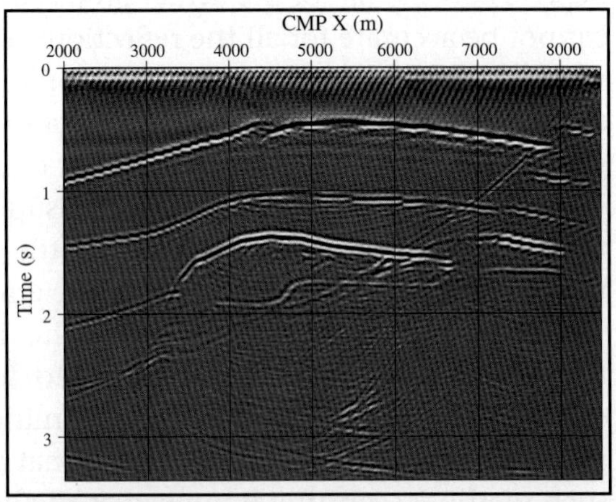

Figure 33. Inline 2D prestack time migration for the same crossline location as the section shown in Figure 32. The dipping faults oriented at 45° to the acquisition direction are grossly mispositioned.

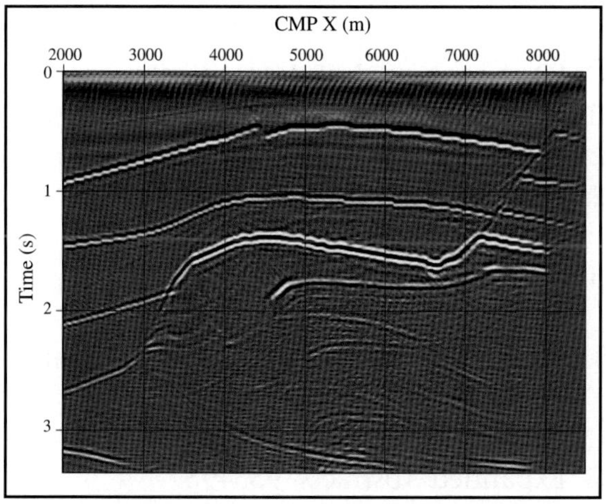

Figure 34. Two-pass 3D prestack time migration for the same crossline location as the sections shown in Figures 32 and 33. Notice the slight overmigration of the dipping fault.

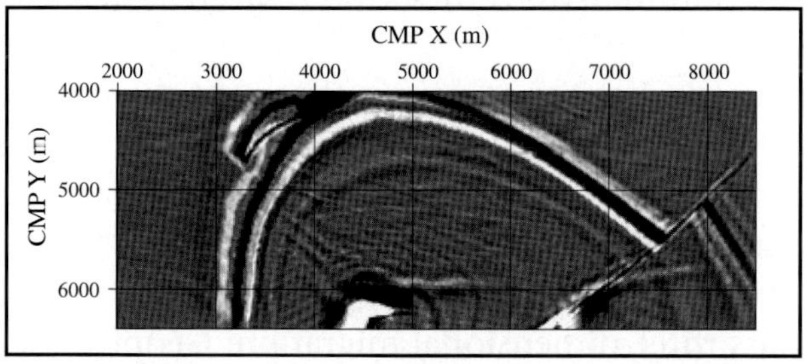

Figure 35. Time slice cut at 1.3 s through the image obtained by full 3D prestack migration. This time slice corresponds to the full 3D prestack migration shown in Figure 32.

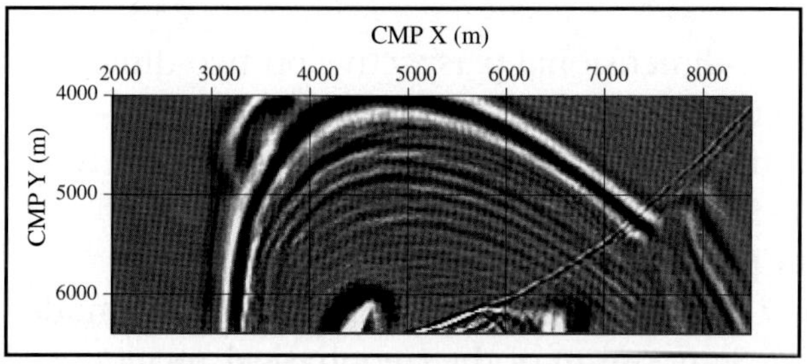

Figure 36. Time slice cut at 1.3 s through the image obtained by 2D inline migrations. This time slice corresponds to the section shown in Figure 33. Notice the gross mispositioning of the dipping fault.

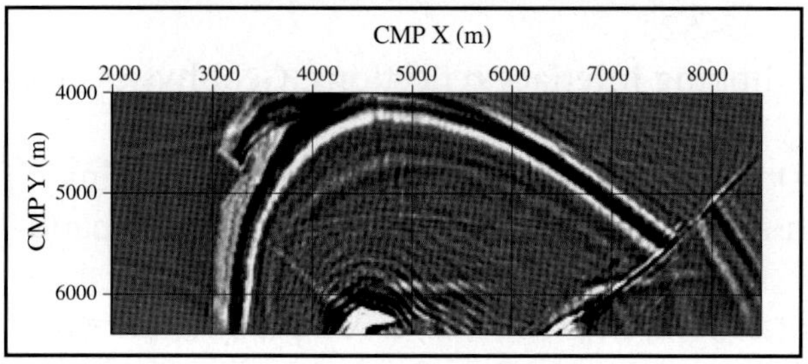

Figure 37. Time slice cut at 1.3 s through the image obtained by two-pass 3D migration. This time slice corresponds to the section shown in Figure 34. Notice the slight overmigration of the dipping fault.

References

Baumstein, A., 2004, 3D inverse shot record DMO: A tool for shot-domain data regularization: 74th Annual International Meeting, SEG, Expanded Abstracts, 2001–2004.

Baumstein, A., and M. T. Hadidi, 2006, 3D surface-related multiple attenuation: Data reconstruction and application to field data: Geophysics, **71,** E25–E33.

Biondi, B., 2006, 3D seismic imaging: SEG Investigations in Geophysics, No. 14.

Biondi, B., and J. Ronen, 1987, Dip moveout in shot profiles: Geophysics, **52**, 1473–1482.

Biondi, B., S. Fomel, and N. Chemingui, 1998, Azimuth moveout for 3-D prestack imaging: Geophysics, **63**, 574–588.

Canning, A., and G. H. F. Gardner, 1992, Feathering correction for 3-D marine data: 62nd Annual International Meeting, SEG, Expanded Abstracts, 955–957.

——, 1993, Two-pass 3D pre-stack depth migration: 63rd Annual International Meeting, SEG, Expanded Abstracts, 892–894.

Deregowski, S. M., and F. Rocca, 1981, Geometrical optics and wave theory of constant offset sections in layered media: Geophysical Prospecting, **29**, 374–406.

Devaux, V., G. H. F. Gardner, and T. Rampersad, 1996, 3-D prestack depth migration by Kirchhoff operator splitting: 66th Annual International Meeting, SEG, Expanded Abstracts, 455– 458.

Dickinson, J. A., 1988, Evaluation of two-pass three-dimensional migration: Geophysics, **53**, 32–49.

Forel, D., and G. H. F. Gardner, 1988, A three-dimensional perspective on two-dimensional dip moveout: Geophysics, **53**, 604–610.

Gibson, B., K. Larner, and S. Levin, 1983, Efficient three-dimensional migration in two steps: Geophysical Prospecting, **31**, 1–33.

Hale, D., 1991, Course notes: Dip moveout processing: SEG.

Hanson, D. W., and S. A. Witney, 1995, 3-D prestack depth migration — Velocity model building and case history: 1995 Spring Symposium of the Geophysical Society of Tulsa, SEG, Seismic Depth Estimation, 27–52.

Levin, F. K., 1971, Apparent velocity from dipping interface reflections: Geophysics, **36**, 510–516.

Matson, K. H., and R. Abma, 2005, Fast 3D surface-related multiple elimination using azimuth moveout for multiples: 75th Annual International Meeting, SEG, Expanded Abstracts, 2064–2067.

Matson, K., S. Michell, R. Abma, E. Shoshitaishvili, M. Williams, I. Ahmed, J. Oldroyd, and R. Fisher, 2004, Advanced subsalt imaging and 3D surface multiple attenuation in Atlantis: A case study: 74th Annual International Meeting, SEG, Expanded Abstracts, 1269–1272.

Ronen, J., 1987, Wave equation trace interpolation: Geophysics, **52**, 973–984.

Verschuur, D. J., A. J. Berkhout, and C. P. A. Wapenaar, 1992, Adaptive surface-related multiple elimination: Geophysics, **57**, 1166–1177.

Chapter 5 Advanced Methods for Prestack Depth Migration

Introduction

The choice of an appropriate migration method is crucial to the success of any imaging project. In Chapters 2 and 3 I introduced the basic concepts underlying the two main types of migration methods; Kirchhoff migration and wavefield-continuation migration. In this chapter I discuss the actual algorithms that are implemented in several advanced depth-migration programs and discuss their respective strength and weaknesses. Because migration technology is evolving rapidly, the specific details of migration algorithms quickly become obsolete. However, the intrinsic properties and characteristics of the different migration methods do not change as quickly as their implementations do. Therefore, in my discussion I often overlook the algorithmic details in favor of a conceptual view of the methods.

The first section of this chapter is dedicated to Kirchhoff depth-migration methods. These methods are still the workhorse of the industry, although they are being slowly supplanted by the more accurate and robust wavefield-continuation methods.

In the second section, I briefly introduce Gaussian-beam migration methods. In many ways, Gaussian-beam methods are in between Kirchhoff methods and wavefield-continuation methods. Gaussian-beam migration provides an often-attractive compromise between the accuracy of wavefield methods and the computational efficiency of Kirchhoff migration. As is the case with Kirchhoff migration, Gaussian-beam migration relies on ray tracing to propagate wavefields. However, it is more accurate in handling multipathing wavefields than Kirchhoff migration is.

The last and most extensive section of this chapter is dedicated to wavefield-continuation migration methods because that is the area of migration technology that is progressing most rapidly and also the one presenting the broadest array of possible algorithmic choices. Straightforward implementation of the prototypical migration methods I introduced in Chapter 3 can be extremely expensive. Consequently, the most commonly used wavefield-continuation migration algorithms rely on insightful optimizations and approximations to improve their computational efficiency. It is important to analyze the assumptions underlying each approach and understand their impact on specific imaging projects.

Kirchhoff Depth-migration Methods

As I discussed in Chapter 2, the accuracy of Kirchhoff depth-migration methods depends strongly on how well the shape of the summation surfaces approximates the actual traveltime delays of the wavefields propagating in the subsurface. Thus, the choice of the method used to evaluate the summation surfaces is the most important factor that distinguishes the different implementations of Kirchhoff migration.

The most efficient method for computing traveltime in a 3D velocity model is the direct solution of the Eikonal equation by a finite-difference method (van Trier and Symes, 1991). However, finite-difference methods have the serious drawback that they can compute only one arrival, which in most cases is the first arrival. When multipathing occurs, single-valued traveltime surfaces might be a poor approximation of the actual ones. Furthermore, the first arrivals often are not the ones carrying the greatest part of the wavefield energy, and thus they are not the arrivals that should contribute the most to the migrated image.

Ray-tracing methods (Cerveny and Pšenčík, 1983) are more demanding computationally than finite-difference solvers are, but, at least in principle, they can follow all the branches of a propagating wavefield. However, implementation of a Kirchhoff migration that uses multivalued summation surfaces computed by ray tracing is an engineering challenge that is not often tackled by production-processing software. Therefore, even when ray tracing is used to compute traveltimes, the summation surfaces often are approximated by use of single-valued surfaces.

The advantage of ray tracing over finite-difference solutions is that ray tracing allows one to be selective in choosing the arrivals one uses for migration. The most common choice is to select the most-energetic arrivals among all the computed arrivals (Audebert et al., 1997). If more than one arrival is used to compute the summation surfaces, the quality of the migration results often improves (Wervelman et al., 2003), although at the price of a substantial increase in computational cost. However, as the example below illustrates, even when one uses multivalued summation surfaces, it can be difficult to represent accurately the traveltimes of a multibranched wavefield.

Figure 1 shows a prototypical situation in which multipathing creates potential problems for Kirchhoff migration because the presence of a low-velocity anomaly causes the wavefield to triplicate. Figure 2 shows the wavefronts that are effectively modeled by four commonly used migration methods. Figure 2a shows the wavefront that is modeled when a first-arrival kinematic computation of the summation surfaces is used. Notice that the slower-traveling but more-energetic components of the wavefront are absent from Figure 2a. Figure 2b shows the wavefront that is modeled when a most-energetic selection crite-

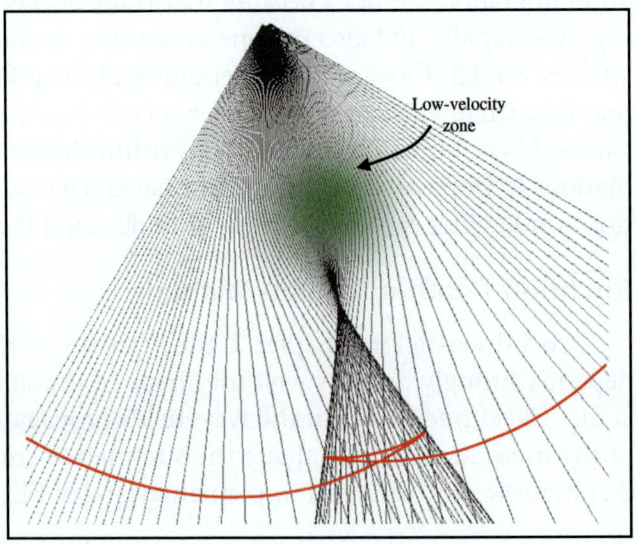

Figure 1. A prototypical situation of wavefront distortion caused by velocity-model complexity. The presence of the low-velocity anomaly causes the wavefield to triplicate. (This image was kindly provided by Ross Hill at Chevron.)

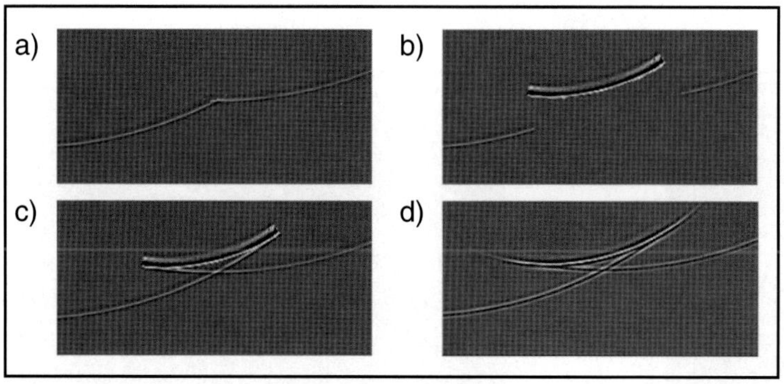

Figure 2. Wavefronts produced by different methods of modeling wave propagation: (a) First-arrival kinematic modeling, (b) ray-trace modeling with most-energetic arrival selection criterion for a single-value summation surface, (c) multiple-arrival ray-trace modeling, and (d) Gaussian-beam modeling. (This image was kindly provided by Ross Hill at Chevron.)

rion is applied to select the wavefronts computed using dynamic ray tracing. The more-energetic arrivals now are included at the expense of less-energetic arrivals. Figure 2c shows the effective wavefront modeled when multivalued summation surfaces are used during migration. All the arrivals are now included during migration. However, the tails of the triplicated wavefront are abruptly cut. These sharp truncations create artificial discontinuities in the wavefield and cause imaging artifacts. Figure 2d shows the wavefront modeled using a Gaussian-beam migration (see next section). This wavefront is similar to the wavefront that would be computed using wavefield-continuation migration. It shows smooth behavior of the amplitudes, even at the tails of the triplicating wavefronts.

Figures 3 and 4 illustrate the choices of Kirchhoff-migration algorithms discussed above when they are applied to a complex and realistic imaging problem: that of imaging the reservoir interval in the Marmousi model (Versteeg and Grau, 1990). Figure 3 shows the depth-migration results using four different methods: (a) shot-profile wavefield-continuation migration, (b) Kirchhoff migration with first-arrival traveltimes, (c) Kirchhoff migration with the most-energetic arrival computed by paraxial ray tracing, and (d) Kirchhoff migration with traveltimes computed using finite-bandwidth waveform modeling as proposed by Nichols (1996). The result from wavefield-continuation migration (Figure 3a) is clearly superior to the other results, and the result from first arrival migration (Figure 3b)is unsatisfactory at the reservoir level ($z \approx 2400$ m). Both migration methods based on the selection of the most-energetic arrivals for evaluating single-value summation surfaces produce satisfactory results, although with significantly stronger artifacts than the wavefield-continuation migration. The migration that takes into account the finite-frequency nature of wavefield propagation to compute the summation surfaces (Figure 3d) is smoother and has fewer artifacts than the one relying on a asymptotic approximation (Figure 3c).

The wavefields shown in Figure 4 display Green functions computed using wavefield-continuation in the Marmousi model and corresponding to a scatterpoint placed in the reservoir interval. The dots superimposed onto the wavefields are traveltimes computed by (a) first-arrival kinematic modeling and (b) the most-energetic arrival, computed using paraxial ray tracing. The traveltimes computed with the first-arrival kinematic modeling follow branches of the wavefield that have fairly low energy, whereas the traveltimes selected on the basis of amplitudes computed by ray tracing correctly follow the most energetic branch of the wavefield at almost all the horizontal locations. The differences in the selection of the traveltime branches account for the large differences that are reflected in the migrated images shown in Figure 3.

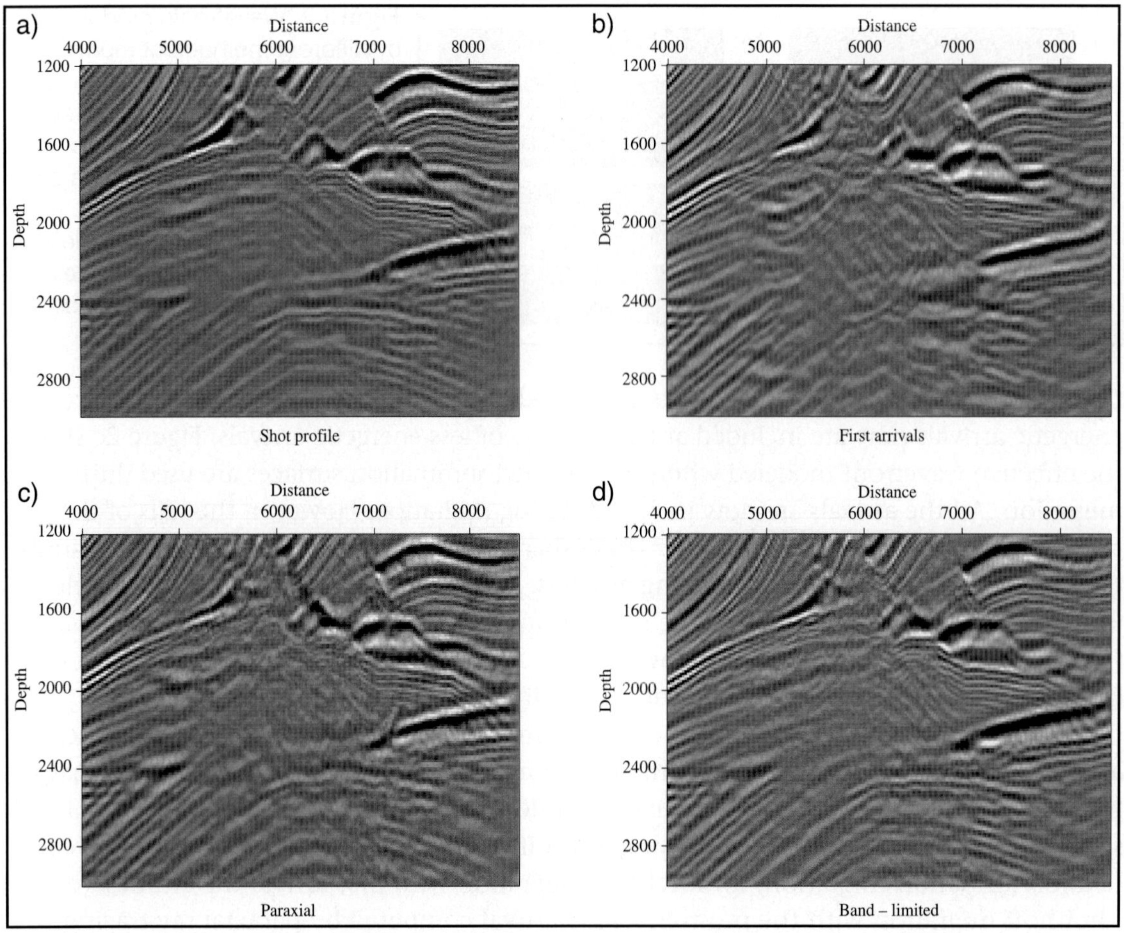

Figure 3. Depth migration results of the Marmousi data set: (a) shot-profile wavefield-continuation migration, (b) Kirchhoff migration with first-arrival traveltimes, (c) Kirchhoff migration with most-energetic arrival computed by paraxial ray tracing, and (d) Kirchhoff migration with traveltimes computed using finite-bandwidth waveform modeling. Figure is adapted from Nichols (1994).

Gaussian-beam Migration Methods

Migration methods based on decomposition of the recoded data and the source function in Gaussian beams (Hill, 1990, 2001; Gray, 2005) are known under the generic name of Gaussian-beam migration (GBM) methods. GBM methods enjoy many of the attractive properties of both Kirchhoff migration methods and wavefield-continuation migration methods. They are more robust with respect to wavefield multipathing than Kirchhoff methods are, but retain some of the computational efficiency and capability of imaging overturned events that are the strength of Kirchhoff methods.

GBM is based on decomposition of the wavefields in Gaussian beams by local slant stacks (Schultz and Claerbout, 1978; Stoffa et al., 1981; Treitel et al., 1982; Temme, 1984) that are weighted using a Gaussian function. Each beam can be propagated around its central ray, and the beam's wavefield u can be evaluated at each spatial location and temporal frequency ω using the expression

$$u(s,n,\omega) = A\exp\left\{i\omega\left[\tau(s) + M(s)n^2\right]\right\}, \tag{5.1}$$

where s and n are ray-centered coordinates (see right sketch in Figure 5), τ is the traveltime along the ray, A is the amplitude along the central ray, and the complex function $M(s)$ is computed by paraxial ray tracing (Hill, 1990). The real part of M controls the beam curvature and the imaginary part of M controls the amplitude decay away from the central ray. In order for the amplitudes not to grow exponentially the imaginary part of M must be always positive. The right sketch in Figure 5 defines the ray-centered coordinate system used to evaluate the wavefield in space by applying equation 5.1. Figure 5a shows a Gaussian beam propagating through the low-velocity anomaly shown in Figure 1. Notice that the beam starts flat at the surface and that its curvature changes sign as the central ray goes through the caustic visible in Figure 1. The superposition of many Gaussian beams builds up the whole wavefield, and is capable of accurately modeling multibranched wavefronts such as the one shown in Figure 2d.

Figure 6 illustrates the ability of GBM to model multibranched wavefronts in a realistic situation. Figure 6a shows a geologic model derived from field data and corresponding to the migration results shown in Figure 7. The proximity of the two salt bodies creates the possibility of severe multipathing of the propagating wavefields. Figure 6b

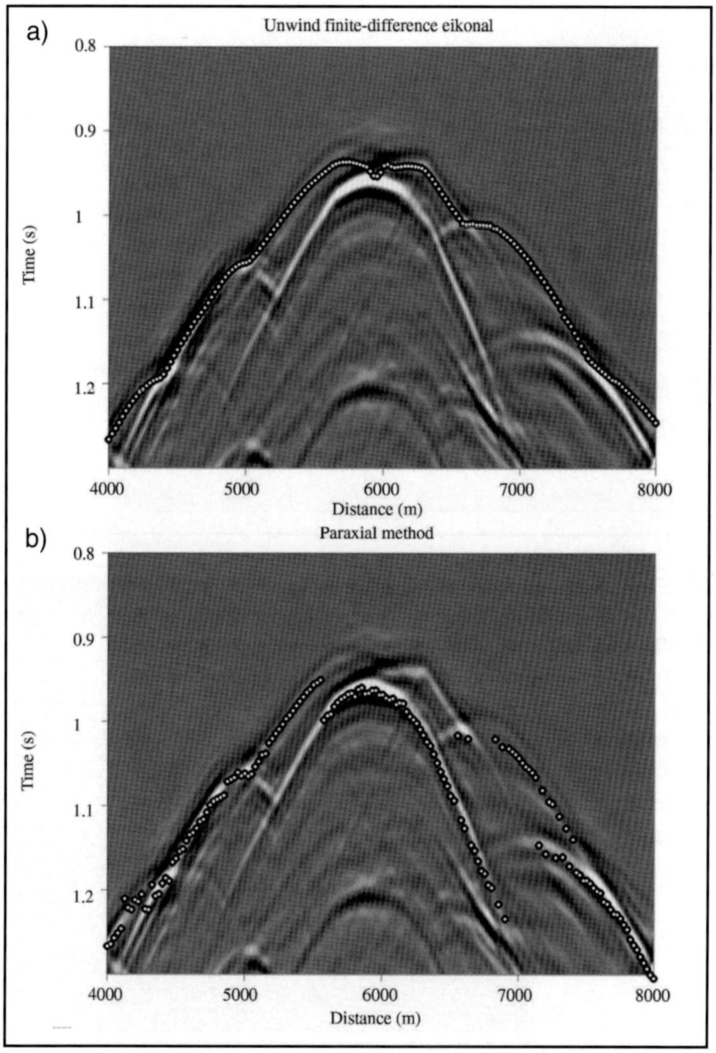

Figure 4. Green functions computed using wavefield-continuation, on which are superimposed the traveltime computed by: (a) first-arrival kinematic modeling and (b) most-energetic arrival computed using paraxial ray tracing. Figure is adapted from Nichols (1994).

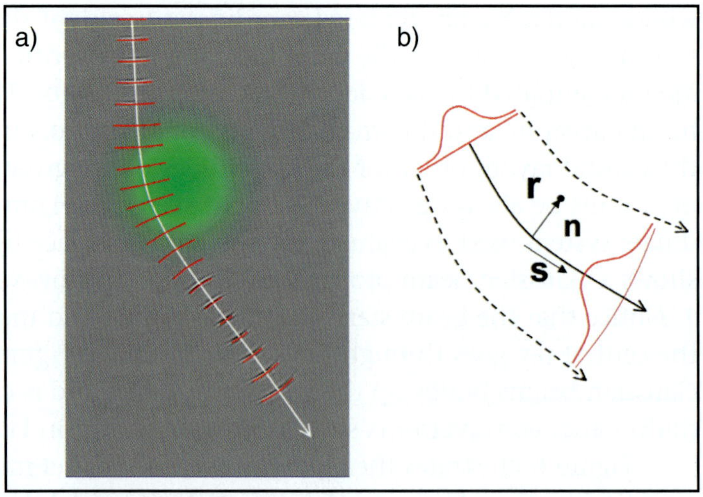

Figure 5. Illustration of wavefield propagation by Gaussian beams. (a) A Gaussian beam propagating through the low-velocity anomaly shown in Figure 1. Notice that the beam starts flat at the surface and that its curvature changes sign as the central ray goes through the caustic visible in Figure 1. (b) A sketch that defines the ray-centered coordinate system used to evaluate the wavefield by applying equation 5.1. (Image (a) was kindly provided by Ross Hill at Chevron.)

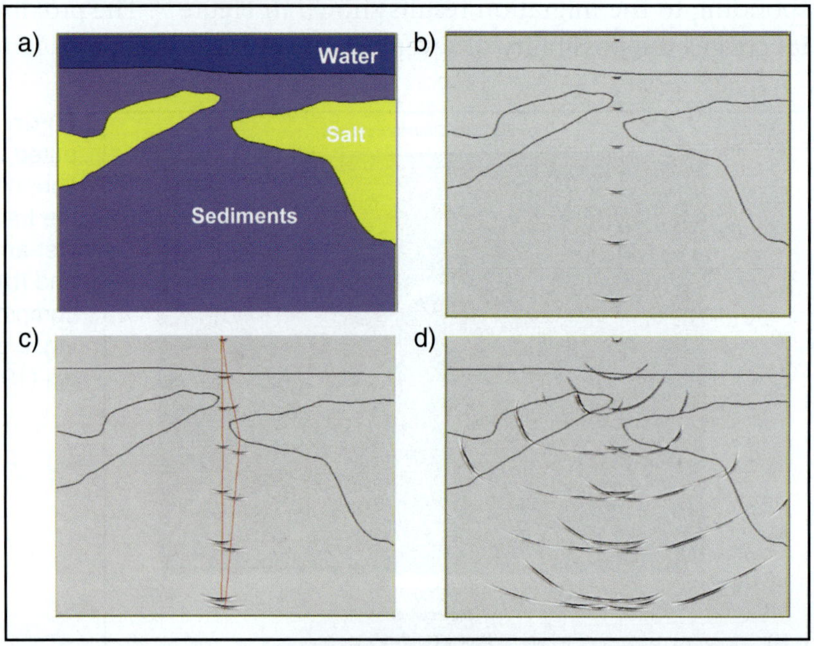

Figure 6. Illustration of the ability of GBM to model multibranched wavefronts in a realistic situation. (a) A geologic model derived from field data. (b)–(d) The superposition of (b) one, (c) two, and (d) several Gaussian beams. The superposition of several beams (d) models the complete wavefronts generated by a point source at the surface. (These images were kindly provided by Ross Hill at Chevron.)

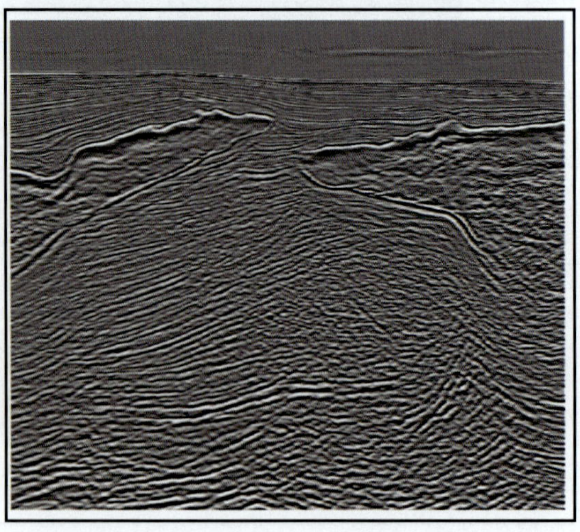

Figure 7. An inline section cut through GBM results obtained from a 3D field data set recorded in the Gulf of Mexico. This image corresponds to the sketch of the geologic model shown in Figure 6a. The migrated image shows continuous reflectors in the central part of the section. The image loses some coherency under the salt edges because of poor illumination. (This image was kindly provided by Ross Hill at Chevron.)

shows an individual Gaussian beam propagating from the surface into the subsurface. One isolated beam cannot model multipathing because a beam wavefield is evaluated by a single-value expansion around a central ray (equation 5.1). Figure 6c shows the superposition of a second beam relative to the original one shown in Figure 6b. The new beam propagates through the salt body on the right, so below the salt the second beam's kinematics significantly depart from the first beam's kinematics. The two beams barely interfere in the sediments below the salt bodies. Finally, Figure 6d shows the superposition of several beams to model the complete wavefronts generated by a point source at the surface. Above the salt body the wavefronts have the expected circular shape. In contrast, below the salt several branches of the wavefront carry significant energy. Poorly illuminated areas (shadow zones) develop under the edge of the salt body on the right and under the flank of the salt body on the left.

The example in Figure 6 illustrates how propagating individual Gaussian beams and summing the results can propagate the wavefield from a point source into the subsurface. A similar procedure can be applied to back-propagate the recorded wavefield into the subsurface. First, the data are decomposed into Gaussian beams, and then the beams are propagated along their central rays. Conceptually, the migrated image then can be created by crosscorrelating the source and receiver wavefields using an imaging condition similar to the ones introduced in Chapter 3 for shot-profile migrations. However, a straightforward implementation of this conceptual migration process would not be the most efficient possible. Gray (2005) discussed several algorithms for shot-profile GBM that achieve different levels of accuracy and computational efficiency. Hill (2001) describes a prestack GBM algorithm that operates on common-offset cubes. His algorithm is both efficient and accurate, but it requires regularization of the recorded data into common-offset cubes prior to migration.

Figure 7 shows an inline section cut through the migration results obtained by applying Hill's algorithm to a 3D field data set recorded in the Gulf of Mexico. Figure 6a is a sketch of the geologic model corresponding to the migrated section shown in Figure 7. The migrated image has continuous reflectors in the central part of the section where the beam modeling in Figure 6d shows severe multipathing of the propagating wavefield. The image loses some coherency under the salt edges because of poor illumination (as suggested by Figure 6d). I discuss the problem of poor subsalt illumination in detail in the last part of Chapter 9.

Wavefield-continuation migration methods

Wavefield-continuation methods can be more accurate for 3D prestack migration than methods based on the Kirchhoff integral are, but their computational cost discourages their use in most imaging projects. Wavefield-propagation algorithms are most efficient when the computational grid is regular and has a horizontal extent similar to the horizontal coverage of the recorded data being imaged. The first difficulty to overcome — although not the most challenging one — is that realistic 3D data-acquisition geometries are not regular (Chapter 1). Thus, one must regularize the data geometry before migration by applying data-regularization algorithms, such as those described in Chapter 9, or methods that achieve similar results.

Wavefield-continuation methods can be extremely inefficient when a mismatch occurs between the geometry of the recorded data and the geometry of the computational grid that the imaging algorithm requires. That problem is difficult to avoid when shot-gather migration is applied, as described in Chapter 3 — particularly when the acquisition geometries have a narrow azimuthal range, such as they do with marine-streamer geometries.

The sketch shown in Figure 8 illustrates the challenges faced by shot-gather migration when one is imaging deep and steeply dipping targets. The sketch represents the geometry of the prestack imaging process for two targets, one shallow and the other deep. Each threesome of rays corresponds to the propagation paths for the events recorded at zero and at maximum offset. The dashed boxes show the boundary of the computational domain required for imaging the events of interest. The figure shows that to image the deep and steeply dipping event, a shot-gather migration would require a computational domain with horizontal extent much wider than the original maximum offset of the recorded data. In practice, the recorded shot gather must be padded all around with an enormous number of zero traces. For example, if padding by a factor of four is needed on either side of the receiver array and along both the inline and the crossline axes, the computational domain will be larger than the original shot gather by a factor of 81 [$(4 + 1 + 4)^2 = 81$]. If computational costs are reduced by use of a narrower computational domain, steeply dipping deep events might be lost.

Several solutions to this computational-efficiency challenge have been proposed. Most of them fall into one of two categories: (1) variations of the source-receiver downward-continuation migration transformed into the midpoint-offset coordinates (Chapter 3), or (2) synthesis of generalized-source gathers with a wide horizontal extent that are obtained by weighted sums of the shot gathers recorded in the field.

The algorithms that belong to the first category can be extremely efficient, in particular for marine-streamer data, because they exploit the fact that the offset axes tend to shrink monotonically with depth. That phenomenon is illustrated in Figure 8. Therefore, the offset axes at the surface need to be only as wide as the data offset axes and can be trimmed down further with depth. Furthermore, the crossline offset of marine-streamer data is narrow at the surface, as is illustrated in Figure 9, and thus the computational

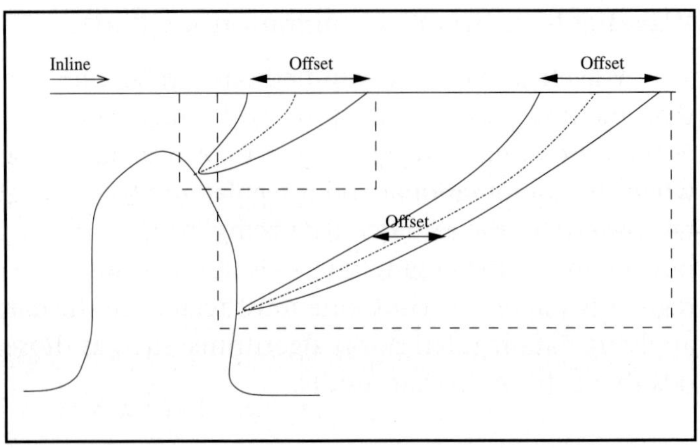

Figure 8. A schematic showing the raypaths, projected along the inline direction, for two prestack events recorded from two steeply dipping reflectors — one shallow and the other deep. The dashed boxes show the boundaries of the minimum-size computational domains required to image the events. The box for the deep events is substantially wider than the inline offset, indicating potential computational inefficiencies when shot-gather migration is applied.

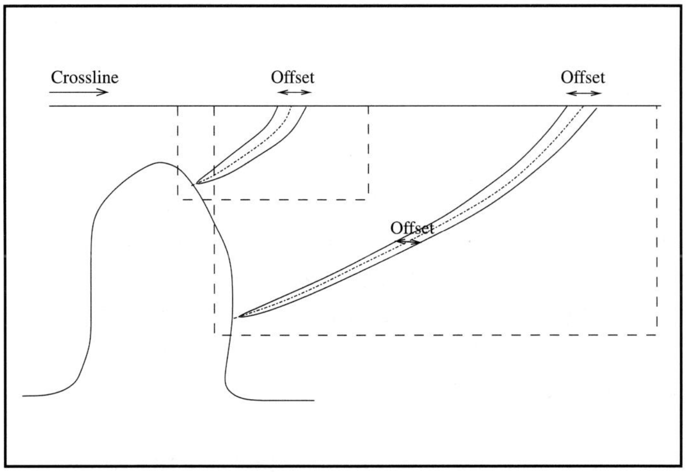

Figure 9. A schematic showing the geometry of the same two prestack events represented in Figure 8, now projected along the crossline direction. The crossline offset is narrow at the surface because a marine-streamer geometry is assumed. If a source-receiver migration transformed into midpoint-offset coordinates is applied to migrate the data, the computational domain can be much smaller than the one required by a shot-gather migration (dashed boxes).

domain can be reduced further along the crossline-offset axis. At the limit, it can be reduced to only one sample, as in the common-azimuth algorithm presented in this chapter in the section titled "Common-azimuth downward continuation and migration." In that section, I present a family of algorithms that achieve high efficiency by taking advantage of the narrow-azimuth nature of streamer data.

Although it is efficient, source-receiver migration is not as flexible as shot-gather migration. The main limitation of source-receiver migration is that it is intrinsically incapable of imaging overturned events because, for such events, at least one depth level always exists at which the horizontal offset between the downward-continued sources and receivers is not defined. In other words, for overturned events, either the source path or the receiver path overturns more shallowly than the other one, thereby making its correct propagation impossible within a source-receiver migration framework, at least in the conventional definition presented in Chapter 3.

In contrast with the methods based on source-receiver migration, the algorithms based on the synthesis of generalized-source gathers enjoy the same flexibility as does the original shot-gather migration. However, the algorithms based on generalized-source gathers are not as efficient computationally as the source-receiver methods, because they cannot (at least as they are understood at this writing) fully exploit either the narrowness of the offset axes (in particular of the crossline axis) or the fact that the offset axes shrink with depth. The next section discusses the general principles underlying all the synthesis of generalized shot gathers and briefly demonstrates how plane-wave migration can be adapted simply to migrate overturned events by propagating the source and receiver wavefields along a tilted Cartesian grid.

Migration of synthesized generalized-source gathers

An effective remedy for the computational inefficiency of shot-gather migration discussed above is to combine the data from several gathers recorded from point sources (shot gathers) and synthesize equivalent gathers from that spatially extended source. I will refer to such synthesized gathers as generalized-source gathers. The generalized-source gathers generated by this process have a wide horizontal extent, at the limit covering the whole survey area, for both the equivalent source function and the synthesized data. Using a shot-gather migration method to image these generalized-source gathers — that is, independently propagating the source and receiver wavefields (equations 3.6 through 3.7) and imaging at depth with the imaging condition in equation 3.8 — avoids some of the inefficiencies one encounters when imaging point-source gathers.

Imaging of the synthesized generalized-source gathers is based on linearity of the acoustic wave equation with respect to the propagating wavefield. Linearity is a property shared by all commonly used approximations of the wave equation, including the one-way wave equation used for downward-continuation migration and the two-way wave equation used for reverse-time migration. Because of linearity, the wavefield obtained at depth by propagating a generalized-source gather is the same as the wavefield one would have obtained by independently propagating each recorded shot gather and then combining the propagated wavefields at depth.

Generally, the combination of wavefields generates cross-talk artifacts in the image. Such artifacts are caused by spurious correlation of source and receiver wavefields related to different shot gathers. However, when a sufficient number of synthesized generalized-source gathers are migrated, the cross-talk artifacts can be attenuated, and at the limit, they vanish. Accuracy of the results obtained by imaging several generalized-source gathers depends on the method used for the combination (i.e., on the algorithm chosen to determine the summation weights) and on the total number of generalized-source gathers synthesized and imaged. Depending on the scheme used for determining the summation weights, the errors can be analyzed as being caused either by cross talk or by an incomplete (aliased) representation of source functions.

Synthesis of the generalized-source gathers is performed by the following sum of N shot gathers, weighted by the complex coefficients $w_{i,j}(\omega)$:

$$\hat{P}^g(\omega,x,y;j)=\sum_{i=1}^{N} w_{i,j}(\omega)P^g(\omega,x,y;\mathbf{s}_i), \tag{5.2}$$

which creates the jth generalized-source gather, $\hat{P}^g(\omega, x, y; j)$. Similarly, the jth generalized-source function, $\hat{P}^s(\omega, x, y; j)$, is created by the following weighted sum of N source functions:

$$\hat{P}^s(\omega,x,y;j)=\sum_{i=1}^{N} w_{i,j}(\omega)P^s(\omega,x,y;\mathbf{s}_i). \tag{5.3}$$

The weighted sums expressed in equations 5.2 and 5.3 can be repeated to create M generalized-source gathers and M generalized-source functions. In the following, I will assume for the sake of simplicity that N is equal to the total number of shot gathers in the survey, but that assumption is not necessary. Equations 5.2 and 5.3 are in the temporal-frequency domain ω, but the same procedure can be performed in the time domain if one substitutes convolution along the time axis for multiplication in the frequency domain.

The summation weights $w_{i,j}(\omega)$ in equations 5.2 and 5.3 are written as general functions of the temporal frequency, ω, the original shot-gather index, i, and the synthesized generalized-source-gather index, j. The choice of the functional form of these weights determines the name and characteristics of the migration method. Two main families of methods exist: the phase-encoding migration methods presented in the next section, and the plane-wave migration methods presented in the section with that name.

- Phase-encoding migration

Phase-encoding migration methods rely on application of different phase shifts to the original shot gathers (i.e., on phase encoding) to decrease the cross-talk artifacts caused by migrating the synthesized gathers (Morton and Ober, 1998; Romero et al., 2000). The summation weights are phase-only complex numbers and are independent from the spatial coordinates. They can be expressed as

$$w_{i,j}(\omega) = \frac{e^{i\phi_{i,j}(\omega)}}{\sqrt{M}}, \tag{5.4}$$

where $\phi_{i,j}(\omega)$ is a real phase function. Because the summation weights are independent from the spatial coordinates, their application to the wavefields commutes with the application of the propagation operators, and the prestack image cube obtained by applying shot-gather migration to the M generalized-source gathers (either by downward continuation or by reverse-time migration) can be written as

$$\begin{aligned} I(z,x,y,x_h,y_h) &= \sum_{\omega}\sum_{j=1}^{M} \hat{P}_z^g(\omega,x+x_h,y+y_h;j)\overline{\hat{P}_z^s(\omega,x-x_h,y-y_h;j)} \\ &= \sum_{\omega}\sum_{i=1}^{N} \hat{w}_{i,i}(\omega) P_z^g(\omega,x+x_h,y+y_h;\mathbf{s}_i)\overline{P_z^s(\omega,x-x_h,y-y_h;\mathbf{s}_i)} \\ &+ \sum_{\omega}\sum_{k=1;k\neq l}^{N}\sum_{l=1}^{N} \hat{w}_{l,k}(\omega) P_z^g(\omega,x+x_h,y+y_h;\mathbf{s}_l)\overline{P_z^s(\omega,x-x_h,y-y_h;\mathbf{s}_k)}, \end{aligned} \tag{5.5}$$

where the subscript z indicates the wavefields propagated from the surface down to depth z, and where

$$\hat{w}_{l,k}(\omega) = \sum_{j=1}^{M} w_{l,j}(\omega)\overline{w}_{k,j}(\omega), \tag{5.6}$$

and consequently

$$\hat{w}_{i,i}(\omega) = \sum_{j=1}^{M} \left| w_{i,j}(\omega) \right|^2. \tag{5.7}$$

When the sum in equation 5.7 is equal to one and when the cross-talk contributions vanish — that is, when the sum in equation 5.6 is equal to zero — the image cube computed by equation 5.5 is equal to the image obtained by conventional shot-gather migration of the original shot gathers (equation 3.8). The choice of the particular form for the weights expressed in equation 5.4 assures that the sum in equation 5.7 is always equal to one.

Several strategies have been proposed to minimize the cross-talk term. Because of the particular form the weights take, the sum in equation 5.6 can be rewritten as

$$\hat{w}_{l,k}(\omega) = \frac{1}{M}\sum_{j=1}^{M} e^{\left[\phi_{l,j}(\omega)-\phi_{k,j}(\omega)\right]}. \tag{5.8}$$

To assure that the sum in equation 5.8 approaches zero as M increases, one can choose the phase function $\phi_{i,j}(\omega)$ to be a random variable equally distributed in the interval $0 - 2\pi$ (Morton and Ober, 1998; Bonomi and Cazzola, 1999; Romero et al., 2000). This choice for the weights is effective when N, the number of shot gathers combined to create each generalized-source gather, is large, but it is less effective when N is small. When N is small, deterministic algorithms for designing the weights can be more effective (Jing et al., 2000; Romero et al., 2000). When random phase shifts are chosen for the summation weights, the cross-talk noise tends to be incoherent and to have a zero mean. It also can be shown that total energy of the cross-talk noise decreases as $1/\sqrt{M}$.

The following example shows that if enough generalized-source gathers are used for migration, the final image can be almost free from cross-talk artifacts. The results were produced using downward-continuation migration of generalized-source gathers that were synthesized using a random (Monte Carlo) phase-encoding scheme; these results have been presented by Cazzola et al. (2004). Figure 10 displays a 3D rendering of salt-pillow and top-of-the-salt surfaces interpreted from the images obtained from a marine data set acquired offshore West Africa. The presence of several salt pillows creates an imaging challenge. Figure 11 compares the image obtained by Kirchhoff prestack depth migration (Figure 11a) with the image obtained by migrating all generalized-source gathers (Figure 11b). This migration was computed using1800 generalized-source gathers synthesized from 75,600 field shot gathers. The total computational cost was estimated to be about 12% of the cost of migrating all the shot gathers independently. The computational saving is about a factor of five smaller than the ratio between the number of field shot gathers and the number of generalized-source gathers because the generalized gathers covered the whole extent of the output image. The Monte Carlo image is clearly superior to the Kirchhoff image under the salt pillow, and is mostly free from cross-talk noise. Figure 12 shows another section that was extracted form the Monte Carlo image cube with superimposed the color-coded velocity model. The velocity model was derived using Kirchhoff migration, and in this case it fits the wavefield-continuation image fairly well.

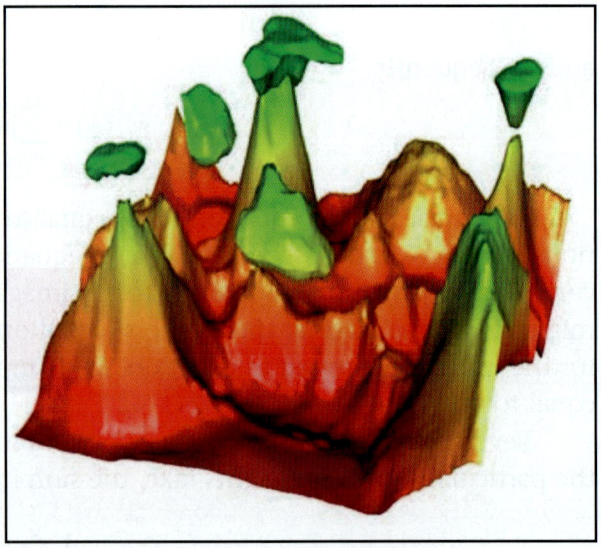

Figure 10. A 3D rendering of salt-pillow and top-of-the-salt surfaces corresponding to the images shown in Figures 11 and 12. The top-of-salt surfaces are color coded according to their depths. The presence of several salt pillows creates an imaging challenge. Figure is from Cazzola et al. (2004).

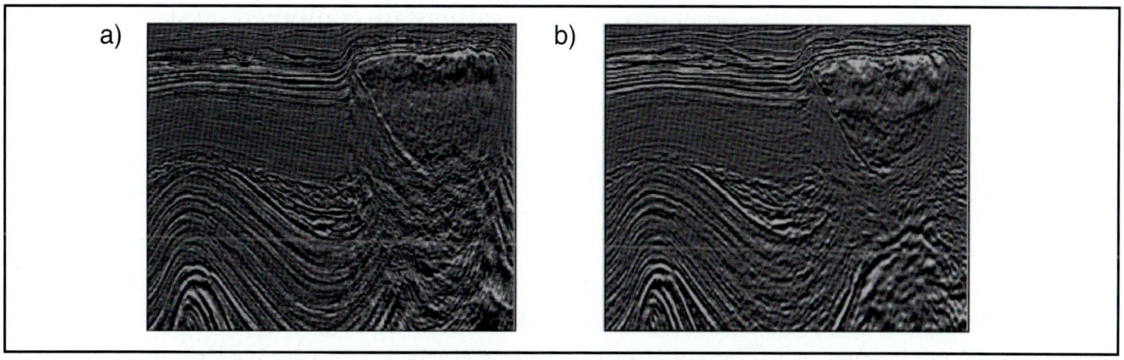

Figure 11. Comparison of (a) a Kirchhoff prestack depth migration with (b) a wavefield-continuation migration of generalized-source gathers using a Monte Carlo phase-encoding scheme. Below the salt pillow the Monte Carlo image is clearly superior to the Kirchhoff image. Figure is from Cazzola et al. (2004).

Figure 12. An inline section extracted form the Monte Carlo image cube onto which is superimposed the color-coded velocity model. The velocity model (derived using Kirchhoff migration) and the Monte Carlo image are consistent. Figure is from Cazzola et al. (2004).

One attractive application of random phase encoding is that it might reduce the computational cost of the iterative migrations needed by migration-velocity-analysis (MVA) procedures (Chapter 7). Many of the methods we apply during MVA for extracting velocity information from the migrated images are based on curvature scans (e.g., on semblance-spectra functions of ρ), which are robust to incoherent noise in the prestack image. Therefore, application of phase encoding with small values of M is likely to be sufficient for MVA purposes.

- Plane-wave migration

The imaging of seismic data after their decomposition into plane-wave components was introduced in the early days of seismic imaging (Schultz and Claerbout, 1978; Stoffa et al., 1981; Treitel et al., 1982; Temme, 1984). However, results of plane-wave depth migrations of real data have been published more recently (Whitmore and Garing, 1993). Plane-wave migration methods have since become popular, thanks to their computational efficiency for wavefield-continuation migration of 3D prestack marine data (Duquet et al., 2001; Notfors et al., 2003). Whitmore (1995) provides an excellent introduction to the basic concepts of plane-wave migration.

Decomposition of prestack data into plane-wave components according to the ray parameters at the source, (p_{x_s}, p_{y_s}), is a particular case of phase encoding whereby the resulting synthesized source functions are plane waves with different values of the ray parameters p_{x_s} and p_{y_s}, and the generalized-source gathers are computed by slant stacks of common-receiver gathers (Liu et al., 2006). To achieve this result, I express the phase function in equation 5.4 as

$$\phi_{i,j}(\omega)=\omega\left[p_{x_s,j}\left(x_0-x_{s,i}\right)+p_{y_s,j}\left(y_0-y_{s,i}\right)\right], \tag{5.9}$$

where x_0 and y_0 are arbitrary horizontal coordinates that are the same for all the values of i and j.

In that case, one also can apply the quantitative analysis of the cross-talk noise developed in equations 5.5. The summation in equation 5.5 can be written as

$$\hat{w}_{l,k}(\omega)=\sum_{j=1}^{M} w_{l,j}(\omega)\overline{w}_{k,j}(\omega)=\frac{1}{M}\sum_{j=1}^{M}\exp\left\{i\omega\left[p_{x_s,j}\left(x_{s,k}-x_{s,l}\right)+p_{y_s,j}\left(y_{s,k}-y_{s,l}\right)\right]\right\}. \tag{5.10}$$

The sum in equation 5.10 vanishes if the ranges of the ray parameters p_{x_s} and p_{y_s} are (1) wide enough with respect to the minimum distances between shots, $(x_{s,k} - x_{s,l})$ and $(y_{s,k} - y_{s,l})$, and (2) sampled densely enough to assure that the oscillatory argument of the summation over j in equation 5.10 is not aliased. Not surprisingly, the criterion for eliminating the cross-talk noise turns into a sampling criterion for a plane-wave representation of a function defined at the original shot locations of the survey. Those sampling criteria are the same ones that one usually applies when one is concerned about the accuracy of plane-wave migration (Whitmore, 1995; Duquet et al., 2001; Notfors et al., 2003).

Conical-wave migration is an interesting variation of plane-wave migration that is cost-effective for 3D marine-streamer data. It was discussed first by Whitmore (1995) and was applied first to 3D data by Duquet et al. (2001). The method takes advantage of both the narrow-azimuth nature of marine data and the large distance between parallel sail lines of typical 3D marine surveys to achieve higher computational efficiency than a full 3D plane-wave migration could achieve (Duquet et al., 2001; Notfors et al., 2003). Each sail line in a 3D survey is assumed to be a common-azimuth line, that is, to have zero crossline offset. The 3D acquisition geometries used in the field are not strictly common-azimuth geometries, but one can transform the data into an ensemble of common-azimuth lines by a preprocessing step, such as the AMO process presented in Chapter 4. The shot gathers in each common-azimuth line then are combined according to the inline source-ray parameter p_{x_s} and are imaged independently into a 3D volume. The method's name derives from the fact that each wavefield produced by the line-source functions for each p_{x_s} propagates as a conical wave in a 3D volume.

- Imaging overturned waves by plane-wave migration in tilted coordinates

The computational advantages of plane-wave migration can be substantial when the inaccuracies caused by limiting the number of plane-wave components do not interfere with the final goals of the imaging project. However, the most promising application of plane-wave migration might be the migration of overturned events by one-way propagation operators.

One-way extrapolation has many advantages over two-way propagation. It is computationally more efficient and less memory intensive because it can be performed easily in the temporal frequency domain. In addition, the size of the depth steps is not limited by stringent stability constraints, as is the time-step size in explicit finite-difference two-way solvers. Furthermore, when one uses one-way propagators, sharp velocity boundaries do not cause the artifacts that are created by reverse-time migration, which I discussed at the beginning of Chapter 3. The main limitation of one-way extrapolators is that they cannot handle events that propagate along both directions of the extrapolation axis, such as overturned events.

Higginbotham et al. (1985) proposed a simple method for using a zero-offset migration with a one-way propagator to image overturned events reflected by a specific interface, such as a salt flank. The idea is to tilt the propagation axis from the conventional depth direction to a direction that better matches the propagation of the events of interest. This idea is simple and powerful, but it is difficult to generalize to prestack migration when the overturned events have no preferred propagation direction, as in complex 3D geology. Creating generalized-source gathers identified by the takeoff angle of the source wavefields at the surface can separate events according to their prevailing propagation direction. Thus, it enables us to determine an optimal tilt angle for imaging each plane-wave component. Both the generalized-source function and the generalized-source gather can be continued effectively by one-way propagators advancing along the tilted direction defined by the takeoff angle at the source (Etgen, 2002; Etgen, personal communication, 2003; Shan and Biondi, 2004) or even at a slightly steeper angle when the velocity is known to increase monotonically with depth.

Figure 13 illustrates this idea for one plane-wave component. It shows that both the source and the receiver legs of an overturned event are more likely to be propagated properly by one-way extrapolation in the tilted coordinate system than by conventional extrapolation along the depth axis. For events not reflected at normal incidence, the propagation direction of the receiver wavefield does not coincide exactly with the propagation direction of the source wavefield. However, the angle between the propagation direction of the receiver wavefield and that of the source wavefield is limited by the aperture angle at the reflection point, which is limited by the maximum offset of the recording geometry.

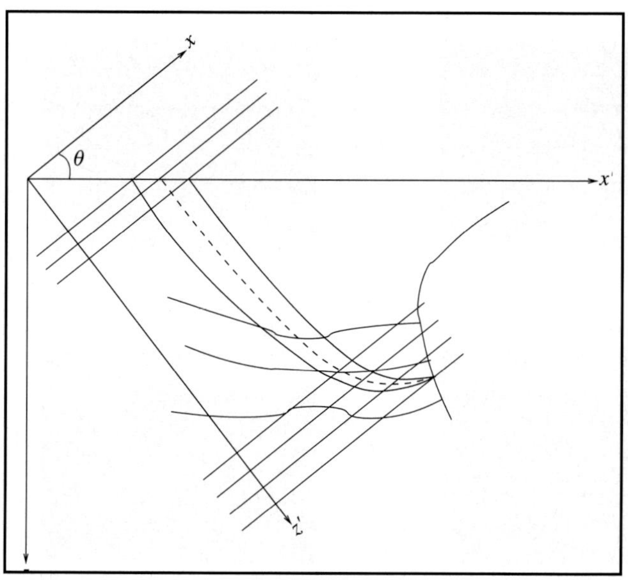

Figure 13. A schematic showing the geometry of an overturned event reflected from a salt flank. The event cannot be imaged properly by a downward-continuation method propagating the wavefields along the depth axis z. In contrast, it can be imaged properly by a plane-wave migration propagating the wavefields along the z' propagation direction, which is rotated by the tilt angle θ with respect to z. The tilt angle can be different for different plane-wave components. The dotted lines represent the horizontal direction in the tilted coordinates. They help to verify visually that neither of the two legs of the event overturns with respect to the z' propagation direction.

Shan and Biondi (2004) described an implementation of this idea and showed that overturned events generated by a vertical salt edge can be imaged using a one-way propagator as well as a reverse-time-migration algorithm. Furthermore, they applied geometric analysis of the kinematics of ADCIGs (angle-domain common-image gathers) of overturned events, presented by Biondi and Shan (2002) and by Biondi and Symes (2004), to define a simple algorithm for computing high-quality ADCIGs for the overturned events using the tilted-coordinate plane-wave migration method. As Biondi and Shan (2002) and Biondi and Symes (2004) discussed, the quality of the ADCIGs obtained from horizontal-offset CIGs degrades progressively as the geologic dip increases, and it degenerates when events overturn. Propagation in tilted coordinates presents a simple solution to this problem. The offset-domain CIGs are computed along the x' directions and then transformed to equivalent CIGs computed in both the x and z directions. The contributions from all the source plane-wave components are merged and then transformed to the angle domain (Shan and Biondi, 2004).

Figures 14 and 15 show the quality of the stacked images and the CIGs that can be obtained by applying a plane-wave migration in tilted coordinates. The data set is a 2D line extracted from a 3D marine survey recorded in the North Sea. A vertical salt edge generates overturned events that cannot be migrated by using downward-continuation migration. Thus, this data set is a good test bed for migration methods aiming to image overturned events.

Figure 14 compares migration results for three prestack migration methods. Figure 14a shows the image obtained by a conventional downward-continuation migration. Figure 14b shows the image obtained by a reverse-time migration (Biondi and Shan, 2002). Figure 14c shows the image obtained by plane-wave migration with tilted coordinates. The fan of rays shown in Figure 14d is traced through the migration velocity model, starting from the vertical salt-edge reflection. Notice that the reverse-time migration image (Figure 14b) is sampled more coarsely than either of the other two images (Figure 14a and c). The vertical salt edge is imaged equally well by the plane-wave migration

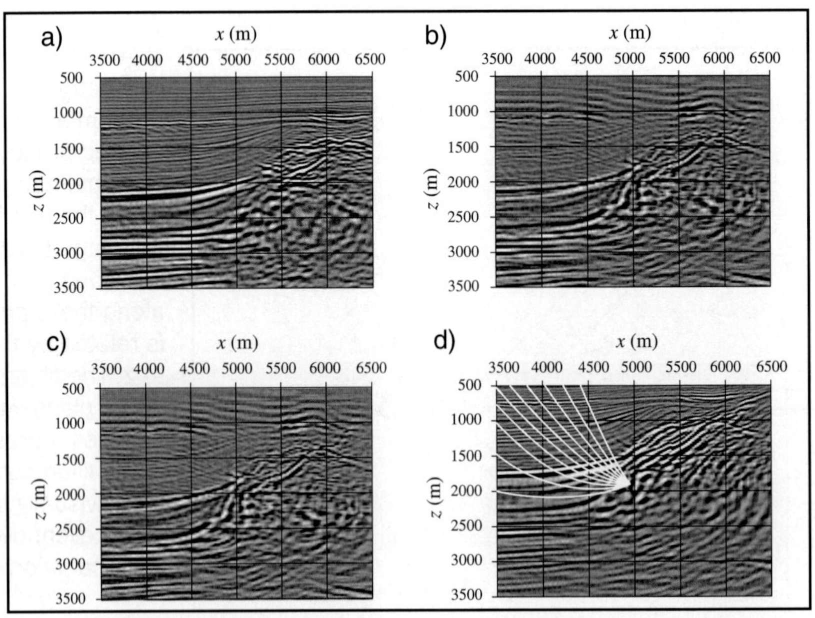

Figure 14. Migration of a North Sea real data set in the proximity of a vertical salt edge: (a) the image from downward-continuation migration; (b) the image from reverse-time migration; (c) the image from plane-wave migration in tilted coordinates; and (d) the same image as in (c), with a superimposed fan of rays corresponding to overturned events reflected from the salt edge. Figure is from Shan and Biondi (2004).

(Figure 14c) and the reverse-time migration (Figure 14b). Although the salt-edge reflectors are discontinuous and attenuated in the downward-continuation migration results (Figure 14a), they are continuous and have strong amplitudes in the plane-wave migration (Figure 14c).

Figure 15 compares the CIGs obtained by reverse-time migration with the CIGs obtained by plane-wave migration, extracted at a surface location, $x = 4000$ m, where the reflectors are dipping only slightly. Figure 15a and b compare the offset-domain CIGs, and Figure 15c and d compare the angle-domain CIGs. The two sets of CIGs are similar, although they differ slightly, primarily because of different spatial sampling between the reverse-time and plane-wave migration images.

Source-receiver migration in midpoint-offset coordinates

The methods presented in the previous section are most efficient when the number of synthesized generalized-source gathers, M, is substantially lower than the total number of shot gathers to be imaged (Cazzola et al., 2004). Otherwise, the computational gain may be modest (Stork and Kapoor, 2004). Whether this condition is fulfilled depends on many parameters, such as the required image accuracy, the structure of the target, and the data-acquisition geometry.

Another effective way to reduce the computational cost of wavefield-continuation migration is to take advantage of the fact that the data always are acquired with offsets shorter than the total horizontal extent of the survey because of practical and economic limitations (Figures 8 and 9). Furthermore, most land and marine data-acquisition geometries have an anisotropic distribution of offset vectors (Chapter 1), with the offset range being substantially wider in one direction (the inline direction) than in the other (the crossline direction).

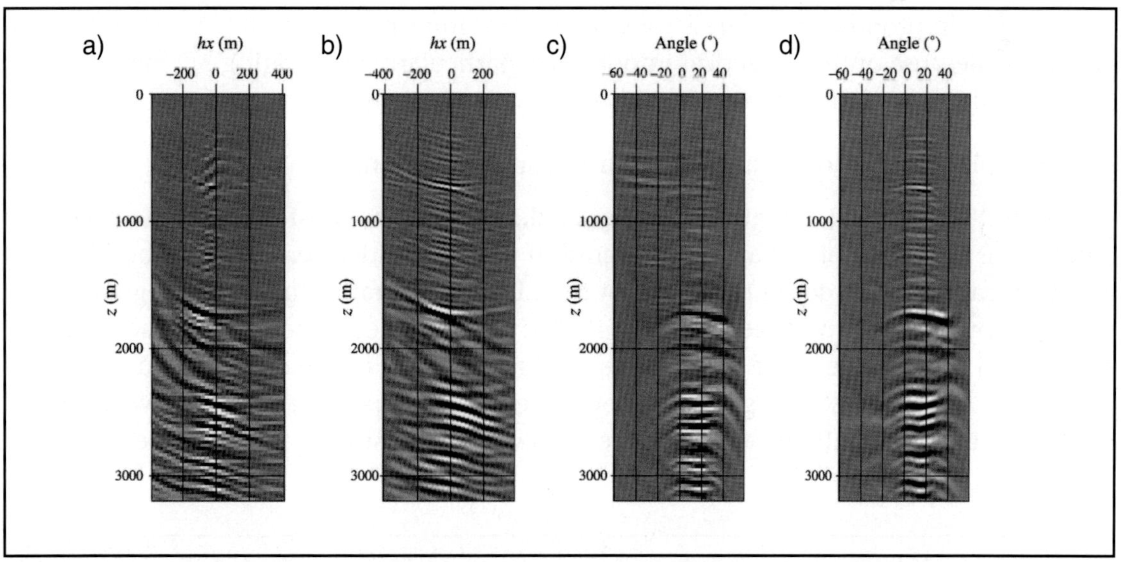

Figure 15. CIGs extracted from the prestack images at x = 4000 m: (a) offset-domain CIGs (ODCIGs) obtained by reverse-time migration, (b) ODCIGs obtained by plane-wave migration in tilted coordinates, (c) ADCIGs obtained by reverse-time migration, (d) ADCIGs obtained by plane-wave migration in tilted coordinates. Figure is from Shan and Biondi (2004).

Source-receiver migration transformed into midpoint-offset coordinates can take advantage of the limited offset ranges in a simple and straightforward way, because the computational domain easily can be restricted to encompass only the existing offsets. From Chapter 3, we know that the whole prestack wavefield expressed in the frequency and wavenumber domain, $P_z(\omega, \mathbf{k_m}, \mathbf{k_h})$, can be downward-continued from depth z to depth $z + \Delta z$ by the phase shift

$$P_{z+\Delta z}(\omega, \mathbf{k_m}, \mathbf{k_h}) = P_z(\omega, \mathbf{k_m}, \mathbf{k_h}) e^{-ik_z \Delta z}, \tag{5.11}$$

where the vertical wavenumber k_z is evaluated using the following midpoint-offset wavenumber DSR (double-square-root) equation

$$k_z = \sqrt{\frac{\omega^2}{v(\mathbf{s},z)^2} - \frac{1}{4}\left[\left(k_{x_m} - k_{x_h}\right)^2 + \left(k_{y_m} - k_{y_h}\right)^2\right]} + \sqrt{\frac{\omega^2}{v(\mathbf{g},z)^2} - \frac{1}{4}\left[\left(k_{x_m} + k_{x_h}\right)^2 + \left(k_{y_m} + k_{y_h}\right)^2\right]}. \tag{5.12}$$

The prestack image is computed by applying the imaging condition of equation 3.13, that is, by evaluating the downward-propagated wavefield at zero time (i.e., summing over frequencies).

The prestack image obtained by source-receiver migration is equivalent to the one obtained by shot-gather migration (Biondi, 2006). Therefore, I will refer to the migration algorithm defined by equations 5.11, 5.12, and 3.13 as full-azimuth migration. Full-azimuth migration can be fairly efficient, but there are two main practical obstacles to its achieving absolute efficiency: (1) The downward continuation must be performed on 4D data sets (constant-frequency slices) and thus requires either a large amount of memory or out-of-core computations, and (2) computational efficiency demands that the crossline-offset range be as narrow as possible while still being sufficiently wide to capture all useful propagation paths and avoid boundary artifacts. Several variations of full-azimuth migration have been proposed to address those issues and improve the method's computational efficiency. Because of their practical importance, particularly for imaging 3D marine data, I dedicate the rest of this chapter to analyzing these methods.

- Offset plane-wave downward continuation and migration

Offset plane-wave migration (Mosher et al., 1997) is the most efficient method among the variations of full-azimuth migration that I consider here. The method is a generalization to 3D depth migration of the 2D time-migration methods originally proposed by Ottolini and Claerbout (1984). It is based on decomposition of each CMP gather into offset plane waves by slant stacks, as a function of the offset ray parameters p_{x_h} and p_{y_h}. After slant stacking, each offset plane-wave component can be downward continued independently by applying the following expression of the DSR operator:

$$k_z = \sqrt{\frac{\omega^2}{v(\mathbf{s},z)^2} - \frac{1}{4}\left[\left(k_{x_m} - \omega p_{x_h}\right)^2 + \left(k_{y_m} - \omega p_{y_h}\right)^2\right]} + \sqrt{\frac{\omega^2}{v(\mathbf{g},z)^2} - \frac{1}{4}\left[\left(k_{x_m} + \omega p_{x_h}\right)^2 + \left(k_{y_m} + \omega p_{y_h}\right)^2\right]}, \tag{5.13}$$

which is obtained by substituting $k_{x_h} = \omega p_{x_h}$ and $k_{y_h} = \omega p_{y_h}$ into equation 5.12.

Independent imaging of the offset plane-wave components solves the practical challenges of full-azimuth migration discussed above — the large memory requirements and the boundary artifacts along narrow offset axes. The number of offset plane waves needed to achieve the desired imaging accuracy is an important parameter influencing computational efficiency. For narrow-azimuth data sets, the number of required offset plane waves is substantially lower than the number of source plane waves needed for the plane-wave migration presented earlier in this chapter.

The price one pays for efficiency gained with the independent downward continuation of the offset plane waves is inaccuracy in depth migration with complex velocity models. The offset plane waves should mix during downward continuation when the velocity varies laterally. Whereas in layered media, the horizontal ray parameters, (p_{x_s}, p_{y_s}, p_{x_g}, p_{y_g}), are constant with depth, in arbitrary media, the horizontal ray parameters vary with depth according to the velocity function along the source and receiver raypaths. The ray parameters change when later inhomogeneities are encountered; consequently, the offset ray parameters should be allowed to change as well. An alternative way to understand this issue is to notice that in laterally varying media, the operators represented by the square roots in the DSR operator are not diagonal in the frequency-wavenumber domain. Instead, they become convolutional operators that mix the wavefield components corresponding to different wavenumbers; that is, they mix different offset plane waves.

- Common-azimuth downward continuation and migration

Almost all marine data sets acquired with streamers have a narrow crossline-offset range. The common-azimuth migration method takes advantage of that characteristic to reduce the dimensionality of the computational domain of full-azimuth migration (Biondi and Palacharla, 1996). It assumes that all the data have a zero crossline offset, that is, all the traces in the data share the same azimuth, hence the term *common-azimuth*. By reducing the crossline-offset axis to only one sample, common-azimuth migration solves both issues of full-azimuth migration discussed above: The constant-frequency slices become smaller (3D instead of 4D), and the problem of boundary artifacts along the crossline-offset axis disappears.

Because the actual data are not common-azimuth data, a preprocessing step is necessary to transform all the data into equivalent common-azimuth data (see Chapter 4, the section titled "AMO transformation to common-azimuth data"). In principle, this preprocessing step might introduce errors, but usually the amount of azimuthal rotation required to convert real geometries into true common-azimuth geometries is very small. Consequently, the errors introduced by the rotation are also small.

Analytical derivation of the common-azimuth downward-continuation operator takes advantage explicitly of the reduced dimensionality of the data space; thus, it derives a continuation operator with only four dimensions. The expression for the common-azimuth operator is derived by substituting the following stationary path (Biondi and Palacharla, 1996):

$$\hat{k}_{y_h} = k_{y_m} \frac{\sqrt{\frac{\omega^2}{v(\mathbf{g},z)^2} - \frac{1}{4}\left(k_{x_m} + k_{x_h}\right)^2} - \sqrt{\frac{\omega^2}{v(\mathbf{s},z)^2} - \frac{1}{4}\left(k_{x_m} + k_{x_h}\right)^2}}{\sqrt{\frac{\omega^2}{v(\mathbf{g},z)^2} - \frac{1}{4}\left(k_{x_m} + k_{x_h}\right)^2} + \sqrt{\frac{\omega^2}{v(\mathbf{s},z)^2} - \frac{1}{4}\left(k_{x_m} + k_{x_h}\right)^2}}, \qquad (5.14)$$

into the expression for the full DSR of equation 5.12; that is,

$$\hat{k}_z = \sqrt{\frac{\omega^2}{v(\mathbf{s},z)^2} - \frac{1}{4}\left[\left(k_{x_m} - k_{x_h}\right)^2 + \left(k_{y_m} - \hat{k}_{y_h}\right)^2\right]} - \sqrt{\frac{\omega^2}{v(\mathbf{g},z)^2} - \frac{1}{4}\left[\left(k_{x_m} + k_{x_h}\right)^2 + \left(k_{y_m} + \hat{k}_{y_h}\right)^2\right]}. \qquad (5.15)$$

Common-azimuth migration can be implemented accurately with mixed-domain (ω-k/ω-x) downward-continuation methods, such as an extended split-step algorithm [Chapter 5 in Biondi (2006)], as long as implementation properly takes into account the fact that the stationary path expressed in equation 5.14 is a function of the interval velocities at the source and receiver locations.

The common-azimuth dispersion relation of equation 5.15 can be recast, after some algebraic manipulations, as a cascade of two dispersion relations. The first is the relation for 2D prestack downward continuation along the inline direction,

$$k_{z_x} = \sqrt{\frac{\omega^2}{v(\mathbf{s},z)^2} - \frac{1}{4}\left(k_{x_m} - k_{x_h}\right)^2} - \sqrt{\frac{\omega^2}{v(\mathbf{g},z)^2} - \frac{1}{4}\left(k_{x_m} + k_{x_h}\right)^2}. \qquad (5.16)$$

The second is the one for 2D zero-offset downward continuation along the crossline axis,

$$\hat{k}_z = \sqrt{k_{z_x}^2 - k_{y_m}^2}. \qquad (5.17)$$

This inline and crossline separation of the common-azimuth dispersion relation is an alternative proof of the full separability of the 3D prestack migration operator derived in Chapter 4. If the propagation velocity in equations 5.16 and 5.17 is assumed to be constant, these dispersion relations can be used directly for cascading two 2D Stolt migrations (Stolt, 1978).

Figures 16 and 17 demonstrate the effectiveness of common-azimuth migration to depth-image 3D data in the presence of a complex velocity function. The data were recorded in the deep waters of the Gulf of Mexico over a complex salt body with a deep canyon on the top and steep flanks on the bottom. The figures show the inline sections extracted from the migrated cubes imaged by common-azimuth migration (Figure 16) and by Kirchhoff migration (Figure 17). The same velocity function was used for both migrations. Notice that below the canyon, the sediments and the salt-bottom reflection are imaged better by the common-azimuth migration than by the Kirchhoff migration. The shape of the salt bottom on the right-hand side is not well determined; thus, both images lose coherency in that area. However, the common-azimuth migration succeeds in imaging several events that could be used for improving the velocity model, using the methods presented in Chapter 7.

Figure 18 shows similar advantages for common-azimuth migration over Kirchhoff migration, but for the SEG-EAGE salt data set, for which the velocity model is known.

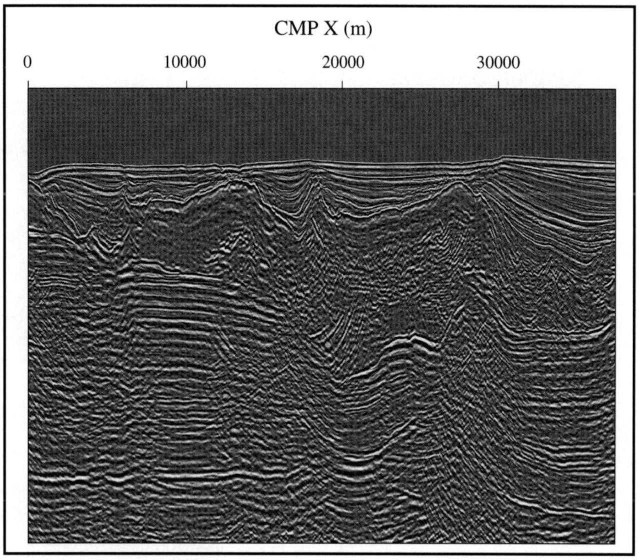

Figure 16. An inline section extracted from a migrated cube imaged by common-azimuth migration.

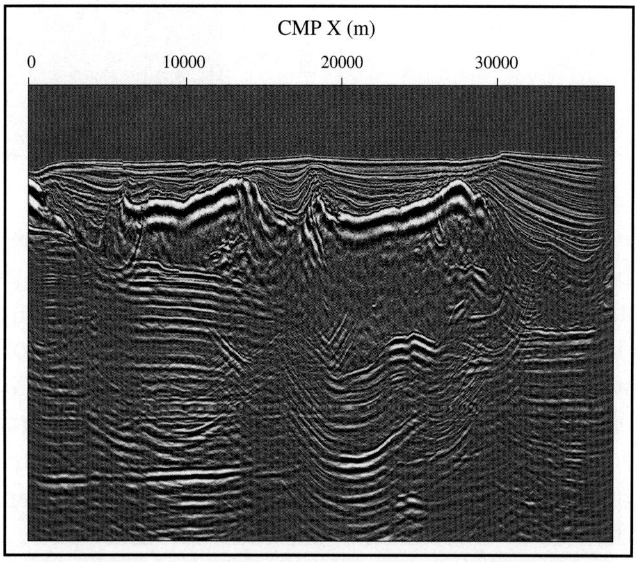

Figure 17. An inline section extracted from a migrated cube imaged by Kirchhoff migration. This section has the same crossline coordinate as the section shown in Figure 16.

Figure 18 compares two inline sections. The structure under the salt is imaged better by common-azimuth migration.

- Narrow-azimuth downward continuation and migration

Narrow-azimuth migration is a generalization of common-azimuth migration, and it is based on the downward continuation of a narrow strip around the zero crossline offset (Biondi, 2003a). Computational cost of such a generalization obviously is proportional to the number of crossline offsets used to represent this narrow strip. Minimum width of the strip needed to image all events properly depends on the reflector geometry and on the velocity model. Maximum sampling needed in the crossline-offset direction depends on the crossline-offset dip spectrum. The common-azimuth relationship in equation 5.14 provides a guideline for the crossline dip spectrum as a function of the reflector geometry.

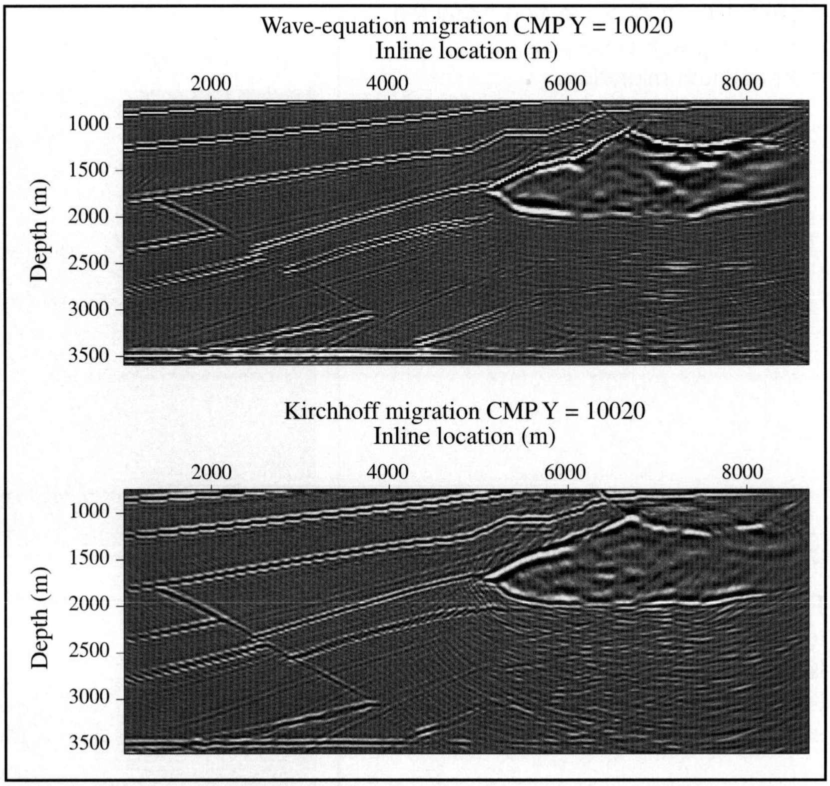

Figure 18. (a) Common-azimuth migration compared with (b) Kirchhoff migration.

To achieve computational efficiency, the crossline-offset range must be as narrow as possible and still be able to capture all the useful propagation paths and avoid boundary artifacts. That goal can be accomplished best by applying two complementary procedures: (1) defining an optimal range of crossline-offset wavenumbers, k_{y_h}, for the downward continuation and (2) windowing the ADCIGs along the azimuth axis to limit the effects of the boundary artifacts on the final migrated image. The information provided by the common-azimuth equation can be exploited to define a range of crossline-offset dips that corresponds to propagation azimuths centered on the zero-azimuth direction defined by the common-azimuth stationary path. Chapter 7 in Biondi (2006) provides a more detailed derivation of narrow-azimuth migration.

- Narrow-azimuth migration of the SEG-EAGE salt model

Common-azimuth migration produces high-quality images of the SEG-EAGE salt data set for most of the reflectors, but it is not as accurate as shot-gather migration in imaging a few reflectors. Figure 19 shows an occurrence of this problem for the inline section taken at y_m = 6000 m. Figure 19a displays the inline section of the velocity model and Figure 19b displays the stacked image obtained by common-azimuth migration. With the exception of the bottom of the salt reflector between x_m = 3000 m and x_m = 4000 m, this image is as accurate as the image obtained by shot-gather migrations performed with a large migration aperture (i.e., with large padding around the recorded shot gather), but this image is obtained at a fraction of the cost. Reflectors that are not imaged well by common-azimuth migration can be imaged well and efficiently by using narrow-azimuth migration.

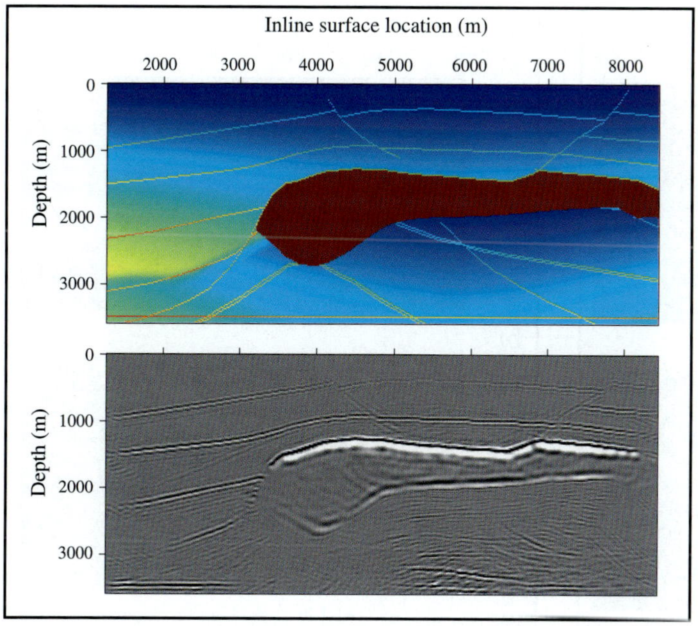

Figure 19. Inline sections taken at y_m = 6000 m of (a) the velocity model and (b) the common-azimuth migration.

To ensure that the differences between migration results are exclusively the result of differences between migration algorithms, I created the data set used for narrow-azimuth migration by padding the original common-azimuth data with seven empty crossline offsets ($N_{y_h} = 8$), and I assumed $\Delta y_h = 160$ m. I obtained the stacked images after windowing in the angle domain. Only the reflection azimuths with $|\phi| \leq 30°$ contributed to the final stack. After windowing, I stacked the ADCIGs over both the azimuth axis and the aperture-angle axis.

Figure 20 compares the results of common-azimuth migration (b) and narrow-azimuth migration (c) in a small window around the reflector of interest. The narrow-azimuth image clearly is superior to the common-azimuth image, and it correctly images all the reflectors within the window, as one can confirm by comparing the migrated image with the velocity model (a). In the narrow-azimuth migration, even the steeply dipping sediment interface that truncates against the left flank of the salt is imaged well.

Similar improvements are visible in depth slices cutting through the bottom of the salt around the inline section shown in Figure 20. Figure 21 compares the slices taken at a depth of 2400 m. Figure 21a shows the velocity model. Figure 21b shows the image obtained by common-azimuth migration. Figure 21c shows the image obtained by narrow-azimuth migration. In this image, the salt-bottom boundary is well defined for all the inline sections.

Analysis of the ADCIGs confirms that narrow-azimuth migration images the reflector with the correct kinematics. Figure 22 shows the stack (left panel) and the ADCIG taken at $x_m = 3860$ m (right panel) obtained by common-azimuth migration. The ADCIG displays an upward moveout that confirms that common-azimuth migration is inaccurate for this reflector. In contrast, in the corresponding ADCIG obtained by narrow-azimuth migration, shown in Figure 23, the moveout is flat. The spurious coherent energy that is visible above the reflector centered at around $\gamma = 20°$ probably is linked to boundary artifacts. The ADCIG and the stacked section are obtained by stacking over the

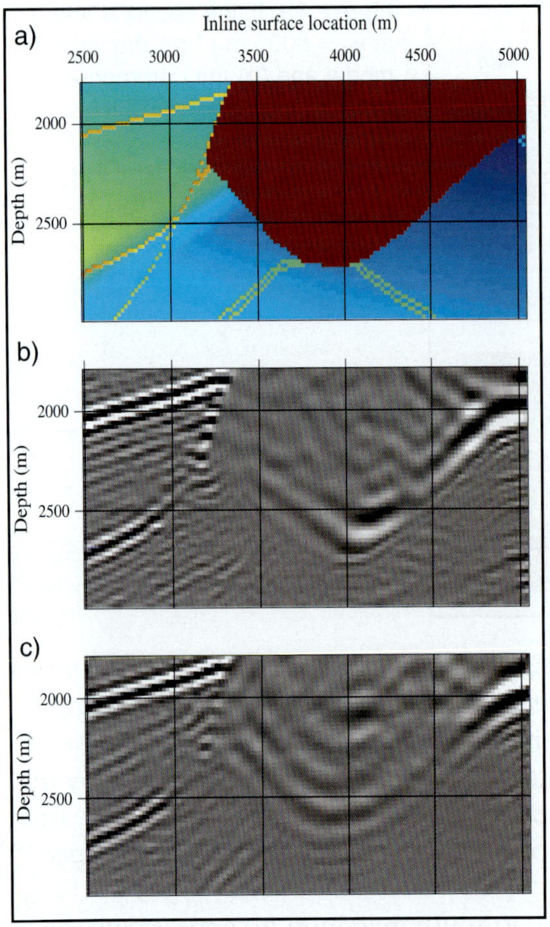

Figure 20. Windows around the reflector of interest of (a) the velocity model, (b) the common-azimuth migration, and (c) the narrow-azimuth migration.

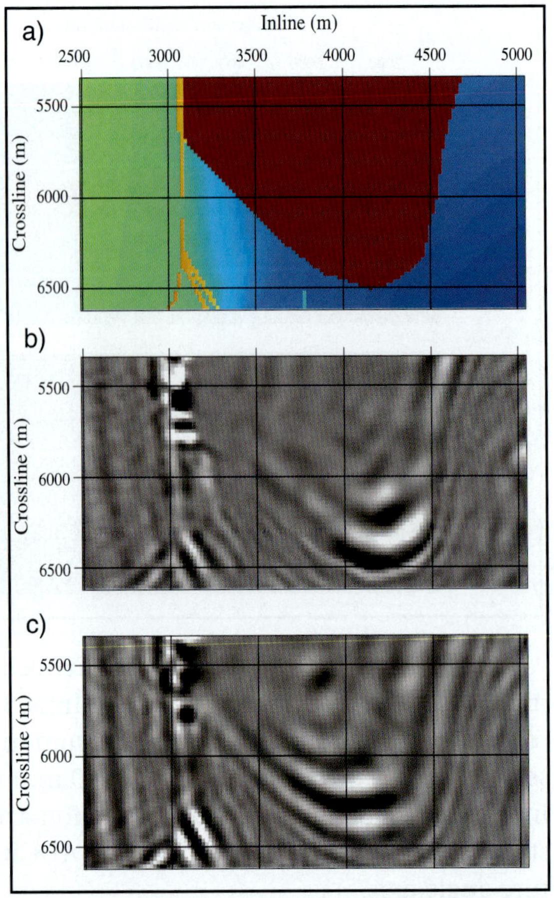

Figure 21. Depth slices taken at z = 2400 m of (a) the velocity model, (b) the common-azimuth migration, and (c) the narrow-azimuth migration.

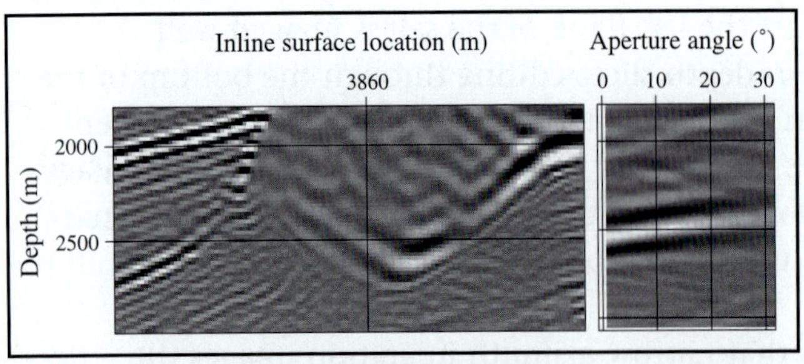

Figure 22. A stacked inline section (left panel) and an ADCIG taken at x_m = 3860 m (right panel), produced by common-azimuth migration.

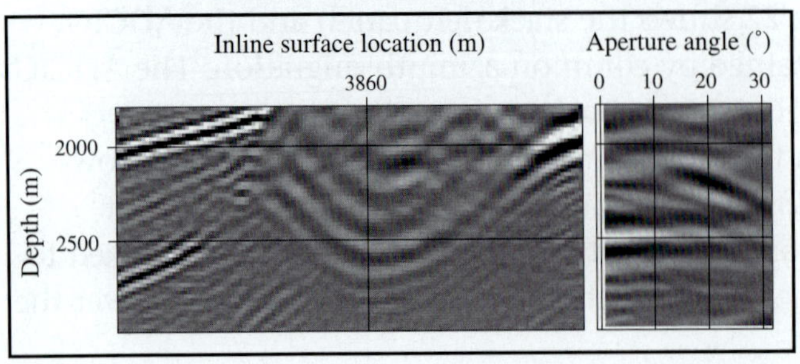

Figure 23. A stacked inline section (left panel) and an ADCIG taken at x_m = 3860 m (right panel), produced by narrow-azimuth migration.

azimuthal range of $-30° \leq \phi \leq 30°$. Stacking over azimuths would strongly attenuate the events with wide aperture angles, unless a proper Jacobian correction is applied to the ADCIGs before stacking (Biondi, 2003b). The ADCIG shown in Figure 23 was obtained after application of the appropriate Jacobian correction.

References

Audebert, F., D. Nichols, T. Rekdal, B. Biondi, D. E. Lumley, and H. Urdaneta, 1997, Imaging complex geologic structure with single-arrival Kirchhoff prestack depth migration: Geophysics, **62**, 1533–1543.

Biondi, B., 2003a, Narrow-azimuth migration of marine streamer data: 73rd Annual International Meeting, SEG, Expanded Abstracts, 897–900.

———, 2003b, Amplitude balancing of 3-D angle-domain common-image gathers: Stanford Exploration Project, Report 114, 45–56.

———, 2006, 3D seismic imaging: SEG Investigations in Geophysics, No. 14.

Biondi, B., and G. Palacharla, 1996, 3-D prestack migration of common-azimuth data: Geophysics, **61**, 1822–1832.

Biondi, B., and G. Shan, 2002, Prestack imaging of overturned reflections by reverse time migration: 72nd Annual International Meeting, SEG, Expanded Abstracts, 1284–1287.

Biondi, B., and W. W. Symes, 2004, Angle-domain common-image gathers for migration velocity analysis by wavefield-continuation imaging: Geophysics: **69**, 1283–1296.

Bonomi, E., and L. Cazzola, 1999, Prestack imaging of compressed seismic data: A Monte Carlo approach: 69th Annual International Meeting, SEG, Expanded Abstracts, 1914–1917.

Cazzola, L., L. Pizzaferri, L. Ratti, G. Cardone, and E. Bonomi, 2004, An example of wavefield depth migration and Monte Carlo imaging in West Africa deep waters: 74th Annual International Meeting, SEG, Expanded Abstracts, 1037–1040.

Cerveny, V., and I. Pšenčík, 1983, Gaussian beam and paraxial ray approximation in three-dimensional inhomogeneous media: Bulletin of the Seismological Society of America, **70**, 47–77.

Duquet, B., P. Lailly, and A. Ehinger, 2001, 3-D plane wave migration of streamer data: 71st Annual International Meeting, SEG, Expanded Abstracts, 1033–1036.

Etgen, J., 2002, Waves, beams and dimensions: An illumination if incoherent view of the future of migration: Presented at the 72nd Annual International Meeting, SEG.

Gray, S. H., 2005, Gaussian beam migration of common-shot records: Geophysics, **70**, S71–S77.

Higginbotham, J. H., Y. Shin, and D. V. Sukup, 1985, Directional depth migration: Geophysics, **50**, 1784– 1796.

Hill, N. R., 1990, Gaussian beam migration: Geophysics, **55**, 1416–1428.

———, 2001, Prestack Gaussian-beam depth migration: Geophysics, **66**, 1240–1250.

Jing, X., C. Finn, T. Dickens, and D. Willen, 2000, Encoding multiple shot gathers in prestack migration: 70th Annual International Meeting, SEG, Expanded Abstracts, 786–789.

Liu, F., D. W. Hanson, N. D. Whitmore, R. S. Day, and R. H. Stolt, 2006, Toward a unified analysis for source plane-wave migration: Geophysics, **71**, S129–S139.

Morton, S. A., and C. C. Ober, 1998, Faster shot-record migrations using phase encoding: 68th Annual International Meeting, SEG, Expanded Abstracts, 1131–1134.

Mosher, C. C., D. J. Foster, and S. Hassanzadeh, 1997, Common angle imaging with offset plane waves: 67th Annual International Meeting, SEG, Expanded Abstracts, 1379–1382.

Nichols, D. E., 1994, Imaging complex structures using band limited Green's functions: Ph.D. thesis, Stanford University.

———, 1996, Maximum energy traveltimes calculated in the seismic frequency band: Geophysics, **61**, 253–263.

Notfors, C., S. Gray, J. Sun, J. Young, Y. Zhang, and L. Chernis, 2003, Delayed-shot 3D prestack depth migration: 73rd Annual International Meeting, SEG, Expanded Abstracts, 1027–1030.

Ottolini, R., and J. F. Claerbout, 1984, The migration of common-midpoint slant stacks: Geophysics, **49**, 237–249.

Romero, L. A., D. C. Ghiglia, C. C. Ober, and S. A. Morton, 2000, Phase encoding of shot records in prestack migration: Geophysics, **65**, 426–436.

Schultz, P. S., and J. F. Claerbout, 1978, Velocity estimation and downward-continuation by wavefront synthesis: Geophysics, **43**, 691–714.

Shan, G., and B. Biondi, 2004, Imaging overturned waves by plane-wave migration in tilted coordinates: 74th Annual International Meeting, SEG, Expanded Abstracts, 969–972.

Stoffa, P. L., P. Buhl, J. B. Diebold, and F. Wenzel, 1981, Direct mapping of seismic data to the domain of intercept time and ray parameter — A plane-wave decomposition: Geophysics, **46**, 255–267.

Stolt, R. H., 1978, Migration by Fourier transform: Geophysics, **43**, 23–48.

Stork, C., and J. Kapoor, 2004, How many P values do you want to migrate for delayed shot wave equation migration?: 74th Annual International Meeting, SEG, Expanded Abstracts, 1041–1044.

Temme, P., 1984, A comparison of common-midpoint single-shot and plane-wave depth migration: Geophysics, **49**, 1896–1907.

Treitel, S., P. R. Gutowski, and D. E. Wagner, 1982, Plane-wave decomposition of seismograms: Geophysics, **47**, 1375–1401.

van Trier, J., and W. W. Symes, 1991, Upwind finite-difference calculation of traveltimes: Geophysics, **56**, 812–821.

Versteeg, R., and G. Grau, 1990, Practical aspects of seismic data inversion, the Marmousi experience: 52nd Annual International Conference and Exhibition, EAEG, Extended Abstracts, 1–194.

Wervelman, R., B. Salomons, and E. van Dedem, 2003, Enhancements in central North Sea subsalt imaging — An example of the use of multi-valued pre-stack depth migration: 65th Meeting, EAGE, Expanded Abstracts, E04.

Whitmore, N. D., 1995, An imaging hierarchy for common-angle plane wave seismograms: Ph.D. dissertation, University of Tulsa.

Whitmore, N. D., and J. D. Garing, 1993, Interval velocity estimation using iterative prestack depth migration in the constant angle domain: The Leading Edge, **12**, 757–762.

Chapter 6 Principles of Velocity Analysis

Introduction

In previous chapters, when I analyzed methods for imaging zero-offset and prestack data, I assumed that the velocity function was known either for rms velocity or for interval velocity. Of course, in reality that is not true, and velocity must be estimated from the seismic data.

Estimation of velocity from seismic data is an ill-posed inverse problem, in the sense that the data do not contain all the necessary information to define a velocity function that varies arbitrarily with depth and along the horizontal directions. Fortunately, in many practical situations, seismic imagers have a priori knowledge of the behavior of the velocity function. That knowledge complements the information contained in the seismic data, and the combination of these two sets of information — the information contained in the data and the a priori knowledge — often is sufficient to constrain the problem adequately.

A typical example is velocity estimation in a horizontally layered medium using the Dix inversion formula. If one assumes that the subsurface is layered horizontally and is isotropic, one can measure the rms velocity function from the data, and one can apply the Dix formula to estimate the interval velocity function. However, in the presence of structural dips and lateral variations in velocity, one must use more sophisticated analytic methods or risk estimating grossly inaccurate velocity functions. The drawback of using methods that are more general than the Dix formula is that velocity estimation becomes poorly constrained as one adds degrees of freedom to the set of admissible solutions.

Velocity plays a dual role in seismic imaging: (1) It affects focusing of the data and (2) it determines positioning of the imaged reflectors in physical space. Consequently, we call focusing velocity the velocity component that affects the data focusing and mapping velocity the velocity component that affects mapping of the reflectors. The V_{rms} function used by Kirchhoff time migration is a well-known example of a focusing-velocity function (Chapter 2). The interval velocity used to perform map migration of time-migrated images is an example of a mapping-velocity function. Focusing and mapping velocities obviously are correlated and, in some cases, they are defined by a single velocity function and are estimated simultaneously. For example, in depth migration, the interval velocity function determines both the focusing and the mapping of the data. However, it is important to distinguish the two velocity components because their estimates have different characteristics. Furthermore, they often are estimated sequentially, with estimation of the focusing velocity preceding estimation of the mapping velocity.

Estimation of the focusing velocity is linked strictly to an imaging operator. For example, estimation of interval velocity using the Dix formula is based on the simplest imaging procedure, that is, on NMO + stack (Chapter 4). That link is the main reason for discussing imaging methods in this book (e.g., stacking, DMO, migration, etc.) before discussing velocity estimation methods.

Focusing velocity is determined nonuniquely from the data; that is, many choices of a velocity function may focus the data equally well. Furthermore, because imaging operators are not linear functions of the focusing velocity and because coherent noise is present in the data (e.g., multiple reflections), several locally optimal solutions may exist. Thus, the convergence to a globally optimal solution is not assured. However, at least in principle, the data contain all the information required to determine a velocity function that focuses the reflections properly.

In contrast, mapping velocity is determined from the seismic data only through its correlation with the focusing velocity. Therefore, the definition of a velocity function that correctly maps reflectors at depth often requires other information (e.g., well data, geologic models) in addition to surface seismic data. In this book, I will cover methods that estimate the mapping velocity from surface seismic data. I will not discuss in detail ways to constrain the inversion for the mapping velocity using information from data other than that contained in the surface seismic data. The problem of constraining the inversion by using nonsurface seismic data is important and challenging, but it is beyond the scope of this book.

Anisotropy in the velocity function (Tsvankin, 2001) is another important challenge that relates strictly to the need for additional data. Anisotropy has the largest effects on the mapping velocity, although in some cases, it also affects the focusing velocity. Thus, estimating velocity by including an anisotropy term is an even more poorly constrained problem than estimating an isotropic velocity model is. The task often is impossible without using additional geophysical or geologic data.

Most velocity-estimation methods are based on measurements of the kinematics of the reflections. An important differentiation among the wide variety of such methods is whether the kinematics are measured directly from the data in the time domain or are measured from the results of migration in the image domain. This chapter presents the methods defined in the data domain, whereas the following chapter presents some of the methods based on migration. When geologic structures are simple and lateral velocity variations are smooth, the data-domain methods usually provide robust and accurate estimates of both the focusing and mapping velocities. However, when geologic structures and/or the velocity function are complex, data-domain methods often fail. Then we need to leverage the power of migration to focus and simplify the data. That will enable us to reliably measure the kinematics of the reflections and converge toward accurate velocity estimates.

Data-domain methods are simple and are well suited for introducing several of the concepts needed for developing the more complex method based on migration. This chapter presents the data-domain methods, in order of increasing complexity of both the methodology and the geologic setting in which the methods are appropriate. I start with the simplest method for inverting stacking velocities, which was introduced by Dix (1955) and which is valid for horizontally layered media. Second, I examine the situations in which the Dix formula can be extended beyond its assumption of a layered earth, and I include examples illustrating the limitations and perils of such extensions. Finally, I introduce the basic concepts of reflection tomography (Bishop et al., 1985;

Stork and Clayton, 1991), which can be applied in the presence of mild structures and velocity variations with wavelengths of the same order as the data's maximum offset.

Flat Reflectors in $V_{\mathrm{rms}}(\tau)$ Media

I define focusing velocity as the velocity function that enables an imaging algorithm to focus the data. Therefore, the focusing velocity needs to describe accurately the relative delays between events reflected from the same discontinuity in the subsurface and recorded at different surface locations. The first step of a velocity-estimation procedure is to measure relative delays of reflections recorded within an ensemble of seismic traces, as a function of the trace source/receiver coordinates. A robust solution to this problem in the presence of complex geology is not trivial.

The simplest method for measuring relative delays in the data is to measure the moveout along the offset axis. When the reflections originate from a flat interface and the medium above the interface is isotropic and homogeneous, the moveout is hyperbolic with offset. The inverse of the velocity is given by the time dip of the hyperbola's asymptote. In that case, the focusing velocity is the velocity parameter that describes the hyperbola over which one can sum the data and produce a well-focused image of the planar interface. This velocity parameter is often called the stacking velocity. Stacking velocities are measured from the data by using stacking-velocity spectra (Taner and Koehler, 1969). A velocity spectrum is computed from a CMP gather by applying an NMO correction with several constant velocities, within a given velocity range. The flatness of the NMO-corrected gathers is measured by using a coherency measure, such as semblance or stacking power. The stacking velocities corresponding to the maxima of the spectra are then picked, either by hand or automatically.

When the subsurface is stratified horizontally, the reflections from flat interfaces still are described well by hyperbolas, but the stacking velocity varies according to the depth of the interface. That is, the stacking velocity is a function of the two-way traveltime, τ_ξ, to the apex of the hyperbolas. The stacking-velocity function is an average velocity, and it cannot be used directly to transform two-way traveltimes into depth. If one knows only the stacking velocity, one cannot image the reflectors at depth properly. According to the terminology introduced above, the stacking-velocity function is the focusing velocity. To convert time to depth, one estimates the mapping-velocity function, that is, the propagation velocity or interval velocity. Often, estimation of interval velocities is a much tougher problem than estimation of average velocities, such as stacking velocity. Fortunately, in a layered medium, the stacking velocity is related to the interval-velocity function by the simple relationship

$$V_{\mathrm{rms}}^2\left(\tau_N\right)=\frac{\sum_{i=1}^{N} v^2\left(\tau_i\right)\Delta\tau_i}{\sum_{i=1}^{N}\Delta\tau_i}, \tag{6.1}$$

where $\Delta\tau_i$ is the time thickness of the ith layer and $\tau_N = \sum_i \Delta\tau_i$ is the total two-way traveltime to the bottom of the Nth layer. Because equation 6.1 expresses the root-mean-square of the interval velocity, the resulting velocity function is called rms velocity. This formula is easily invertible, and the interval velocity in each layer $v(\tau)$ can be estimated from the rms velocity function, $V_{\rm rms}(\tau)$, by applying the following formula, also known as the Dix formula (Dix, 1955):

$$v^2(\tau_N) = \frac{\tau_i V_{\rm rms}^2(\tau_i) - \tau_{i-1} V_{\rm rms}^2(\tau_{i-1})}{\Delta\tau_i} . \quad (6.2)$$

Once the interval velocity is known as a function of the vertical traveltime τ_ξ, the time-to-depth conversion is defined by the transformation

$$z_j = \frac{1}{2} \sum_{i=1}^{j} v(\tau_i) \Delta\tau_i , \quad (6.3)$$

where z_j is the depth at the bottom of the *j*th layer.

In horizontally layered media, the rms-velocity function can be measured directly from the data as the stacking-velocity function, because in a layered medium, the two velocities are approximately equivalent. Therefore, in such simple media, estimation of the mapping velocity from the data is a well-posed problem, although in practice, the recursive inversion of equation 6.1 by the Dix formula in equation 6.2 often can be unstable unless the rms-velocity function has been smoothed properly. In contrast, in more complex media, the stacking velocity is not equal to the rms velocity, and estimation of an interval-velocity function from an average-velocity function is far from unique. In the next section, I explore some of the issues arising in complex media, with the help of examples from the SEG-EAGE salt data set.

Dipping Reflectors in $V_{\rm rms}(\tau)$ Media

The next step toward more complex and realistic media is to consider media that have depth-dependent velocity [$V_{\rm rms}(\tau)$ media] but that contain dipping reflectors. In Chapter 4, I showed that even in a homogeneous medium, the stacking velocity measured from dipping reflectors is not equal to the medium's velocity. Stacking velocities are biased by a factor proportional to the inverse of the cosine of the apparent dip (equation 4.6). In a homogeneous medium, this velocity bias is removed after application of DMO (Chapter 4). Stacking velocities are biased similarly in a medium with depth-varying velocity, and DMO approximately corrects that bias. Therefore, the quality of stacking-velocity analysis greatly improves after application of DMO. Stacking velocities measured from the data after DMO often are called DMO stacking velocities. However, although DMO removes from the stacking-velocity spectra the bias caused by dips, it does not remove all the sources of errors and ambiguities in the velocity analysis. In particular, it does not resolve ambiguities related to the fact that the zero-offset time of reflection from a dipping reflector is larger than the two-way vertical traveltime from the same dipping reflector to the surface. The following example illustrates this issue.

Figure 1 shows an inline section of the DMO-stacked cube of the SEG-EAGE salt data set. Above the salt body, the velocity function increases linearly with depth, but it varies slowly along the horizontal directions. Therefore, the data for reflections that are above the salt can be considered to be recorded in a $V_{rms}(\tau)$ medium. Figure 2 shows a depth slice of the velocity model taken at 1.5 km. The S1-S1′ vertical line in Figure 2 indicates the location of the DMO-stack section shown in Figure 1. I concentrate our attention on the reflections from the salt-sediment interface along the fault plane indi-

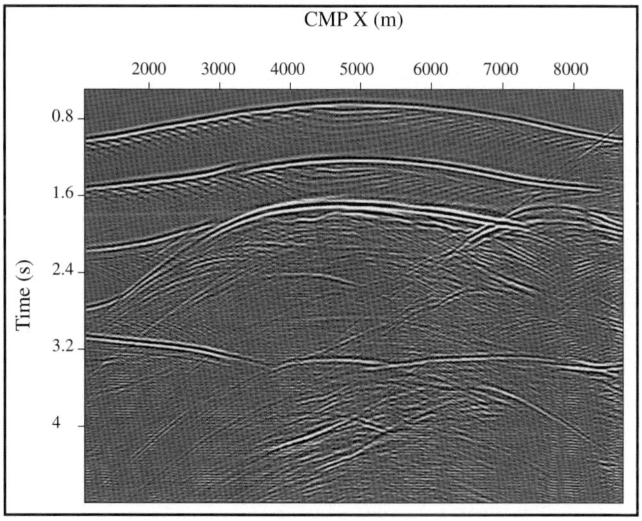

Figure 1. The S1-S1′ inline section of DMO-stacked cube of the SEG-EAGE salt data set.

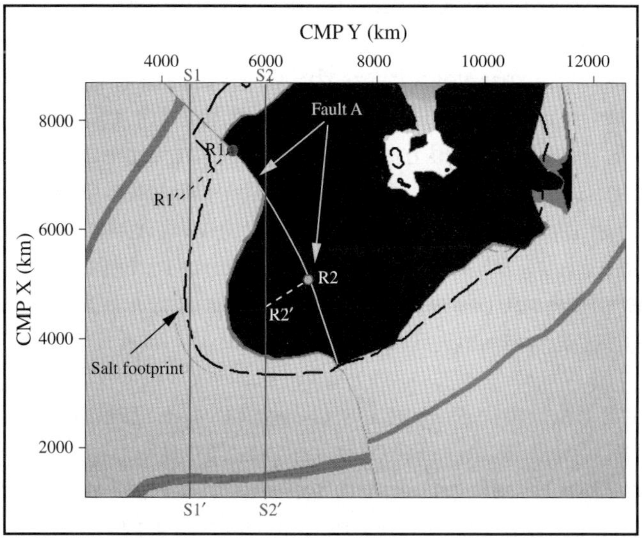

Figure 2. A depth slice of the velocity model for the SEG-EAGE salt data set. The two vertical lines (S1-S1′ and S2-S2′) superimposed onto the velocity function indicate the locations of the two data lines used for the examples discussed in this section and the next one (on dipping reflectors). The dashed lines R1-R1′ and R2-R2′ approximately show the trace of the rays sketched in Figures 3 and 8, respectively. The yellow line marked as Fault A shows the trace of a major faulting plane, visible in the 3D rendering of the model shown in Appendix B, Figure 1.

cated by Fault A in Figure 2. These top-of-salt reflections correspond to the dipping event located around CMP X = 6.5 km in the stacked section S1-S1′. Notice that the dipping event arrives later than the almost flat event corresponding to another reflection from the top of the salt.

The raypaths of these two top-of-salt events are shown in the schematics of Figure 3; the raypaths are superimposed onto an inline section of the velocity model. This inline section was taken at the crossline location that corresponds approximately to the reflection point for the dipping event, indicated as R1 in Figures 2 and 3. The raypath of the zero-offset reflection from the fault plane corresponds to the R1-R1′ lines in the figures. Notice that for the dipping event, the reflection point of the nonzero-offset reflections is moved updip with respect to the zero-offset reflection point (R1). This well-known effect is called reflection-point dispersal. DMO corrects reflection-point dispersal by moving the dipping energy along the midpoint axes and horizontally aligning nonzero-offset events with zero-offset events.

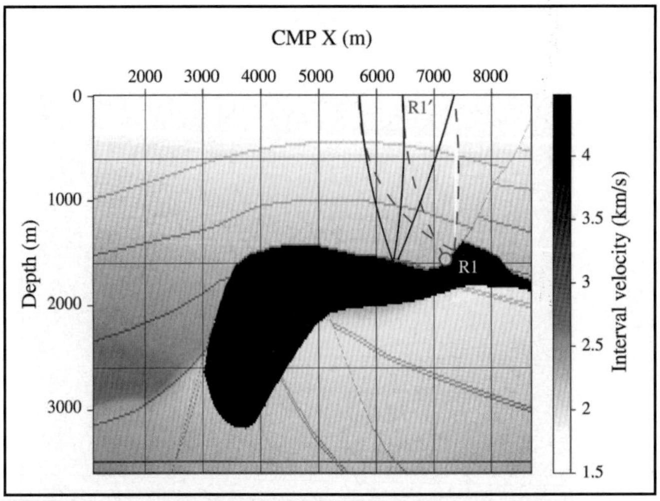

Figure 3. An inline section of the velocity model for the SEG-EAGE salt data set. This inline section was taken at the crossline location approximately corresponding to the reflection point for the dipping event, indicated as R1 in Figure 2.

Figure 4 shows the CMP gather located at CMP X = 6.6 km after simple binning and spatial interpolation. Panel (a) shows the original CMP gather. Panel (b) shows the CMP gather after application of NMO with the rms velocity at the CMP location. Panel (c) shows the velocity spectrum for the gather in (a). The moveout for the fault-plane reflection (zero-offset time is about 2 s) is overcorrected in (b) because of the dip effect. In addition, the moveout for the other top-of-salt reflection (1.8 s) is overcorrected slightly in (b) because of the dip effects. Figure 5 shows the same CMP gather shown in Figure 4 but after 3D DMO has been applied to the data. DMO has made the events in Figure 5b flatter than the events in Figure 4b. The moveout of the fault-plane reflection after DMO is now pointing slightly down because its V_{rms} velocity is lower than the V_{rms} velocity of the medium at its zero-offset time.

Comparing Figure 5c with Figure 4c, one can see that DMO reduces the stacking velocity of the fault-plane reflection by about 250 m/s. Further, because DMO corrects the reflection-point dispersal, in the post-DMO CMP gather, the reflections from the same event originated at the same point in the subsurface, independently of their offsets. However, the velocity spectra after DMO still are misleading because of the timing of the fault-plane reflection. This reflection arrives later than the other top-of-salt reflection, but it has a slightly lower velocity because the reflection point is shallower. Such an inversion of the stacking-velocity function does not correspond to a velocity inversion in the true rms-velocity function. Figure 6 shows the true rms velocity corresponding to the inline section shown in Figure 1.

Notice that at the CMP of interest in Figure 6, the rms velocity is not influenced by the presence of the salt body. That is because the crossline dip of the fault plane causes the acquisition plane S1-S1′ to be displaced with respect to reflection point R1. If one assumed the stacking-velocity function measured from the velocity spectrum in Figure 5c were equal to the rms-velocity function and then converted it to interval velocity by inverting equation 6.1, the resulting interval-velocity function would be inaccurate. In particular, it erroneously would show a low-velocity layer between 1.8 s and 2 s, because the measured rms velocity decreases with time. When structural dips are mild, carefully picking the stacking-velocity spectra and vertically smoothing the rms-velocity function can mitigate the negative effects of this phenomenon. However, when lateral variations

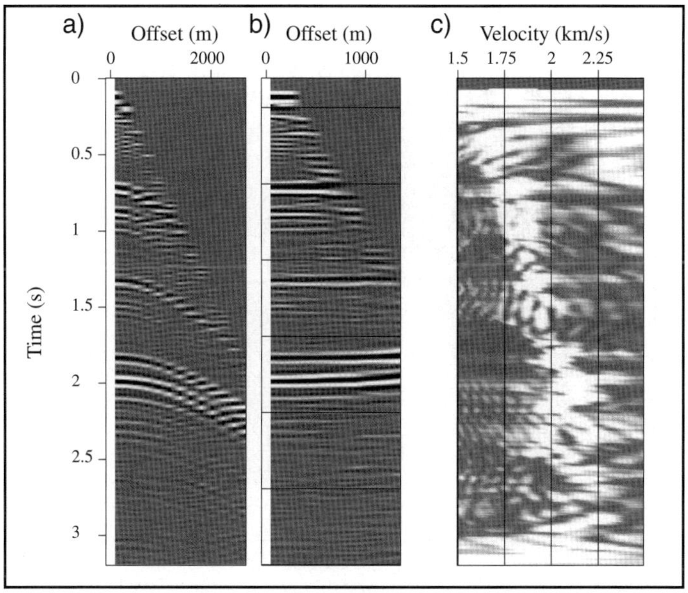

Figure 4. A CMP gather taken at CMP X = 6.6 km in the inline section shown in Figure 1. (a) The original CMP gather, (b) CMP gather after NMO, and (c) velocity spectrum for the gather in (a).

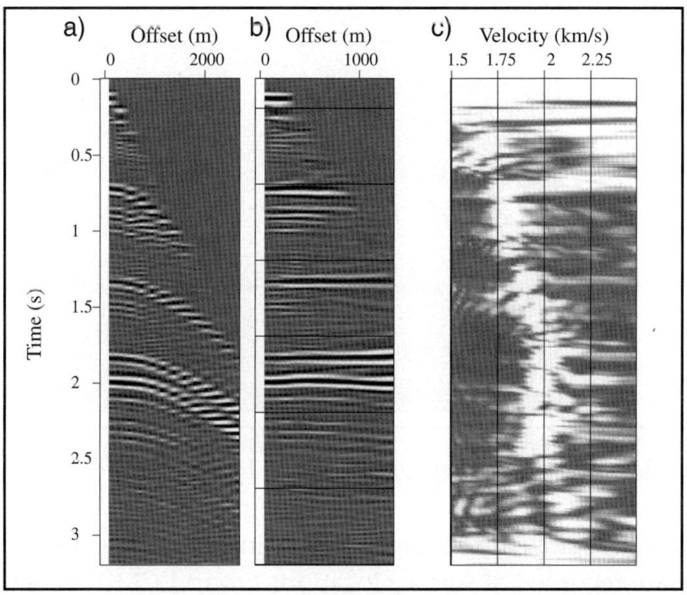

Figure 5. CMP gather after DMO, taken at CMP X = 6.6 km in the inline section shown in Figure 1. (a) Original CMP gather, (b) the CMP gather after NMO, and (c) the velocity spectrum for the gather in (a).

in velocity are present along with dipping reflectors, the dangers of estimating velocity in the unmigrated domain are more significant, as the next example demonstrates.

Dipping Reflectors in Smooth $V_{rms}(\tau, x, y)$ Media

The problems induced by mispositioning of the stacking velocity are even more troublesome when the medium is laterally inhomogeneous [i.e., in $V_{rms}(\tau, x, y)$ media]. Zero-offset arrivals from dipping reflectors not only are delayed but also are shifted horizontally, and when the velocity varies laterally, such horizontal shifts add further ambiguity to interpretation of the stacking-velocity measurements.

These problems of reflector delay and horizontal shifting are illustrated by the following example. Figure 7 shows an inline section of a DMO-stacked cube taken closer to

Figure 6. An inline section of the rms velocity function corresponding to the stack section shown in Figure 1.

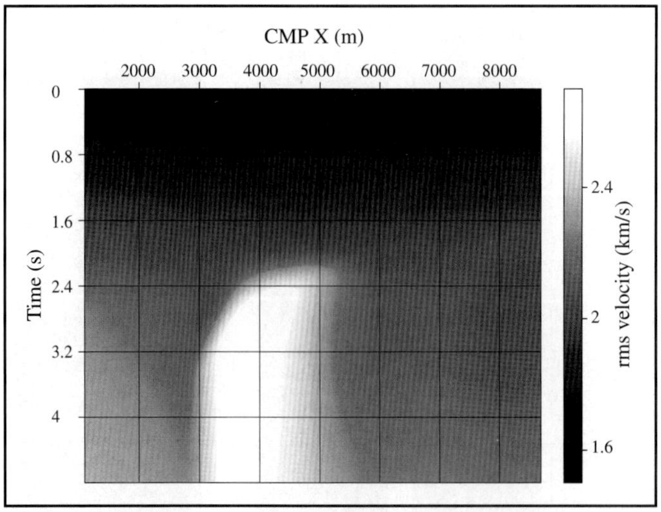

Figure 7. The S2-S2′ inline section of the DMO-stacked cube of the SEG-EAGE salt data set.

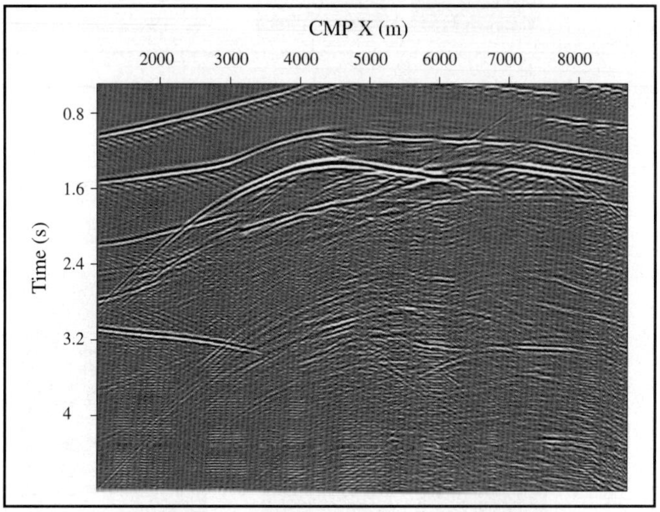

the salt body than was the inline section shown in Figure 1 (S1-S1′). The S2-S2′ vertical line in Figure 2 indicates the location of this inline section. As in the previous example, I concentrate here on the fault-plane reflections at the salt-sediment interface. The zero-offset reflections from that interface are recorded at surface locations around CMP X = 4.5 km. Those reflections interfere with the reflections from the bottom of the salt body, as shown by the raypath schematics superimposed onto the velocity model in Figure 8.

Figure 9 shows the CMP gather located at CMP X = 4.5 km after simple binning and spatial interpolation. Figure 10 shows the same CMP gather as in Figure 9 but after 3D DMO has been applied to the data. The CMP gathers before and after DMO are similar. In both gathers, the fault-plane reflections (zero-offset time is about 1.75 s) have been grossly undercorrected after NMO [panel (b)]. In addition, both velocity spectra show the incongruity between the velocity of the fault-plane reflections (slow) and the bottom of the salt reflections (fast). That incongruity can be explained easily by comparing the stacked section shown in Figure 7 with the corresponding rms-velocity function shown in Figure 11. The fault-plane reflections are recorded at the surface at a traveltime and surface location where the rms velocity is very high because of the presence of the salt

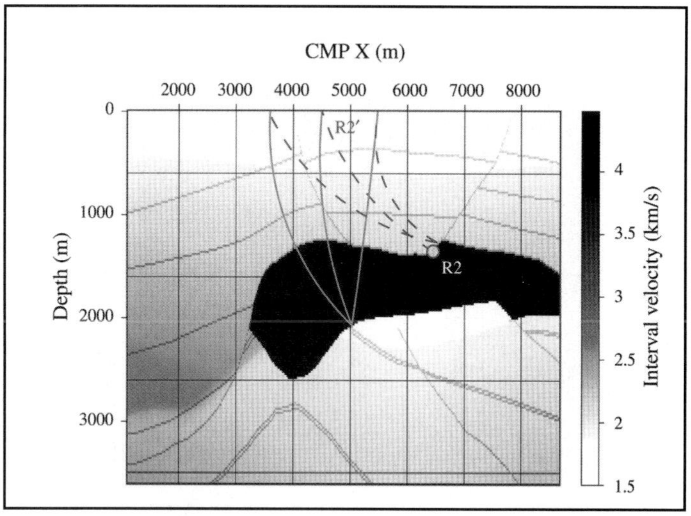

Figure 8. An inline section of the velocity model for the SEG-EAGE salt data set. This inline section was taken at the crossline location approximately corresponding to the reflection point for the dipping event, indicated as R2 in Figure 2.

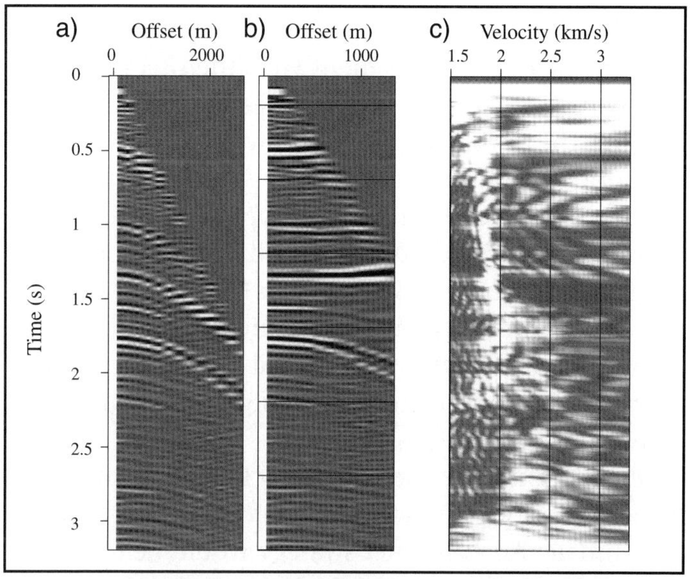

Figure 9. A CMP gather taken at CMP X = 4.5 km in the inline section shown in Figure 7. (a) The original CMP gather, (b) the CMP gather after NMO, and (c) the velocity spectrum for the gather in (a).

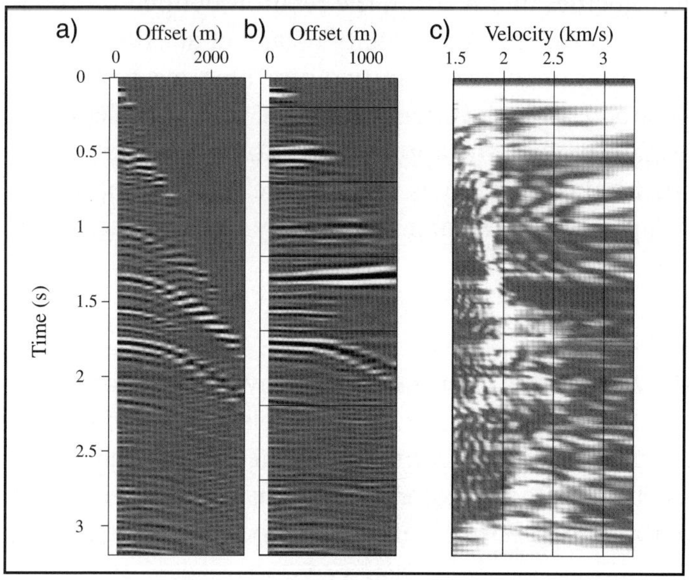

Figure 10. A CMP gather after DMO, taken at CMP X = 4.5 km in the inline section shown in Figure 7. (a) The original CMP gather, (b) the CMP gather after NMO, and (c) the velocity spectrum for the gather in (a).

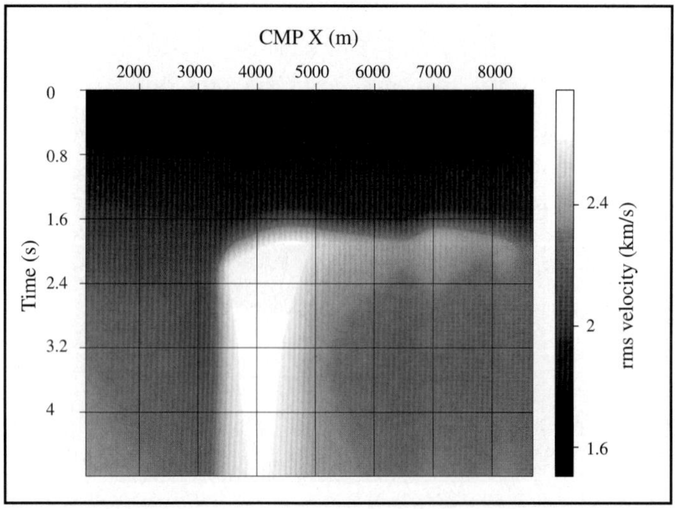

Figure 11. An inline section of the rms-velocity function corresponding to the stacked section shown in Figure 7.

body, and thus they are undercorrected by NMO. In contrast, the salt-bottom reflections travel almost vertically, so their stacking velocity is in agreement with the rms velocity.

Although the fault-plane and salt-bottom reflections are recorded at zero offset at similar times and at the same midpoint, they originated at a distant location in the subsurface, and thus they propagated along different paths and they sample different regions of the velocity function. If the stacking velocity from the dipping reflector were used for estimating the rms velocity, large errors would be introduced in the velocity estimation. The origin of the problem is that although DMO corrects for the effects of the reflector dips, it does not move the reflections to the time and lateral position of the reflection point. Events that reflect from different locations in the subsurface can cross in the unmigrated data. Only full migration (in contrast to partial migration) completely corrects these shifts and resolves the resulting ambiguities.

As was the case for simple vertical variation, here too careful picking and smoothing (both vertical and horizontal) of the rms-velocity function can mitigate the problem of introducing errors into the velocity estimate. An even safer strategy would be to estimate interval velocity from stacking velocities by inverting the relation defined by equation 6.1, using a smoothness constraint on the interval velocity (Clapp et al., 1998). If these simple precautions do not work, one can choose from two possible ways to proceed in the velocity-estimation process. If lateral velocity variations are the main problem, a tomographic method in the data domain may be the appropriate choice; the next section introduces such methods. If structural complexity is present, the velocity analysis methods based on migration, which I present in Chapter 7, are likely to be the best solution.

Traveltime Reflection Tomography

The Dix formula is based on the assumption that the earth is layered horizontally. When lateral velocity variations are sufficiently strong, traveltime delays cannot be modeled analytically as well as they are modeled in a horizontally stratified earth. The relationship between interval velocity and rms velocity expressed in equation 6.1 becomes inaccurate. Consequently, it is inappropriate to estimate interval velocity by inverting equation 6.1 — that is, by applying the Dix formula in equation 6.2. Chapter 10 in Biondi (2006) presents a simple graphical analysis of the error introduced by the Dix formula in the presence of lateral velocity variations. In these cases, estimation of interval velocity must be based on more accurate forward modeling of the traveltimes and an explicit fitting of the modeled traveltimes to measured traveltimes. Methods that estimate interval velocity by fitting modeled traveltimes to measured traveltimes using an inversion procedure generally are categorized as reflection-tomography methods.

An almost infinite number of reflection-tomography methods exists, but all are based on the same principle and face similar challenges. In this section, I review the basic principles of reflection tomography and its main shortcomings. In its simplest form, reflection tomography assumes that traveltimes of reflections are readily available. This is not true in practice, and actually obtaining accurate and robust measures of reflected traveltimes is one of the main challenges of reflection tomography.

Picking individual traveltimes directly from the data is simple conceptually and was the favorite among the earliest reflection-tomography methods (Bishop et al., 1985; Stork and Clayton, 1991). However, picking a traveltime is time-consuming and prone to error, even with relatively simple, high-quality data. Instead of picking individual traveltimes, it is more robust if one picks traveltimes and traveltime derivatives (or ray parameters) from ray-parameter spectra computed by using local slant stacks (Sword, 1987; Biondi, 1992; Billette et al., 1997). The algorithmic details for inverting these picks differ from the details for inverting individual traveltimes, but all these methods share the same fundamental principles. At the limit, the tomographic principles can be applied to invert stacking velocities measured from velocity spectra, because velocity spectra are even less noisy and easier to pick than local slant stacks are (Toldi, 1989). In general, a trade-off exists between theoretical resolution and data uncertainty among the data-domain tomographic methods. Tomography based on the picks of individual traveltimes has a higher theoretical resolution than does tomography based on the picks of average parameters, such as stacking velocity. On the other hand, loss of resolution caused by increased uncertainty in the picks may outweigh the theoretical advantages.

Inversion of the information extracted by a local stacking operator, such as local slant stacks (Sword, 1987; Billette et al., 1997) or beam stacks (Biondi, 1992), is a potentially useful compromise in this trade-off. Recently, such inversion has been revived by development of the common-reflection-point-stacking (CRS) methodology (Perroud et al., 1999) and of the multifocusing imaging method (Gurevich and Landa, 2002). The parameter picked from the spectra obtained from these local stacking operators can be inverted easily for interval velocity, as Duveneck and Hubral (2002) have shown.

Formalization of traveltime tomography as an inverse problem

Traveltime tomography estimates the interval-velocity model by minimizing the differences between the traveltimes measured from the data, (t_{obs}), and the traveltimes computed using a forward-modeling operator, (t_m). Forward modeling of the reflected traveltimes usually is accomplished by solving the Eikonal equation, either by ray tracing or by direct solutions using Eikonal solvers. Chapter 2 has more details on numerically solving the Eikonal equation.

Reflection tomography is distinguished from transmission tomography by the important fact that its traveltimes are functions not only of the velocity model but also of the reflectors' geometry. The reflectors' geometry is itself a complex function of the velocity model and usually is obtained by interpreting the migrated image. Such indirect dependence of the traveltimes on the velocity model through the reflectors' geometry increases the modeling operator's degree of nonlinearity with respect to the unknown parameters (velocity function) and consequently increases the difficulties of reflection tomography.

In mathematical terms, I can formulate the velocity-estimation problem as the minimization of the nonquadratic objective function

$$J_s(\mathbf{s}) = \|\mathbf{t}_m(\mathbf{s},\mathbf{I}) - \mathbf{t}_{\text{obs}}\|_2 , \tag{6.4}$$

as a function of the slowness vector **s**. I have explicitly highlighted in my notation the dependence of the vector of modeled traveltimes, $\mathbf{t}_m$, on a vector of parameters, **I**, that describes the reflectors' geometry.

The minimization problem expressed in equation 6.4 is poorly constrained. In most practical situations, many different slowness vectors minimize the objective function equally well, within the accuracy of the data. To constrain the solution and avoid divergence during numerical solution of the optimization problem, a model-regularization term usually is added to the simple objective function in equation 6.4. The objective function of the regularized problem is

$$J_{\text{reg}}(\mathbf{s}) = \|\mathbf{t}_m(\mathbf{s},\mathbf{I}) - \mathbf{t}_{\text{obs}}\|_2 + \mathbf{s}'\mathbf{C}_s^{-1}\mathbf{s} , \tag{6.5}$$

where $\mathbf{C}_s$ is an estimate of the covariance of the slowness function. The rapidly varying components of the velocity functions are determined poorly from reflection data, because they have little influence on the kinematics of the reflections. Therefore, **s** often is constrained to be a smooth function, and consequently $\mathbf{C}_s^{-1}$ is set to be a roughening operator. In Chapter 7, the section on improving the convergence of tomographic MVA discusses how the model covariance can be estimated from the data to improve the geologic likelihood of the final velocity model. An alternative approach for constraining the velocity function is to define it by using smooth basis functions, such as cubic splines. In some areas (e.g., the North Sea), it also is advantageous to use a geologically driven sparse parameterization of the velocity function. In Chapter 7, I also will discuss the relative advantages and disadvantages of a gridded parameterization of the velocity function compared with those of a sparse parameterization.

The optimization problem in equation 6.5 usually is solved by applying gradient-based optimization methods (Claerbout, 2004). However, the reflection traveltimes are not a linear function of the slowness model because of ray bending and the movement of the estimated reflectors. To apply gradient-based optimization algorithms, one must linearize the problem repeatedly around the current best estimate of the slowness model, ${}_i\hat{\mathbf{s}}$, and reflector geometry, ${}_i\hat{\mathbf{I}}$, where the left subscript i indicates the iteration number. The first linearization is based on the a priori estimate ${}_0\mathbf{s}$. Therefore, given the high degree of nonlinearity of the modeling operator, the final solution depends strongly on the choice of ${}_0\mathbf{s}$.

The objective function for the linearized problem at the ith iteration can be written as a function of the slowness perturbation ${}_i\Delta\mathbf{s} = {}_i\hat{\mathbf{s}} - {}_{i-1}\hat{\mathbf{s}}$:

$$
\begin{aligned}
J_{\text{reg}}\left({}_i\Delta\mathbf{s}\right) &= \left\|\left({}_{i-1}\hat{\mathbf{t}}_m + {}_i\Delta\mathbf{t}_m\right) - \mathbf{t}_{\text{obs}}\right\|_2 + \left({}_{i-1}\hat{\mathbf{s}} + {}_i\Delta\mathbf{s}\right)' \mathbf{C}_s^{-1}\left({}_{i-1}\hat{\mathbf{s}} + {}_i\Delta\mathbf{s}\right) \\
&= \left\|{}_i\mathbf{L}\,{}_i\Delta\mathbf{s} - \left(\mathbf{t}_{\text{obs}} - {}_{i-1}\hat{\mathbf{t}}_m\right)\right\|_2 + \left({}_{i-1}\hat{\mathbf{s}} + {}_i\Delta\mathbf{s}\right)' \mathbf{C}_s^{-1}\left({}_{i-1}\hat{\mathbf{s}} + {}_i\Delta\mathbf{s}\right),
\end{aligned}
\tag{6.6}
$$

where ${}_{i-1}\hat{\mathbf{t}}_m$ are the modeled traveltimes given the slowness function, ${}_{i-1}\hat{\mathbf{s}}$, and ${}_i\Delta\mathbf{t}_m$ are the perturbations in the modeled traveltimes caused by the perturbations, ${}_i\Delta\mathbf{s}$, in the slowness function. These perturbations (${}_i\Delta\mathbf{s}$ and ${}_i\Delta\mathbf{t}_m$) are linked by the operator ${}_i\mathbf{L}$, which is evaluated by linearizing the forward-modeling functional at $\mathbf{s} = {}_{i-1}\hat{\mathbf{s}}$. The next section presents the principles for evaluating the linear operator ${}_i\mathbf{L}$.

Evaluation of the linearized reflection-tomography operator

When we evaluate the linear operator ${}_i\mathbf{L}$, it is important to take into account that the modeled traveltimes, ${}_i\hat{\mathbf{t}}_m$, are functions of the current estimate of the slowness model, ${}_i\hat{\mathbf{s}}$, both directly and indirectly, through the current estimate of the reflectors' geometry, ${}_i\hat{\mathbf{I}}$ (Etgen, 1990; van Trier, 1990; Stork and Clayton, 1991). The operator ${}_i\mathbf{L}$ is made of two distinct components: ${}_i\mathbf{L}_s$, which models the perturbations in reflected traveltimes, ${}_i\Delta_s\mathbf{t}_m$, at fixed reflectors' geometry, and the chain operator ${}_i\mathbf{L}_I = {}_i\mathbf{G}_i\mathbf{H}$, which models the perturbations in reflected traveltimes, ${}_i\Delta_I\mathbf{t}_m$, caused by perturbations in the reflectors' geometry, $\Delta_i\hat{\mathbf{I}}$. The total traveltime perturbations, ${}_i\Delta\mathbf{t}_m$, are equal to the sum of these two terms, as

$$
{}_i\Delta\mathbf{t}_m = {}_i\Delta_s\mathbf{t}_m + {}_i\Delta_I\mathbf{t}_m = {}_i\mathbf{L}_s\,{}_i\Delta\mathbf{s} + {}_i\mathbf{L}_I\,{}_i\Delta\mathbf{s} = \left({}_i\mathbf{L}_s + {}_i\mathbf{G}_i\mathbf{H}\right){}_i\Delta\mathbf{s}\,. \tag{6.7}
$$

The schematic in Figure 12 shows the raypaths used for evaluating the two components of ${}_i\mathbf{L}$ that correspond to one element in the traveltime vectors. The source ray (ray$_\downarrow$) and the receiver ray (ray$_\uparrow$) are used to evaluate ${}_i\mathbf{L}_s$, whereas the normal-incidence ray (ray$_\perp$) is used to evaluate ${}_i\mathbf{H}$.

To numerically evaluate ${}_i\mathbf{L}_s$ and ${}_i\mathbf{H}$, we apply the Fermat principle to all of the relevant rays. We can apply Fermat principle because in each case, we can assume that the

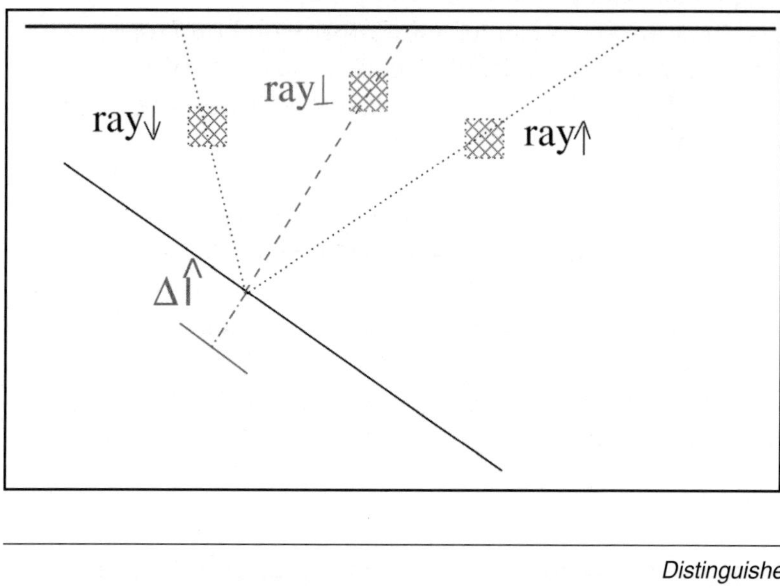

Figure 12. Raypaths used for evaluating the two components (${}_i\mathbf{L}_s$ and ${}_i\mathbf{H}$) of the linearized tomographic operator, ${}_i\mathbf{L}$. The source and receiver rays (ray$_\downarrow$ and ray$_\uparrow$, respectively) are used for ${}_i\mathbf{L}_s$. The normal-incidence ray (ray$_\perp$) is used for ${}_i\mathbf{H}$. $\Delta\hat{\mathbf{I}}$ is the reflector movement caused by velocity perturbations along the normal ray.

end points of the rays are fixed. According to the Fermat principle, the first-order perturbations in the traveltimes between two fixed points are given by the integral of the slowness perturbations along the ray connecting the two points. The slowness perturbations corresponding to the first tomographic operator (${}_{i}\mathbf{L}_{s}$) are represented by the blue boxes in Figure 12, whereas the slowness perturbations corresponding to the second tomographic operator (${}_{i}\mathbf{H}$) are represented by the red box in the same figure. The reflector movement caused by velocity perturbations along the normal ray is $\Delta\hat{\mathbf{I}}$. Application of the adjoint of ${}_{i}\mathbf{L}$ to traveltime residuals spreads (i.e., back-projects) these traveltime residuals along the raypaths into the slowness model. Thus, the adjoint of ${}_{i}\mathbf{L}$ is called the back-projection operator.

The two components of the tomography operator have opposite signs: A positive perturbation in slowness contributes both to an increase in traveltimes through operator ${}_{i}\mathbf{L}$ and to a decrease in traveltimes through ${}_{i}\mathbf{H}$ (Biondi, 2006). The total traveltime perturbations (${}_{i}\Delta\boldsymbol{t}_{m} = {}_{i}\Delta_{s}\boldsymbol{t}_{m} + {}_{i}\Delta_{l}\boldsymbol{t}_{m}$) usually decrease as the aperture angle decreases. They are exactly zero for normal-incidence events, because for such events the source and receiver rays coincide with the normal-incidence ray. Because the positions of the reflectors are unknown, the errors in traveltimes caused by velocity errors cannot be measured at normal incidence. Therefore, from reflection data, we obtain no velocity information in absolute traveltimes but only in the relative delays between events reflected from the same reflector at different angles. That limitation causes the well-known depth-velocity ambiguity, which is another way to say that the mapping-velocity component is determined poorly from reflection data.

References

Billette, F., G. Lambaré, and P. Podvin, 1997, Velocity macromodel estimation by stereotomography: 67th Annual International Meeting, SEG, Expanded Abstracts, 1889–1892.

Biondi, B., 1992, Velocity estimation by beam stack: Geophysics, **57**, 1034–1047.

———, 2006, 3D seismic imaging: SEG Investigations in Geophysics, No. 14.

Bishop, T. N., K. P. Bube, R. T. Cutler, R. T. Langan, P. L. Love, J. R. Resnick, R. T. Shuey, D. A. Spindler, and H. W. Wyld, 1985, Tomographic determination of velocity and depth in laterally varying media: Geophysics, **50**, 903–923.

Claerbout, J. F., 2004, Image estimation by example: Stanford Exploration Project, http://sepwww.stanford.edu/sep/prof/index. html, accessed May 26, 2004.

Clapp, R. G., P. Sava, and J. F. Claerbout, 1998, Interval velocity estimation with a null-space: Stanford Exploration Project, Report 97, 147–156.

Dix, C. H., 1955, Seismic velocities from surface measurements: Geophysics, **20**, 68–86.

Duveneck, E., and P. Hubral, 2002, Tomographic velocity model inversion using kinematic wavefield attributes: 72nd Annual International Meeting, SEG, Expanded Abstracts, 862–865.

Etgen, J., 1990, Residual prestack migration and interval velocity estimation: Ph.D. thesis, Stanford University.

Gurevich, B., and E. Landa, 2002, Multifocusing imaging with controlled reflection-point dispersal: Geophysics, **67**, 1586–1592.

Perroud, H., P. Hubral, and G. Hocht, 1999, Common-reflection-point stacking in laterally inhomogeneous media: Geophysical Prospecting, **47**, 1–24.

Stork, C., and R. W. Clayton, 1991, Linear aspects of tomographic velocity analysis: Geophysics, **56**, 483–495.

Sword, C. H., 1987, Tomographic determination of interval velocities from reflection seismic data: The method of controlled directional reception: Ph.D. thesis, Stanford University.

Taner, M. T., and F. Koehler, 1969, Velocity spectra — Digital computer derivation and applications of velocity functions: Geophysics, **34**, 859–881.

Toldi, J. L., 1989, Velocity analysis without picking: Geophysics, **54**, 191–199.

Tsvankin, I., 2001, Seismic signatures and analysis of reflection data in anisotropic media: Elsevier Science.

van Trier, J., 1990, Tomographic determination of structural velocities from depth migrated seismic data: Ph.D. thesis, Stanford University.

Chapter 7 Migration Velocity Analysis

Introduction

In the presence of complex structures and/or strong lateral velocity variations, extracting velocity information in the data space is both inaccurate and time-consuming. In such situations, the image space is a more appropriate domain for extracting kinematic information because migration focuses and greatly simplifies the events. Even when the migration velocity is far from the true velocity, incomplete focusing of the reflections is a step in the right direction. Velocity-estimation methods that use the focusing capabilities of migration to extract kinematic information more reliably are known commonly as migration velocity analysis (MVA) methods.

MVA is an iterative process in which each iteration comprises two distinct steps: (1) The data are imaged by prestack migration and (2) the velocity function is updated on the basis of the migration results. The global convergence of this process is a first-order concern. The relationship between the velocity model and the focusing quality of the image is highly nonlinear, and consequently, the objective function optimized during MVA is nonconvex and has several local minima. The quality of the starting velocity model is crucial in assuring global convergence. The starting model usually is estimated by using one of the data-domain velocity-estimation methods presented in Chapter 6. Therefore, development of MVA methods has not made the data-domain methods obsolete. On the contrary, data-domain velocity-estimation methods are a necessary preamble to MVA.

The first challenge of MVA is extraction of kinematic information from the migrated image cube. In the data domain, the accuracy of the velocity function is determined by measuring the coherency of the data in CMP gathers along the offset axes, after application of NMO. In the image domain, the accuracy of the velocity function is determined by measuring the focusing of the reflections in the migrated image. The coherence of the image in common-image gathers (CIGs) after migration is the criterion most commonly used to judge the focusing. When offset-domain CIGs (ODCIGs) are generated (such as the ones generated by a Kirchhoff common-offset prestack migration), coherency is measured along the offset axes. When angle-domain CIGs (ADCIGs) are generated (such as the ones generated by a wavefield-continuation migration), coherency is measured along the aperture-angle (γ) and reflection-azimuth (ϕ) axes. Departure from flatness of the events in the CIGs is called residual moveout, or RMO.

As is true in the data-domain case, it often is more robust in the migrated domain to measure the residual moveout as a function of only one parameter or a few parameters instead of measuring the depth errors for each offset or angle. The choice of parameterization of the residual moveout is important for obtaining accurate measurements. In an upcoming section on RMO analysis, I discuss possible choices for the RMO function, concentrating on derivation of a functional expression for the RMO function in ADCIGs.

Extracting kinematic information in the migrated domain instead of in the data domain has another advantage. In the migrated domain, one can evaluate both the focusing of the reflections in the physical space (midpoint and depth) and the residual moveout along the offset (or angle) axes. The additional information provided by the focusing can be useful in the presence of numerous strongly diffracted events, such as those created by a rugose top-salt interface or when the recorded range of reflection-aperture angles is too narrow to provide sufficient velocity resolution. The residual focusing required for optimal imaging of the data can be measured by applying residual prestack migration and creating velocity spectra analogous to the RMO spectra. The upcoming section titled "Residual prestack migration" presents the basic concepts of this method and gives examples of its application.

When the goal of MVA is simply to estimate an accurate rms velocity function for prestack time migration (see the next section), the measurements of RMO parameters are applied directly to update the V_{rms} function. When the goal of MVA is more ambitious and an interval-velocity function is sought to perform depth migration, the kinematics errors measured from the migrated gathers must be converted into interval-velocity updates. Several methods exist that accomplish this conversion. The simplest ones are based on generalizing to the migrated domain of the Dix formula, which was presented in the discussion on unmigrated domain in Chapter 6. The RMO functions picked from the migrated gathers are interpreted to be errors in the rms velocity function. These errors are inverted then into interval-velocity updates along the vertical direction, using an updating procedure based on the Dix formula. An upcoming section presents the basic concepts underlying vertical-updating methods and discusses the limitations of this approach.

In complex situations, tomography often is needed for inverting the measured depth errors into interval-velocity updates. The basic principles presented in Chapter 6 in the discussion of traveltime reflection tomography are directly applicable to tomographic migration velocity analysis, after the depth errors are converted into time errors. The section on tomographic migration velocity analysis in this chapter will present the theory behind those methods.

Tomographic MVA is used in the most challenging imaging projects, in which simpler methods fail. However, even tomographic methods may be challenged by complex problems because of the intrinsic limitations of tomographic velocity estimation — nonlinearity of the problem and nonuniqueness of the solution. Several methods have been proposed to address these limitations and ensure that the estimation process converges to a velocity function that is consistent with a priori geologic models of the subsurface. I review some of these advanced methods at the end of this chapter.

Time-migration Velocity Analysis

The next two major sections after this one present examples in which the zero-offset events are shifted vertically and horizontally with respect to the reflectors' true positions, thereby causing ambiguities in interpretation of the stacking velocity analysis. Those ambiguities cause severe problems, not only when one tries to invert stacking velocities into interval velocities but also when one attempts the simpler task of defining a V_{rms} function that can be used for a prestack time migration. One can unravel the ambiguities in the data domain successfully by performing the velocity analysis in the migrated domain. Gardner et al. (1974) and Sattlegger (1975) first recognized that many of the problems of conventional velocity analysis in CMP gathers can be addressed by comparing the images obtained by migration of different data offsets. The following example illustrates this concept by analyzing the migration results of the SEG-EAGE salt data set along one line.

The location of the migrated line analyzed in the next set of figures is indicated in Figure 1 by the M-M′ vertical line superimposed on the depth slice (z = 1.5 km) of the velocity function. We concentrate our attention on the reflections from the salt-sediment interface along the fault plane indicated by Fault A in Figure 1.

Figure 2 shows the migrated section in a window taken around the image of the fault-plane interface. This section is the result of migrating the data within an offset

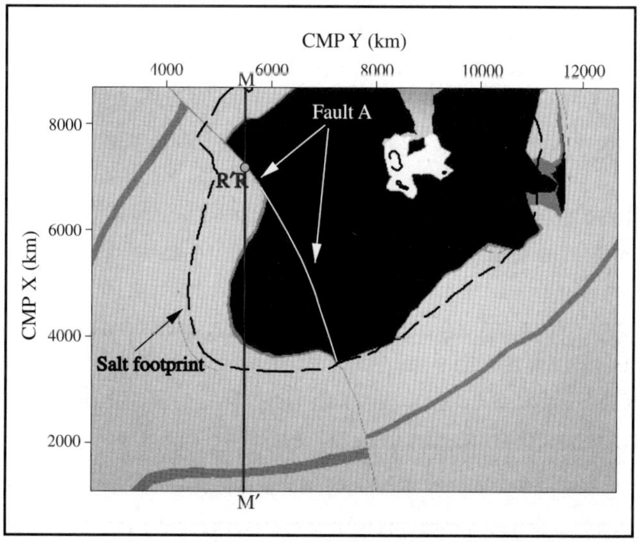

Figure 1. A depth slice of the velocity model for the SEG-EAGE salt data set. The vertical line (M-M′) superimposed onto the velocity function indicates the location of the migrated section shown in the following figures.

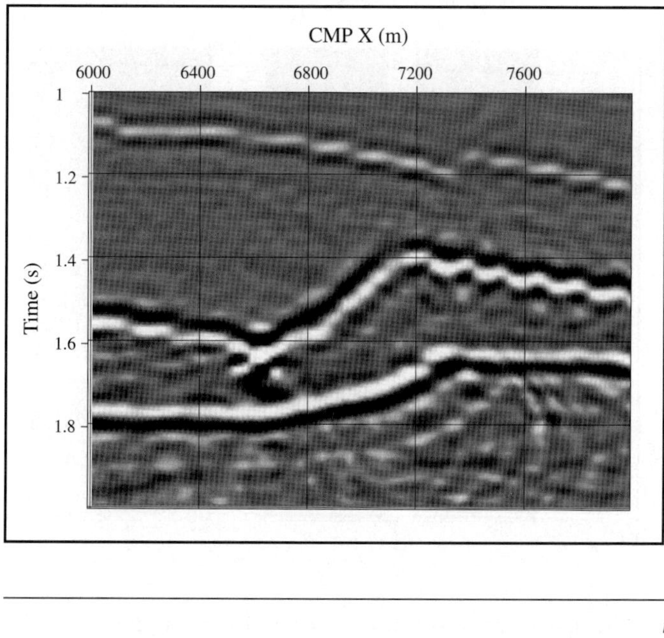

Figure 2. Migration of the fault-plane interface with the correct velocity. This section is the result of migrating the data within an offset range 240 m wide and centered at 500 m.

range 240 m wide and centered at 500 m. Figure 3a shows the ODCIG at CMP X = 7.02 km. As is true with gathers after NMO, if the velocity function used here to image the data is correct, reflections should be flat along the offset axis of the CIGs. In the ODCIG shown in Figure 3a, the fault-plane reflections have been flattened correctly by migration, as have the reflections from the interface directly above it. In contrast, the migration velocity is too slow for the salt-bottom reflections, and thus their moveout curves point upward.

One can use the ODCIGs to improve the migration-velocity function by performing a moveout analysis. CIGs have an advantage over CMP gathers after DMO. With CIGs, the events have been moved, at least approximately, to their proper locations in the subsurface. Therefore, the events in a CIG are less likely to have traveled through different regions in the subsurface and thereby to have caused inconsistency in the moveout analysis.

I shall follow a straightforward procedure to update the velocity function using the CIGs. First, I apply an inverse NMO correction to the gathers, according to the rms velocity used to migrate the data. Figure 3b shows the results of inverse NMO applied to the CIGs. Second, I compute conventional velocity spectra for the CIGs after inverse NMO. Figure 3c shows the velocity spectra computed from the gather in Figure 3b. Third, I use the stacking velocities measured from the velocity spectra to generate a new rms-velocity function. I use the new function to migrate the data again. If needed, I can repeat the process to further improve the velocity.

The procedure outlined above is illustrated better when applied to CIGs obtained with an inaccurate velocity. Figure 4 shows the migration of the same data as in Figure 2, but here with the velocity function having been scaled uniformly by a factor of 0.92. The data have been focused incompletely and show clear signs of undermigration. In particular, residual diffraction hyperbolas are visible below the point at which the fault plane terminates against the salt body (CMP X = 6.6 km). Also notice that the shape of the fault plane is distorted in the undermigrated image.

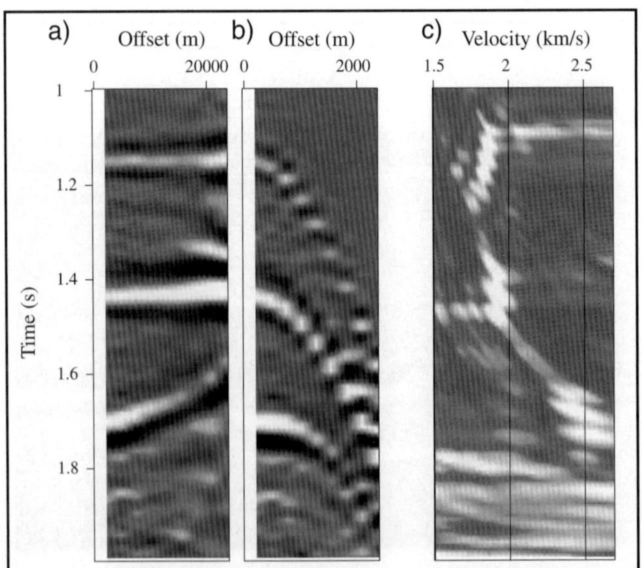

Figure 3. An ODCIG after migration with the correct velocity function. This CIG is located at CMP X = 7.02 km in the migrated section shown in Figure 2: (a) the original CIG, (b) the CIG after inverse NMO, and (c) the velocity spectrum of CIG after inverse NMO.

Figure 5a shows the ODCIG at CMP X = 7.02 km, corresponding to the migrated section shown in Figure 4. Figure 5b shows the results of inverse NMO applied to the CIG, and Figure 5c shows the velocity spectra computed from the CIG shown in Figure 5b. Comparing the velocity spectra in Figure 5c with the spectra in Figure 3c, one notices that the values of the V_{rms} at the semblance peaks corresponding to the two shallower reflections are approximately the same. As expected, the zero-offset time of the dipping event (fault reflection) is slightly lower in the CIG obtained using the lower migration velocity. However, the velocity spectra computed from the CIGs obtained with the inaccurate velocity would yield a fairly accurate V_{rms} function.

Similar results can be obtained when the migration velocity is higher than the true velocity, that is, in presence of overmigration. In that case, however, the mispositioning of dipping reflectors is even more apparent. Figure 6 shows the migration results when the correct velocity function has been scaled by a factor of 1.08. In that case, the imaged

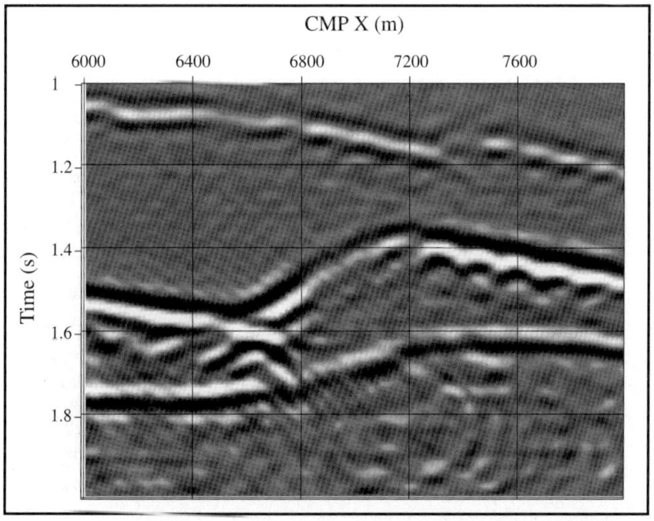

Figure 4. Migration of the fault-plane interface, with the velocity scaled by a factor of 0.92. This section is the result of migrating the data within an offset range 240 m wide and centered at 500 m.

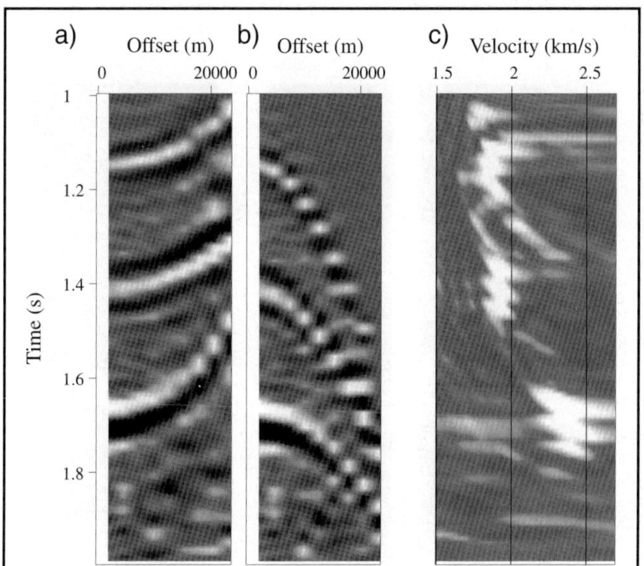

Figure 5. An ODCIG after migration with the velocity scaled by a factor of 0.92. This CIG is located at CMP X = 7.02 km in the migrated section shown in Figure 2: (a) the original CIG, (b) the CIG after inverse NMO, and (c) the velocity spectrum of CIG after inverse NMO.

dipping reflector is too deep and too steep. Typical overmigration smiles also are visible in the section. These smiles are caused by overmigration of the diffractions generated by the steplike discontinuity in the top of the salt. Now the moveouts in the CIG (Figure 7) are undercorrected and point downward. Notice the appearance of a spurious event in the CIG at about 1.4 s, which could be interpreted erroneously as a real event. This event is the result of the overmigration smiles visible in the inline section (Figure 6).

In this section, I have used the cascade of inverse NMO and forward NMO to measure the RMO in the CIGs after migration. This methodology can be applied to the results of time migration because they are defined in the time domain. However, for analyzing depth-migrated images, one needs to derive new expressions for the RMO function and to express the RMO curves as functions of depth instead of time. That task is accomplished in the next section.

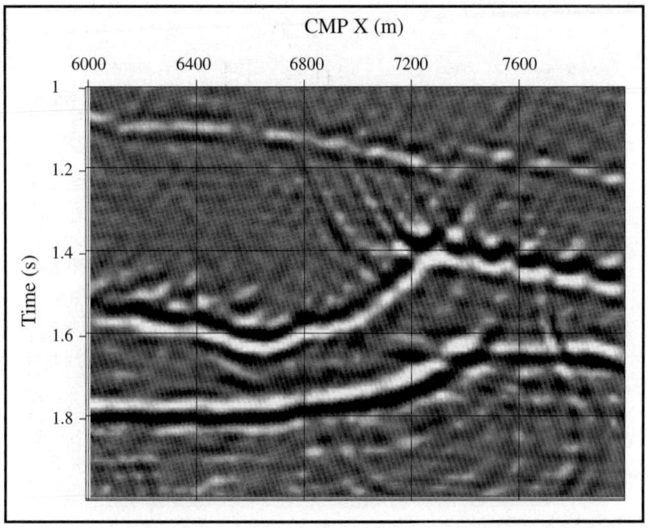

Figure 6. Migration of the fault-plane interface with the velocity scaled by a factor of 1.08. This section is the result of migrating the data within an offset range 240 m wide and centered at 500 m.

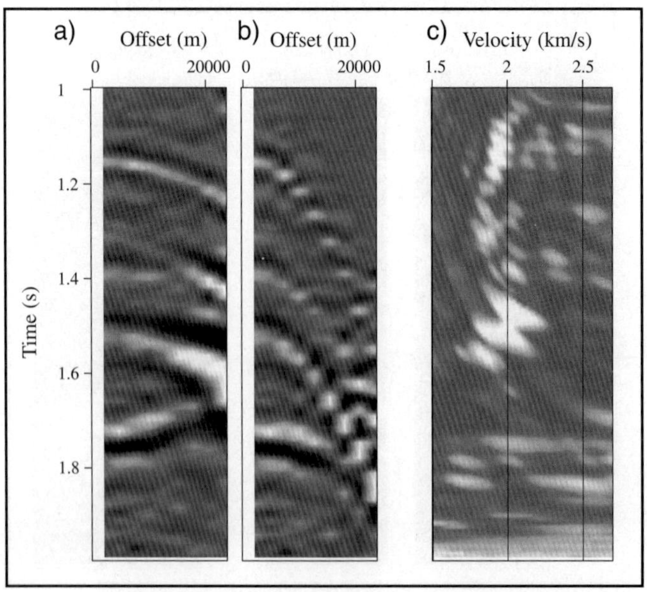

Figure 7. An ODCIG after migration with the velocity scaled by a factor of 1.08. This CIG is located at CMP X = 7.02 km in the migrated section shown in Figure 2: (a) the original CIG, (b) the CIG after inverse NMO, and (c) the velocity spectrum of CIG after inverse NMO.

Extracting Velocity Information from Prestack Images

The MVA methods I present later in this chapter update the interval-velocity function by using the velocity information contained in prestack partial images. Most MVA methods are based on measurements of residual moveouts in CIGs (see the next section). However, the focusing of the reflections in physical space (midpoint and depth) also can provide useful velocity information. Residual prestack migration, presented later in this chapter, can be used to analyze focusing of the reflections in physical space and to analyze the residual moveout along the offset (or angle) axes. Indeed, residual prestack migration is a generalization of RMO, and between these two end members exists a continuum of possible approximations that are potentially useful.

Residual moveout (RMO) analysis

Residual moveout analysis of migrated CIGs is analogous to stacking-velocity analysis of CMP gathers in the data domain. Researchers have developed methods for performing this analysis on both offset-domain CIGs and angle-domain CIGs (see Chapters 2 and 3). Because ODCIGs are computed easily by Kirchhoff migration, offset-domain RMO-analysis methods were developed first (al-Yahya, 1989; Etgen, 1990). Several methods for RMO analysis after migration in the angle domain also have been proposed (Ottolini and Claerbout, 1984; Whitmore and Garing, 1993; Jiao et al., 2002). However, application of RMO analysis of ADCIGs parameterized as a function of the reflection opening angles is a recent development (Biondi and Sava, 1999; Clapp and Biondi, 2000; Mosher et al., 2001).

Regardless of the domain in which prestack partial images are defined, the RMO function usually is parameterized by one single parameter. As I discussed in the traveltime tomography section in Chapter 6, the more parameters one uses to define an RMO function, the noisier the measurements of these parameters are likely to be. Therefore, a trade-off exists between data uncertainty and theoretical resolution. The most common choice is to parameterize the RMO function with only one parameter that describes the average of the migration-velocity error. That parameter often is chosen to be the ratio, ρ, between the average of the true migration slowness, s, and the average of the current migration slowness, s_ρ; that is, $\rho = s_\rho / s$.

In this section, I focus on the methods that parameterize the RMO function with only one parameter, but the reader should be aware that more flexible parameterizations are possible. At the limit, the depth errors could be picked for each value of the offsets and reflection angles (van Trier, 1990). Lafond and Levander (1993) proposed an intermediate solution and defined a two-parameter functional form for the RMO function. An interesting solution, which preserves the information contained in the full prestack partial images but avoids data picking, is the differential semblance optimization (DSO) method introduced by Symes and Carazzone (1991). DSO minimizes the inconsistency between the images at contiguous offsets (or aperture angles) for all the values of the offset (or angle).

- RMO functions in two dimensions

I start by discussing RMO functions in two dimensions and then generalize them to three dimensions. The simplest case is when the reflectors are flat and the RMO is defined as a function of the data's absolute offset, $|\mathbf{h}|$. The residual moveout, $\mathbf{z}_{\mathrm{RMO}}$, with respect to the zero-offset depth $z_{\rho 0_\rho} = z_\rho(|\mathbf{h}| = 0) = z/\rho$, is

$$\mathbf{z}_{\mathrm{RMO}} = z_{\rho 0}\left(\sqrt{1+\frac{1-\rho^2}{\rho^2}\frac{|\mathbf{h}|^2}{z_{\rho 0}^2}}\right)\mathbf{z}\,, \tag{7.1}$$

where $\mathbf{z}$ is the unit vector in the depth direction (Al-Yahya, 1989). Equation 7.1 can be applied directly to measure the value of ρ from ODCIGs by computing ρ-spectra and picking the semblance peaks in the ρ-spectra. However, equation 7.1 is nonlinear with respect to ρ, and its generalization to dipping events is challenging to apply (al-Yahya, 1989).

To overcome such difficulties, one can derive an RMO function that is valid for arbitrary reflectors' dips but is linear with respect to ρ = l. The linearization of equation 7.1 around ρ = 1 yields the RMO expression

$$\Delta\mathbf{z}_{\mathrm{RMO}} \approx -(\rho-1)\frac{|\mathbf{h}|^2}{z_{\rho 0}}\mathbf{z}\,, \tag{7.2}$$

which is linear in ρ but parabolic in offset. Notice that the addition of Δ in the notation indicates that it is a linearized expression. Equation 7.2 is much simpler than equation 7.1; it also is fairly accurate and often is applied in practice.

To generalize equation 7.2 to dipping reflectors, I switch to the angle domain, because the derivation and the expressions are simpler there than in the offset domain. An angle-domain RMO function is immediately applicable to ADCIGs computed either by integral migration methods or by wavefield-continuation methods (Chapter 3). Liu and Bleistein (1995) and Meng and Bleistein (2001) discuss application of angle-domain relationships to the ODCIGs obtained by Kirchhoff migration.

The sketch in Figure 8 represents the geometry of an ADCIG in two dimensions. It is analogous to Figure 30 in Chapter 2, but it illustrates the case in which the migration velocity is inaccurate (too low). Biondi and Symes (2004) demonstrated that in the angle domain, to the first order, the movement of a dipping event is along the normal

Figure 8. Geometry of a 2D ADCIG for a single event migrated with the wrong velocity. The geologic dip angle is labeled as α_x, and the aperture angle as γ. The velocity error causes the image point to shift from $\bar{\mathbf{I}}$ to $\mathbf{I}_\gamma$. Compare this figure with Figure 30 in Chapter 2.

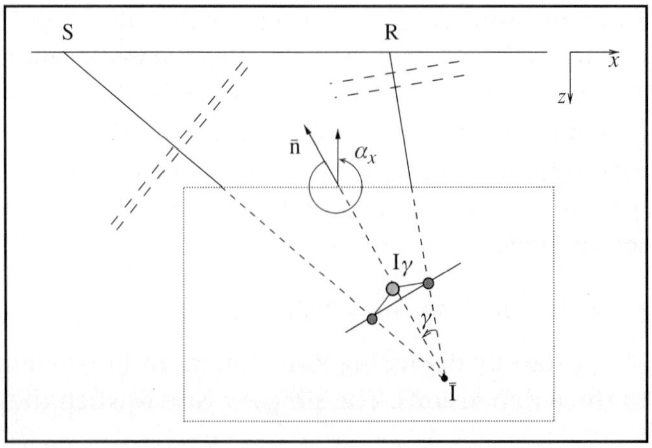

n. Therefore, the RMO expression is for the relative shift along the normal direction ($\Delta \mathbf{n}_{\mathrm{RMO}}$) rather than for the relative shift along the vertical direction ($\Delta \mathbf{z}_{\mathrm{RMO}}$).

The linearized 2D RMO function in the angle domain is

$$\Delta \mathbf{n}_{\mathrm{RMO}} = z_{\rho 0} \frac{(\rho - 1)}{\cos \alpha_x} \frac{\sin^2 \gamma}{\left(\cos^2 \alpha_x - \sin^2 \gamma\right)} \mathbf{n}. \tag{7.3}$$

According to equation 7.3, the RMO shift increases as a function of the apparent geologic dip $|\alpha_x|$. The intuitive explanation for this behavior is that the specular rays become longer as the apparent geologic dip increases, and consequently, the effects of the slowness scaling increase. In the special case of flat reflectors ($\alpha_x = 0$), equation 7.3 simplifies into

$$\Delta \mathbf{n}_{\mathrm{RMO}} = z_{\rho 0} (\rho - 1) \tan^2 \gamma . \mathbf{n}, \tag{7.4}$$

which is the angle-domain equivalent of equation 7.2. Notice that the two equations have opposite signs because the normal vector **n** in Figure 8 is defined as being directed upward.

- RMO functions in three dimensions

Two-dimensional analysis can be extended to three dimensions by following the same geometric considerations I used in Chapter 3 to generalize ADCIGs from 2D to 3D (Biondi and Tisserant, 2004). Figure 9 graphically illustrates this idea and shows the geometry of an angle-domain CIG for a single event migrated with the wrong velocity (low, in this case). However, an important conceptual difference exists between the 2D case and the 3D case. In the 2D case, one can assume that the source and receiver rays cross, even when the data were migrated with the wrong velocity. The cross will occur below the imaging point when the migration velocity is too low and above the imaging point in the opposite case. In the 3D case, this assumption is violated easily, because two rays are not always coplanar. Fortunately, this difficulty can be resolved by a plane-wave interpretation of 3D ADCIGs, as presented in Biondi (2006).

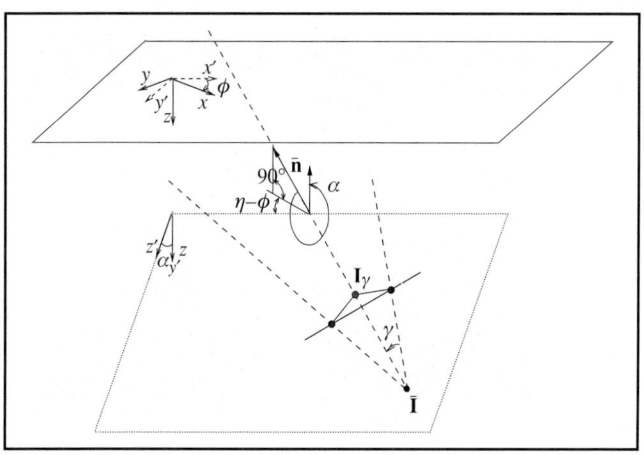

Figure 9. Geometry of an ADCIG for a single event migrated with the wrong velocity. The velocity error causes the image point to shift from $\bar{\mathbf{I}}$ to $\mathbf{I}_\gamma$. Compare this figure with Figure 21 in Chapter 3.

The final result for the RMO function in 3D is

$$\Delta \mathbf{n}_{\mathrm{RMO}} = z_{\rho 0} \frac{(\rho - 1)}{\cos\alpha} \frac{\sin^2\gamma}{\left(1 - \sin^2\alpha\cos^2(\eta - \phi) - \sin^2\gamma\right)} \mathbf{n}, \tag{7.5}$$

where α is the geologic dip, η is the azimuth angle of the normal to the geologic dip, and ϕ is the azimuth angle of the reflected event, as in Figure 9.

In practice, it is simpler computationally to apply the RMO function at a fixed surface location, that is, as a vertical shift ($\Delta \mathbf{z}_{\mathrm{RMO}}$) rather than as a normal shift ($\Delta \mathbf{n}_{\mathrm{RMO}}$). To derive the $\Delta \mathbf{z}_{\mathrm{RMO}}$ functions from the $\Delta \mathbf{n}_{\mathrm{RMO}}$ functions, one merely applies the relationship $\Delta \mathbf{z}_{\mathrm{RMO}} = \Delta \mathbf{n}_{\mathrm{RMO}} \cos\alpha$. In the presence of dipping events, the vertical RMO correction might not be as accurate as the normal RMO correction because of the image-point dispersal phenomenon. After migration with an inaccurate velocity, events with different aperture angles are imaged at different horizontal locations, even if they originated at the same reflecting point in the subsurface.

- Example of 3D RMO analysis

The following example illustrates application of the 3D RMO corrections derived above and the effects of their respective approximations. Figures 10–13 show ADCIGs obtained from migrating the same synthetic common-azimuth data set used in Chapter 3 to illustrate the concept of 3D ADCIGs. The data set contains five planes, dipping at 0°, 15°, 30°, 45°, and 60° toward increasing x and y, as shown in Chapter 3, Figure 26. The azimuth of the planes is 45° with respect to the direction of acquisition. In contrast to the ADCIGs shown in Chapter 3, which were obtained by using the correct migration velocity, the migration velocity was too slow in the present case. The data were migrated with a velocity function 3% slower than the correct one; that is, the slowness function was scaled by 1.03 at every depth level.

All of the following figures show slices of the ADCIGs at one fixed horizontal location, with $x = y = 700$ m. At that location, there are four reflectors. Starting from the shallowest reflector, their respective dips are 0°, 15°, 30°, and 45°. Figure 10 shows two of these ADCIGs, for two reflection azimuths, $\phi = 0°$ (panel a) and $\phi = 7°$ (panel b). As expected, the ADCIG smiles upward because the migration velocity is too slow.

Figure 10. ADCIGs as functions of depth (z) and aperture angle (γ) for two reflection azimuths and at constant horizontal location (x = y = 700 m): (a) $\phi = 0°$ and (b) $\phi = 7°$.

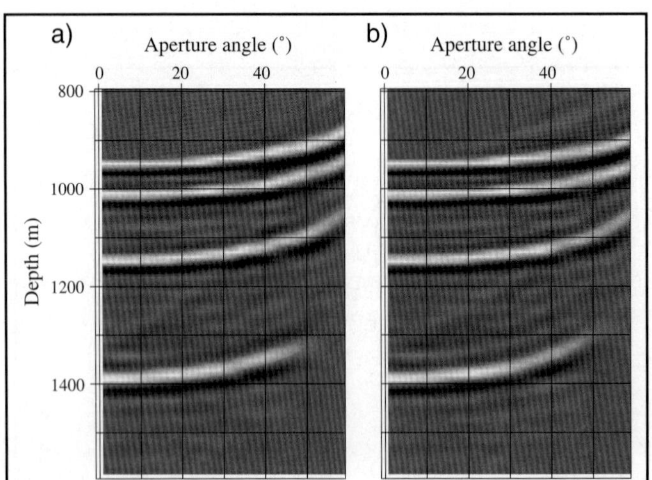

The three panels in Figure 11 display the image as a function of depth (z) and reflection azimuth (ϕ), at a constant horizontal location, for three aperture angles. The aperture angles are (a) $\gamma = 4°$, (b) $\gamma = 20°$, and (c) $\gamma = 40°$. The curvature of the events as a function of the azimuth increases with the aperture angle, indicating that the RMO depends on the reflection azimuth. However, as was the case with the correct-velocity ADCIGs shown in Chapter 3, Figure 29, this dependence occurs not because the general expression of the RMO function (equation 7.3) depends on azimuth but because the common-azimuth data illuminate only a narrow range of reflection azimuths at wide aperture angles.

The apparent reflection azimuth corresponds to the apex of the curves shown in Figure 11. From the figure, it is evident that the apparent azimuth increases with the reflectors' dip. On one end, the apparent azimuth of the flat reflector (top) is zero, and on the other end, the apparent azimuth of the steepest reflector is about 7°. A priori, the reflection azimuth is an unknown. Therefore, to avoid ambiguities, the RMO analysis of narrow-azimuth data, such as the ones in this example, should be performed on ADCIGs obtained by averaging over a range of reflection azimuths. Such stacking over azimuths would strongly attenuate the events with wide aperture angles, unless a proper Jacobian correction is applied to the ADCIGs before stacking (Biondi, 2003). The ADCIG shown in Figure 12a was obtained by averaging the ADCIGs over the azimuthal range $-20° \leq \phi \leq 20°$, after application of the appropriate Jacobian correction.

The panels in Figure 12b–d show ρ-spectra computed using three RMO functions. The panels are centered at $\rho = 1.3$, which corresponds to the actual scaling of the migration velocity. Figure 12b shows the ρ-spectrum computed using the flat reflectors' approximation (equation 7.4). As the reflector dip increases, the semblance peaks shift toward the right; that is, the value of ρ is overestimated.

Figure 12c shows application of the full 3D RMO function (equation 7.5) with the correct values of the geologic dip α but ignoring the relative rotation between the azimuth of the reflectors and the azimuth of the reflections (i.e., assuming $\eta = \phi$). In this case, the RMO function used for computing the ρ-spectrum has a larger curvature than the actual RMO in the ADCIG has. Therefore, the semblance peaks shift toward the left, and the ρ of the dipping reflectors is underestimated.

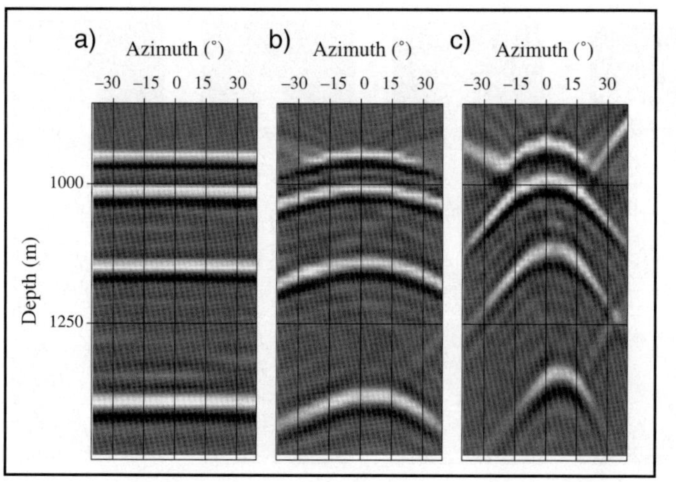

Figure 11. ADCIGs as functions of depth (z) and reflection azimuth (ϕ) for three aperture angles and at a constant horizontal location (x = y = 700 m): (a) $\gamma = 4°$, (b) $\gamma = 20°$, and (c) $\gamma = 40°$.

Finally, Figure 12d shows application of the full 3D RMO function with the correct value of η (45°) and approximately the correct value of ϕ (0°). The semblance peaks are in approximately the right positions. A residual error is caused by the fact that the background velocity increases linearly with depth, rather than being constant as the derivation of equation 7.5 assumes. Because of ray bending, the true rays are shorter than the straight rays assumed by the derivation of equation 7.5. Consequently, the traveltime errors accumulated along the true rays are lower than those predicted along straight rays, and ρ is underestimated slightly.

As I remarked above, the azimuth of the reflections is not known a priori. It is likely to fall within a range centered on the data-acquisition azimuth. Generally, it depends on the velocity model in the overburden as well as on the orientations of the data azimuth and of the geologic dips.

I will now describe a simple procedure for estimating the reflection azimuth by exploiting its effects on the RMO in the ADCIGs. The ability to estimate the reflection azimuth directly from the ADCIGs is attractive. It can improve the RMO analysis, because equation 7.5 depends on the difference between the reflection azimuth and the azimuth angle of the normal to the geologic dip ($\eta - \phi$). Furthermore, the accuracy of a tomographic updating of the interval-velocity function (discussed later in this chapter) would benefit from the knowledge of the reflection azimuth.

Estimation of the reflection azimuth starts with computation of ρ-spectra, such as the one shown in Figure 12b-d. Once the parameters of the RMO function are determined, a ϕ-spectrum can be computed by applying the fixed RMO function to all the ADCIGs with azimuths within a range of interest and then computing the semblance along the γ-axis of the moved-out gathers. Figure 13 shows an example of a ϕ-spectrum. For each reflector, the value of the apparent reflection azimuth corresponds to the semblance peaks in the panel. As expected, the semblance peak for the flat reflector is at ϕ = 0, whereas the semblance peaks shift to the right as the reflectors' dips increase. The semblance peak for the most steeply dipping reflector (45°) is at about 7°, which is consistent with the position of the apex of the curve shown in Figure 11.

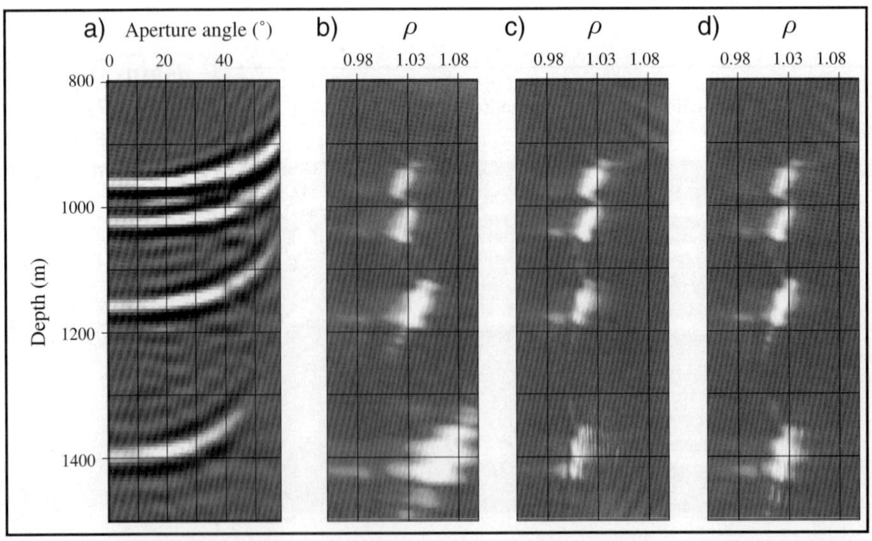

Figure 12. (a) An ADCIG and (b)-(d) corresponding ρ-spectra computed using the 3D RMO function in equation 7.5 with different sets of parameters: (b) α = 0°, (c) η = ϕ, and (d) η = 45°, ϕ = 0°.

- RMO analysis for anisotropic media

The application of RMO analysis after depth migration can be extended to the updating of anisotropic velocity functions after anisotropic depth migration (Sarkar and Tsvankin, 2004). In the anisotropic case, the RMO function becomes more complicated. More crucially, the process of updating the interval velocity becomes even more poorly constrained by the information contained solely in surface-seismic data.

Figures 14 and 15 show an example of the RMO caused by errors in anisotropic parameters in ADCIGs that are computed after wavefield-continuation prestack migration. Figure 14 shows the comparison of anisotropic prestack depth migration (panel a) with isotropic prestack migration (panel b) that we saw in Figure 22 of Chapter 2 using the anisotropic parameters displayed in Figure 21, Chapter 2. Figure 15 shows two examples of ADCIGs computed from both the anisotropic and the isotropic migration results. The CIGs shown in Figure 15a and 15b are taken at a surface location of 3725 m (left vertical black line in Figure 14), and the CIGs shown in Figure 15c and 15d are taken at a surface location of 11,625 m (right vertical black line in Figure 14). The ADCIGs obtained by anisotropic migration (panels a and c) are uniformly flatter than the ADCIGs obtained by isotropic migration (panels b and d). The ADCIGs obtained by isotropic migration display the typical hockey-stick behavior commonly seen, in the presence of anisotropy, in CMP gathers after isotropic NMO (Thomsen, 2002).

The RMO functions after anisotropic depth migration can be analyzed quantitatively by generalizing the concepts employed to derive the RMO expressions presented in equations 7.3 to 7.5 (Biondi, 2005). An important difference between the anisotropic case and the isotropic one is that in the case of anisotropy one needs to distinguish between phase and group angles and velocities. See Tsvankin (2001) for an introduction to the concepts of group and phase quantities in the context of anisotropy. Hereafter I will use the subscript G to denote group quantities and the subscript P to denote phase quantities. For flat reflectors (equation 7.4) in the simple case in which all the anisotropic parameters are scaled uniformly by ρ, the RMO function becomes the following (Biondi, 2005):

$$\Delta \mathbf{z}_{\text{RMO}} = z_{\rho 0}\left(\rho - 1\right)\tan\gamma_P \tan\gamma_G \mathbf{z}. \tag{7.6}$$

In the more general case, the anisotropic-velocity parameters are scaled independently. For a VTI medium, the scaling parameters are ρ_{V_V} for the vertical velocity, ρ_{V_H} for the horizontal velocity, and ρ_{V_N} for the NMO velocity. In the VTI case, the RMO functions are expressed as the following sum of three terms:

$$\Delta \mathbf{z}_{\text{RMO}} = {}_{V_V}\Delta \mathbf{z}_{\text{RMO}} + {}_{V_H}\Delta \mathbf{z}_{\text{RMO}} + {}_{V_N}\Delta \mathbf{z}_{\text{RMO}}, \tag{7.7}$$

where

$${}_{V_V}\Delta \mathbf{z}_{\text{RMO}} = -z_{\rho 0}\left(\rho_{V_V} - 1\right)\left[\frac{1}{S_G\left(\gamma_G\right)}\frac{\partial S_G}{\partial \rho_{V_V}}\left(1 + \tan\gamma_G \tan\gamma_P\right) + 1\right]\mathbf{z},$$

$${}_{V_H}\Delta \mathbf{z}_{\text{RMO}} = -z_{\rho 0}\left(\rho_{V_H} - 1\right)\frac{1}{S_G\left(\gamma_G\right)}\frac{\partial S_G}{\partial \rho_{V_H}}\left(1 + \tan\gamma_G \tan\gamma_P\right)\mathbf{z}, \tag{7.8}$$

and

$${}_{V_N}\Delta \mathbf{z}_{\text{RMO}} = -z_{\rho 0}\left(\rho_{V_N} - 1\right)\frac{1}{S_G\left(\gamma_G\right)}\frac{\partial S_G}{\partial \rho_{V_N}}\left(1 + \tan\gamma_G \tan\gamma_P\right)\mathbf{z}.$$

Notice the dependence on the local group velocity at the reflector. The different form for the vertical velocity component compared with that from the other two relates to the fact that at normal incidence, only $\partial S_G/\partial \rho_{V_V}$ is different from zero and equal to $-S_G(\gamma_G = 0)$.

Figure 16 shows application of the RMO functions expressed in equations 7.7 and 7.8 to predict the RMO functions measured from the ADCIG extracted from the field-data

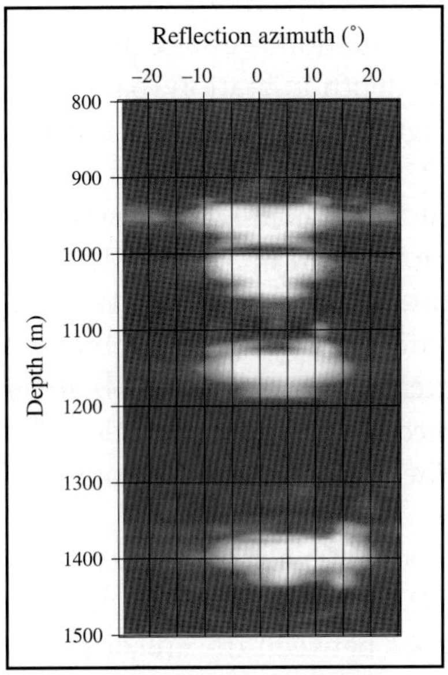

Figure 13. A semblance plot used to estimate the apparent reflection azimuth of reflections. For each reflector, the position of the semblance peaks corresponds to the estimated reflection azimuth.

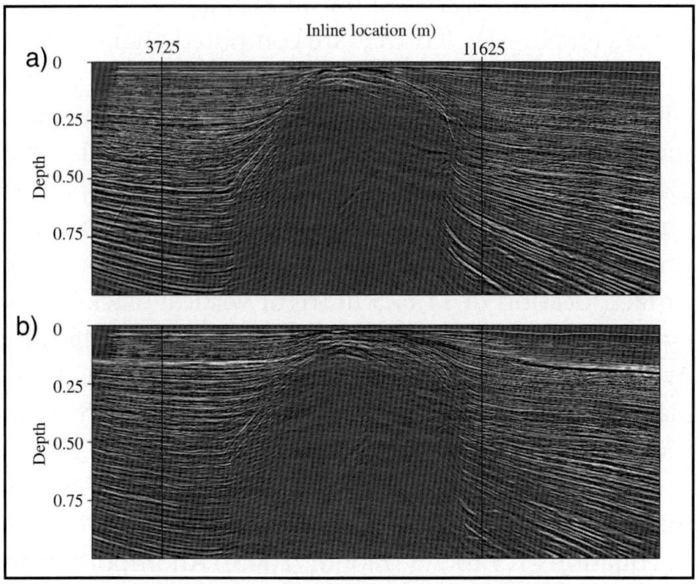

Figure 14. The same comparison of anisotropic prestack depth migration (panel a) with isotropic prestack migration (panel b) as that shown in Figure 22, Chapter 2 using the anisotropic parameters displayed in Figure 21, Chapter 2. The vertical lines superimposed onto the images indicate the locations of the ADCIGs shown the following figures.

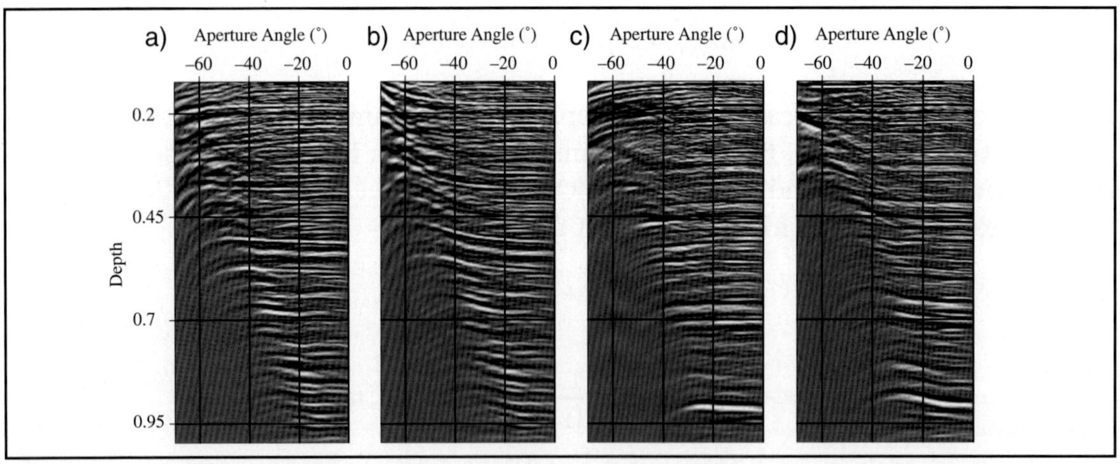

Figure 15. ADCIGs computed by anisotropic migration [(a) and (c)] and isotropic migration [(b) and (d)]. The ADCIGs shown in panels (a) and (b) are taken at the surface location of 3725 m. The ADCIGs shown in panels (c) and (d) are taken at the surface location of 11,625 m.

example. Figure 16c shows the ADCIG obtained after anisotropic migration using the following average parameters below the seafloor: V_V = 1750 m/s, ε = 0.11, δ = 0.04, and η = 0.065. Figure 16d shows the ADCIG obtained after isotropic migration using V_V = 1750 m/s. The shallow reflection of interest is flat in Figure 16c, whereas it is smiling upward in Figure 16d. For comparison, Figure 16a and 16b show a detail from Figure 15a and 15b, enlarged here to match the scale of Figure 16c and 16d. The curve superimposed onto both Figure 16b and 16d was computed using the generalized RMO functions expressed in equations 7.7 and 7.8. The computed RMO function (green line) perfectly overlaps the event in the ADCIG in Figure 16d. In contrast, the computed RMO function (red line) overestimates the moveout in the ADCIG obtained by migrating the data using the original isotropic model (Figure 16b). The cause of this discrepancy is the ray bending induced by the vertical gradient in the original heterogeneous anisotropic model.

Residual prestack migration

The RMO analysis presented in the previous section provides information on the migration-velocity errors by measuring the consistency of the prestack partial images as a function of the reflection angles (or data offsets). This CIG-flatness criterion has the advantage of being independent from the structure of the subsurface. However, it is not the only criterion for judging the quality of the migration velocity. Focusing the reflections in the physical space (midpoint and depth) often carries additional valuable velocity information. That information can be extracted from the prestack partial images by analyzing the results of residual prestack migration. This process is more interpretative and structure dependent than the RMO analysis is. Therefore, it is more challenging to automate and is used more rarely. In areas where diffracted events are numerous (e.g., rugose salt bodies), analysis of residual focusing of diffractions can provide useful information for updating interval velocities (Sava et al., 2005).

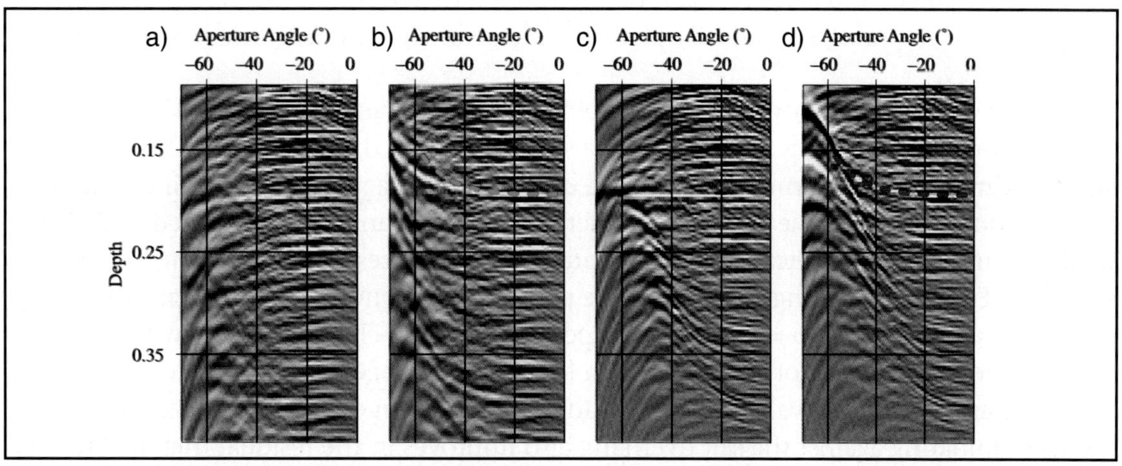

Figure 16. ADCIGs taken at the surface location of 3725 m and with the layer below the seafloor being: (a) anisotropic and heterogeneous, (b) isotropic and heterogeneous, (c) anisotropic and homogeneous (V_V = 1750 m/s, ε = 0.11, δ = 0.04, and η = 0.065), and (d) isotropic and homogeneous (V_V = 1750 m/s). The RMO curve that is superimposed onto panels (b) and (d) is computed using equations 7.7 and 7.8.

Residual migration has the added advantage over residual moveout of further reducing the effects of the image-point dispersal (discussed in a previous section) between events imaged at the same physical location but with different aperture angles. Removal of the image-point dispersal can improve the accuracy of the semblance spectra computed after residual migration, compared with the ones computed after residual moveout. One major disadvantage of residual migration is computational complexity. Residual migration of large 3D prestack partial images can be both computationally intensive and a data-handling challenge.

Rocca and Salvador (1982) introduced the concept of residual migration for zero-offset time migration. As is the case for conventional full migration, residual migration can be performed either by applying an integral operator or by downward continuation (Rothman et al., 1985; Etgen, 1990; Fomel, 2003). The CRP-scan methodology developed by Audebert et al. (1996) is another method that is related to residual migration and applicable to Kirchhoff-like migrations. However, the most computationally efficient residual migration methods are based on the frequency-wavenumber domain algorithm first proposed by Stolt (1978) to perform constant-velocity migration. Both the prestack residual-time-migration method proposed by Stolt (1996) and the prestack residual-depth-migration method proposed by Sava (2003) are based on the Stolt migration algorithm. In this section, I show results obtained by the residual-migration method presented by Sava (2003).

Prestack residual depth migration is conveniently parameterized using the same parameter used for RMO analysis, that is, the ratio, ρ, of the average of the true migration slowness, s, to the average of the actual migration slowness, s_ρ. To perform a residual-migration analysis, one can perform residual prestack migration for several values of the ρ parameters and then compute ρ-spectra by semblance analysis over the aperture-angle axis. The semblance peaks in the ρ-spectra indicate the values of ρ that improve focusing the most. Alternatively or as a complement to the semblance analysis, the stacks over the aperture angles and azimuths can be interpreted to determine the values of ρ that correspond to the most geologically plausible image or the one with the best-focused diffracted energy.

Figures 17 and 18 show an example of this structural analysis. Figure 17 shows an inline section taken from a 3D marine data set recorded in the North Sea. The image shows clear signs of undermigration. Some unfocused diffracted energy from the tip of the salt body exists, and the sediments under the salt overhang are not imaged well. The thick box in Figure 17 indicates the area corresponding to residual migration results shown in Figure 18. This figure shows nine panels corresponding to different values of ρ ranging from $\rho = 1.08$ to $\rho = 1.00$. The shape of the salt tip changes with ρ when steeply dipping energy that was not well focused in the original image comes into focus and then defocuses again at higher values of ρ. Focusing of the bottom of the salt and of the sediments terminating against the salt overhang also improves in the residual migrated results.

To gain intuition into how residual prestack migration differs from residual normal, Figure 19 shows the results of applying residual moveout to the North Sea prestack partial images. Application of residual moveout improves the coherency and the power of the stack, but it does not correct the misfocusing of the salt tip or improve the image of the sediments below the salt overhang as much as residual migration does.

Vertical Interval-velocity Updates from Measured Average-velocity Errors

Information on the average-velocity errors measured using the methodology presented in the previous section can be used to estimate interval-velocity updates in an iterative velocity-estimation scheme. The three steps of migration, RMO analysis, and

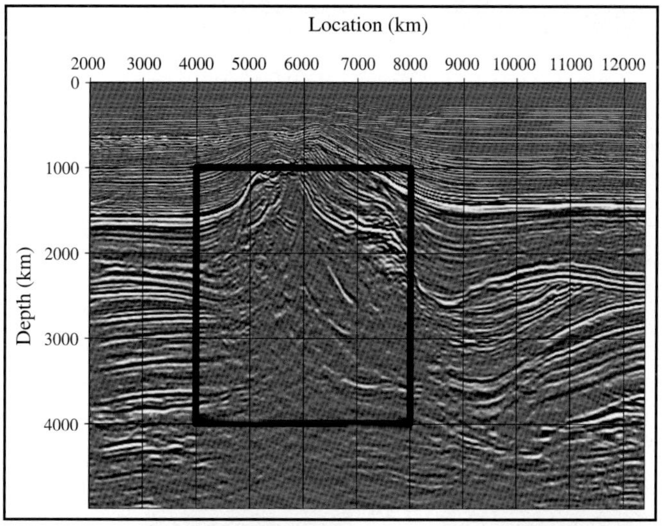

Figure 17. Migrated image of a North Sea data set. The thick box indicates the area corresponding to the residual migration and residual moveout results, respectively, shown in Figures 18 and 19. Figure is from Sava (2003).

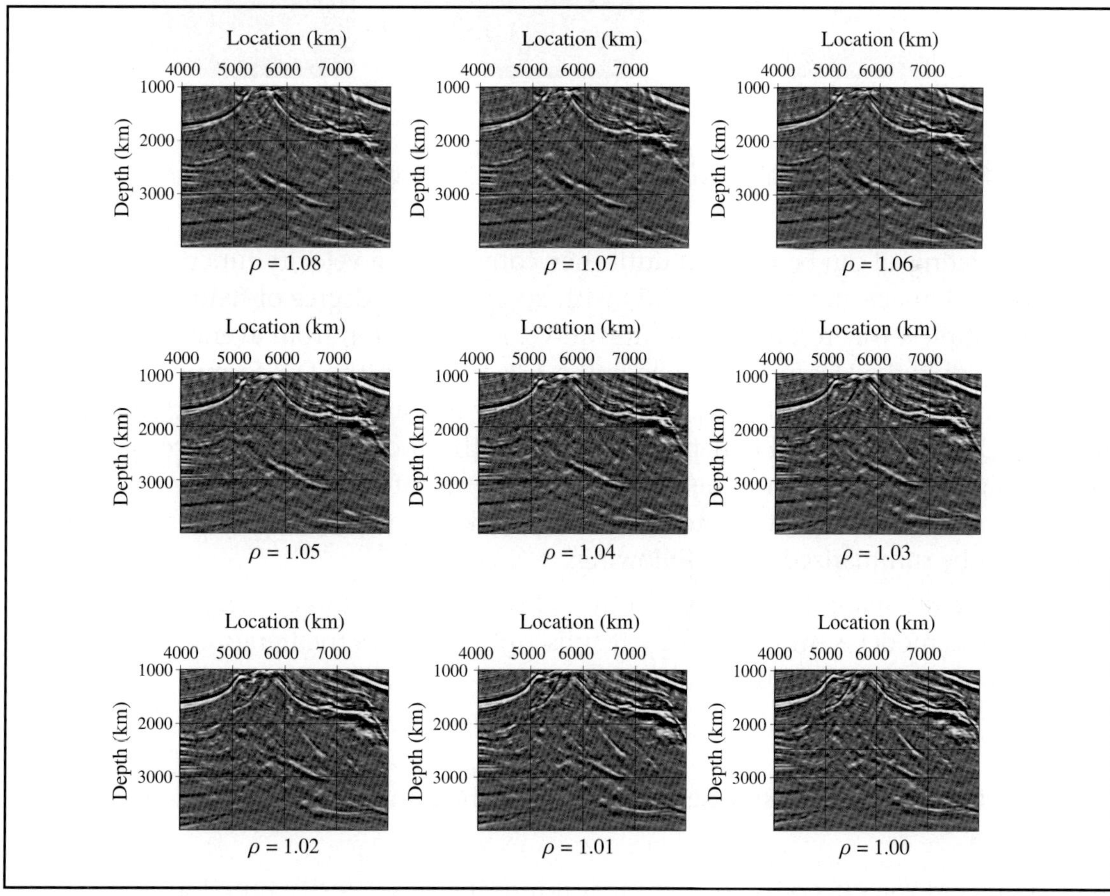

Figure 18. Stacks after residual migrations for $1 \leq \rho \leq 1.08$. Figure is adapted from Sava (2003).

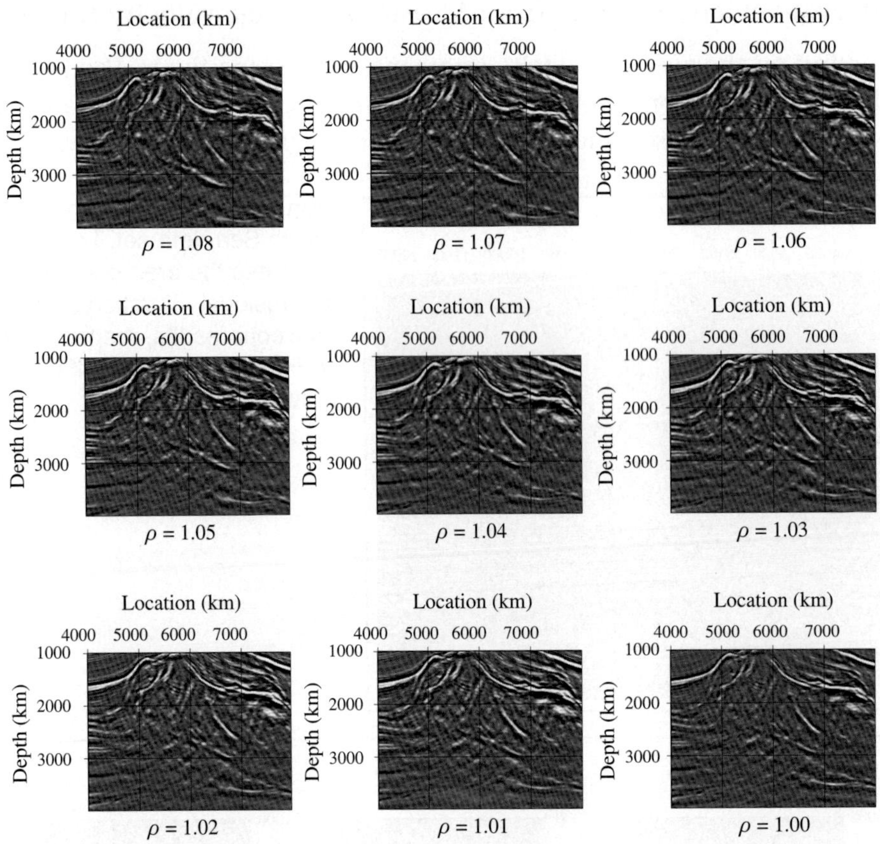

Figure 19. Stacks after residual moveout for $1 \le \rho \le 1.08$. Figure is adapted from Sava (2003).

velocity updating all can be repeated until they converge to a velocity function that yields a well-focused image and migrated CIGs with an acceptable degree of flatness.

The simplest way to update the interval-velocity function from average-velocity errors is to apply the Dix formula in equation 6.2. However, the Dix formula is expressed in two-way traveltimes, τ, whereas the velocity-error measurements obtained from depth-migrated CIGs are expressed in depth. To reconcile this inconsistency, one converts the current migration velocity and the error measurements into time, performs the updates, and then converts the updated interval velocity back into depth. The main steps of this process can be summarized as the following:

1) Evaluate the mapping between depth and two-way traveltime, using the current slowness model ${}_i\hat{s}$ (in which the left subscript i indicates the iteration number) in the following relationship:

$$\tau(z) = 2\int_0^z {}_i\hat{s}\left(z'_\xi\right)dz'_\xi . \tag{7.9}$$

2) Transform the current slowness model ${}_i\hat{s}$ and the error measurements ρ from depth to time.

3) Transform the interval-velocity function into the rms-velocity function ${}_i\hat{V}_{\mathrm{rms}}(\tau)$, using equation 6.1.

4) Compute the new rms-velocity function ${}_{i+1}\hat{V}_{\text{rms}}(\tau) = {}_i\hat{V}_{\text{rms}}(\tau) \,/\, \rho(\tau)$.

5) Apply the Dix formula to convert the new ${}_{i+1}\hat{V}_{\text{rms}}(\tau)$ into the new interval-velocity function.

6) Recompute the time-to-depth mapping function, using the new interval-slowness function ${}_{i+1}\hat{s}$ in the inverse of equation 7.9, that is,

$$z(\tau) = \frac{1}{2}\int_0^{\tau} \frac{d\tau'}{{}_{i+1}\hat{s}(\tau')}. \tag{7.10}$$

7) Map the new interval slowness function from time to depth, using the new mapping function.

This process may seem circuitous compared with updating the interval velocity directly in depth by applying a depth version of the Dix formula. However, by updating the interval velocity in time and then using the new interval-velocity function to map it back to depth, one repositions the whole interval-velocity function according to the updated (and supposedly more accurate) time-to-depth mapping. I discuss this issue in more detail in the next section, because it is important for tomographic MVA methods also (Kosloff et al., 1996; Biondi et al., 1998; Clapp, 2001).

An iterative process that includes the updating scheme outlined above or a similar updating scheme often is referred to as a Deregowski loop (Deregowski, 1990). The Deregowski loop is adequate for depth migration when the velocity function does not vary rapidly in the lateral directions and the reflectors' geometry is not too complex. However, vertical back-projection of the residual-moveout analysis using the Dix formula is based on two assumptions that are not necessarily fulfilled in depth-migration problems. First, one assumes that the velocity does not change rapidly within the offset span of the data. Second, one assumes that the vertical direction used for updating is a good approximation of the true normal to the reflectors.

The first assumption is similar to the assumption one makes when one inverts stacking velocity using the Dix formula and encounters the same problems in the presence of strong lateral velocity variations. In such cases, the assumption that interval velocities and average velocities are linked by the simple relation in equation 6.1 is incorrect, and using the Dix formula may cause problems similar to the ones encountered when applying the Dix formula to the inversion of stacking velocities in the presence of strong lateral variations in velocity.

The schematic in Figure 20 illustrates the problems caused by the second assumption. In the Deregowski loop, the interval velocity is updated along the vertical at the same horizontal location of the CIG used for the analysis. As is apparent from the figure, when the reflectors are dipping, this vertical line is a poor approximation of the true zero-offset rays. Furthermore, the zero-offset ray from each reflector may be distant from the zero-offset rays of the other reflectors. Consequently, average velocities measured from reflectors with different dips may sample different areas of the velocity field. When velocity varies laterally, the measured average-velocity errors can be inconsistent among reflectors. This inconsistency may cause problems when the measured errors in average velocities are used for interval-velocity updates. That problem could be addressed by inverting the average-velocity errors along the normal rays instead of along vertical lines.

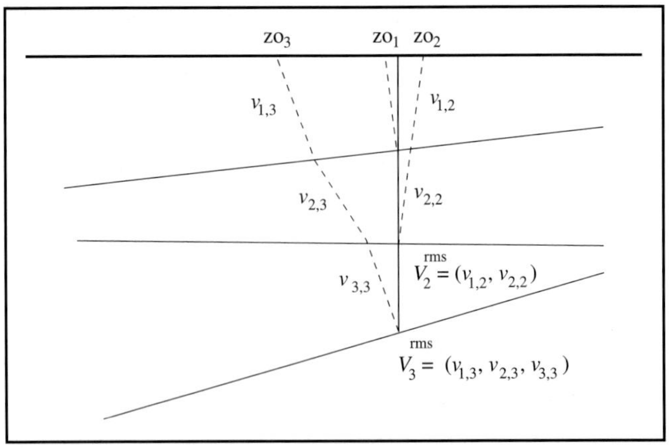

Figure 20. A schematic showing how the average velocities V_{rms} measured from two dipping reflectors may sample different interval velocities in the same layers, because of the different dips of the reflectors. This inconsistency may cause problems when the average velocities are inverted into interval velocities along the same vertical line.

However, normal rays may cross, and subsequent inconsistencies in the updates must be resolved by a tomographic inversion method (Audebert et al., 1997). Therefore, the simplicity of the Deregowski loop — a primary reason for its widespread use — is lost when normal-ray updates are employed.

The accuracy and convergence of the Deregowski loop may be improved by applying it in a layer-stripping fashion — that is, by estimating the velocity layer by layer, starting from the shallow layers and moving down in depth. The two assumptions discussed above are more valid when vertical updating is performed on only a thin layer above the reflectors. Figure 21 graphically shows this concept. The vertical update assumes that the velocity perturbations are centered on the vertical ray (gridded area), whereas the measured effects are caused by perturbations in a wider area around the zero-offset ray (solid area). The two regions overlap significantly only in the layer just above the reflector.

In the presence of steep dips and strong lateral variations, a layer-stripping procedure must proceed slowly, with relatively thin layers, because the errors introduced by the Deregowski updating increase with the thickness of the layer. A layer-stripping procedure has the more substantial limitation of estimating the velocity one layer at a time and assuming that the velocities for all the layers above are correct. Therefore, a layer's velocity is estimated by using the information contained in the reflections from the interfaces just below it. In reality, the focusing and the moveout of deep reflectors are influenced strongly by shallow velocity anomalies; thus, they contain valuable information on those anomalies. Layer stripping neglects that information. Even worse, if the velocity function frozen in the shallow layers contains errors, those errors propagate downward into the estimated velocities for all the layers below, often preventing the process from converging to a satisfactory solution.

Readers interested in examples of application of the Deregowski loop can find several excellent case studies in Yilmaz (2001) and in numerous other articles. The Deregowski loop combined with layer stripping often is used because of its simplicity and because it can be driven by the interpreter's geologic understanding. However, in complex media, the method has substantial drawbacks and may require more iterations for the process to converge than does the more accurate and global approach of tomographic MVA, which is presented in the next section.

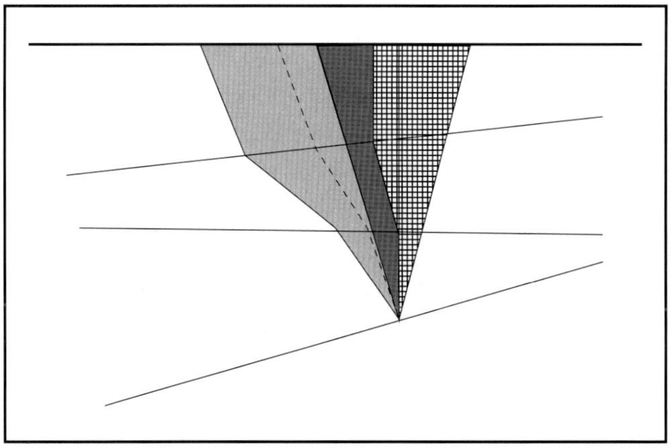

Figure 21. A schematic showing that the error introduced by layer stripping increases with the thickness of the layer. The amount of overlap between the area of influence around the vertical ray (gridded) and the true normal ray (solid) is higher when only the layer immediately above the reflector is considered than when all the layers are considered.

Historically, vertical velocity updating for prestack depth migration was defined for the wavefield-continuation methods, particularly for source-receiver migration (Doherty and Claerbout, 1976; Yilmaz and Chambers, 1984; MacKay and Abma, 1993), prior to being defined for Kirchhoff-migration methods (Deregowski, 1990). When one performs source-receiver migration, migration-velocity errors prevent the prestack wavefield from focusing at zero time and zero offset. However, it still focuses, at least approximately, at a time different from zero. Measurements of that focusing-time error (hence the generic name of focusing analysis, for these methods) provide information on the average-velocity errors that is similar to the information that measurements of the RMO parameter ρ provide for the Deregowski loop described in this section.

Thus, the time shifts of the foci can be inverted into interval-velocity updates along the vertical line. Such focusing-analysis methods are a valid alternative to RMO analysis, when simple vertical updating is performed. However, focusing analysis does not allow tomographic updating as readily as does the RMO analysis of ADCIGs described above.

Tomographic Migration Velocity Analysis

The simple migration-velocity-analysis procedure presented in the previous section may yield biased estimates or it may even diverge when the velocity function varies rapidly in the horizontal directions and the reflectors' structure is complex. To solve such challenging imaging problems, one must update the migration-velocity function using a full-fledged tomographic method.

Tomographic MVA methods are based on the conceptual framework that I developed in Chapter 6 for traveltime tomography. The linear tomographic operators that are inverted to estimate velocity updates actually are equivalent in the two cases (Stork, 1992). The crucial difference between MVA tomography and traveltime tomography is that in MVA tomography, traveltime errors are measured indirectly from the migrated images and not from the data. In complex areas, extracting kinematic information in the data space is practically impossible, whereas the RMO analysis of CIG gathers not only is possible, it is robust, even if the current migration velocity is not particularly close to the true velocity.

The first formulations of tomographic MVA were based on kinematic information extracted from ODCIGs obtained by Kirchhoff migration (Etgen, 1990; van Trier, 1990; Stork, 1992). More recently, the MVA method has been adapted to invert the velocity information extracted from ADCIGs obtained by either Kirchhoff migration (Brandsberg-Dahl et al., 1999) or wavefield-continuation migration (Biondi and Sava, 1999; Clapp and Biondi, 2000). In this section, I present the ray-based tomographic operator that is used to invert the kinematic information contained in ADCIGs. The mathematical formalism for deriving the MVA tomographic operator in the offset domain is very similar.

Objective function of tomographic migration velocity analysis

The goal of MVA is to improve the quality of the migrated images. In particular, one aims at flattening the migrated CIGs or, in other words, zeroing the RMO function measured from the CIGs, $\tilde{\mathbf{n}}_{\mathrm{RMO}}$. To achieve this goal, one minimizes the objective function

$$J_{\mathrm{reg}}(\mathbf{s}) = \left\| \tilde{\mathbf{n}}_{\mathrm{RMO}}(\mathbf{s}) \right\|_2 + \mathbf{s}' \mathbf{C}_s^{-1} \mathbf{s}, \tag{7.11}$$

which includes a slowness regularization term based on the estimated covariance of the slowness function, $\mathbf{C}_s$. The objective function expressed in equation 7.11 is highly nonquadratic. As was true in the case of traveltime tomography (Chapter 6), this nonquadratic optimization problem usually is solved by applying gradient-based optimization methods (Claerbout, 2004), so it needs to be linearized repeatedly.

The measured RMO function must be extracted from the migrated images each time a new migration is performed. Thus, it is a function of the nonlinear iteration index, *i*, and it is expressed as ${}_i\tilde{\mathbf{n}}_{\mathrm{RMO}}$. In practice, ${}_i\tilde{\mathbf{n}}_{\mathrm{RMO}}$ is measured from a set of interpreted reflectors and thus is defined as a function of the reflection angles (or offsets) along the picked reflectors. At each point along the reflectors, the RMO function can be picked directly from the migrated CIGs for each value of the reflection angle(s) or offset(s) (van Trier, 1990). However, a more robust way to measure this RMO function is to use either the RMO or the residual-migration analysis described earlier in this chapter. When one applies simple RMO analysis, the RMO function is determined directly by the ρ values picked from ρ-spectra (Etgen, 1990).

At each nonlinear iteration, the tomographic operator, ${}_i\mathbf{M}$, is obtained by linearizing the relation between the RMO function and the velocity around the current velocity model, ${}_{i-1}\hat{\mathbf{s}}$, as I will describe in the next section. The linearized operator is used to compute the perturbations, $\widehat{{}_i\Delta\mathbf{n}_{\mathrm{RMO}}}$, in the modeled RMO function caused by perturbations, ${}_i\Delta\mathbf{s}$, in the slowness function. The estimate of the covariance of the slowness function, ${}_i\mathbf{C}_s$, also should vary with the iterations, because it ought to be based on structural interpretation of the migrated image, as discussed in an upcoming section.

Comparing the objective function expressed in equation 7.11 with its homologous expression for traveltime tomography, equation 6.5, one notices that in equation 7.11, no observed data vector ($\mathbf{t}_{\mathrm{obs}}$ in traveltime tomography) exists as an explicit target for the modeled data vector ($\mathbf{t}_m$ in traveltime tomography, $\tilde{\mathbf{n}}_{\mathrm{RMO}}$ in tomographic MVA). Although it does not explicitly appear in the objective function, the implicit target of tomographic MVA is a null RMO function. In other words, at each nonlinear iteration, the objective function of ray-based MVA aims to drive the measured RMO function for the interpreted reflections toward the constant global target of a null RMO function, regardless of whether the measured RMO function is far from being flat. At each nonlinear iteration, the velocity updates are estimated by solving a quadratic optimization problem that minimizes the difference between ${}_i\widehat{\Delta \mathbf{n}}_{\mathrm{RMO}} = \mathbf{0} - {}_i\tilde{\mathbf{n}}_{\mathrm{RMO}}$ and the modeled perturbations, ${}_i\widehat{\Delta \mathbf{n}}_{\mathrm{RMO}}$, in the RMO function.

The objective function for the velocity-updating problem at the ith iteration is written as

$$\begin{aligned} J_{\mathrm{reg}}\left({}_i\Delta \mathbf{s}\right) &= \left\| {}_i\widehat{\Delta \mathbf{n}}_{\mathrm{RMO}} - {}_i\widetilde{\Delta \mathbf{n}}_{\mathrm{RMO}} \right\|_2 + \left({}_{i-1}\hat{\mathbf{s}} + {}_i\Delta \mathbf{s}\right)' {}_i\mathbf{C}_{\mathrm{s}}^{-1}\left({}_{i-1}\hat{\mathbf{s}} + {}_i\Delta \mathbf{s}\right) \\ &= \left\| {}_i\mathbf{M}\,{}_i\Delta \mathbf{s} - {}_i\widetilde{\Delta \mathbf{n}}_{\mathrm{RMO}} \right\|_2 + \left({}_{i-1}\hat{\mathbf{s}} + {}_i\Delta \mathbf{s}\right)' {}_i\mathbf{C}_{\mathrm{s}}^{-1}\left({}_{i-1}\hat{\mathbf{s}} + {}_i\Delta \mathbf{s}\right) . \end{aligned} \tag{7.12}$$

Minimization of this objective function provides an estimate of the slowness perturbations (${}_i\Delta \mathbf{s}$) that cause the kinematics of the next prestack image (${}_{i+1}\tilde{\mathbf{n}}_{\mathbf{RMO}}$) to be as close as possible, in a least-squares sense, to the kinematics of a target image that has flat CIGs.

As I noted above, all the terms in the linearized objective function expressed in equation 7.12 change at each nonlinear iteration. Such variations can be substantial, because both the measured RMO function (${}_i\tilde{\mathbf{n}}_{\mathbf{RMO}}$) and the estimate of the slowness covariance (${}_i\mathbf{C}_{\mathrm{s}}$) result from an interpretative process. Furthermore, even the domain of the objective function's definition changes at each iteration, when reflectors are added (and subtracted) from the interpretation because they come in and out of focus.

Differential semblance optimization (DSO) (Symes and Carazzone, 1991) is another viable approach for flattening the RMO function that, unlike the conventional MVA method described in this section, relies not on the definition of a distant target but only on local measurements of nonflatness in the CIGs. The basic principle of DSO is similar to that of conventional MVA; that is, the quality of a velocity model is judged by measuring the departure from flatness of the migrated CIGs. However, instead of minimizing the RMO of the image at each angle (or offset) with respect to the normal-incidence (zero-offset) image, the DSO method minimizes the relative differences in imaging depth between contiguous angles (or offsets). Mulder and ten Kroode (2002) demonstrated that minimizing the DSO objective function can improve the migration velocity in a field-data case. DSO can be effective with complex imaging problems, but more tests are necessary.

Evaluation of the linearized tomographic-MVA operator

The linearized tomographic operator ${}_i\mathbf{M}$ can be evaluated with a method that is similar to the procedure outlined in Chapter 6 for evaluating the traveltime-tomography operator ${}_i\mathbf{L}$. Biondi (2006) actually shows how direct derivation of the individual components of ${}_i\mathbf{M}$ yields a result that is equivalent to the scaling of the individual components of ${}_i\mathbf{L}$ by the inverse of the cosine of the aperture angle γ multiplied by the local slowness at the reflector; that is:

$${}_i\widehat{\Delta \mathbf{n}}_{\mathrm{RMO}} = \frac{{}_i\Delta t_m}{2\,{}_{i-1}\hat{s}\cos\gamma}\mathbf{n}, \tag{7.13}$$

where ${}_i\Delta t_m$ are the traveltime perturbations computed using equation 6.7.

That result demonstrates that the tomographic operator is the same (ignoring a scaling factor $2\,{}_{i-1}\hat{s}\cos\gamma$), regardless of whether one applies tomography in the data domain or in the migrated domain. In the first case, one subtracts from the traveltime perturbations the contribution of the reflector movement at normal incidence. In the second case, one directly matches the difference between the image movements at an arbitrary offset and the image movements at normal incidence. A direct consequence of the equivalence between the tomographic operators is that traveltime tomography and tomographic MVA suffer from similar problems related to the nonuniqueness of the solution. An upcoming section discusses methods for addressing this problem, but first I recapitulate the whole process of tomographic MVA with the help of two examples.

Examples of 3D tomographic MVA

All the steps performed in a tomographic MVA procedure are illustrated best with a case study. The data used in this example are the same North Sea data used in the residual-migration example shown earlier in this chapter. The results are taken from Clapp (2001). The data set is challenging because of the presence of a shallow chalk layer and of a salt body piercing through the chalk.

The first step of a tomographic MVA is to estimate a starting velocity model close enough to the true model that the process has the chance of converging in a limited number of iterations. In the present case, the starting velocity model was estimated using a layer-based parameterization of the model (see the discussion in the next section, on model parameterization). Thus, the model displays sharp discontinuities at the salt and chalk boundaries but smooth gradients inside the layers. Figure 22 shows three orthogonal sections cut through the starting model.

The second step is to perform a prestack depth migration and create prestack partial images. A wavefield-continuation migration is applied in this case, because of the difficulties of imaging the subsalt areas with Kirchhoff migration. The common-azimuth algorithm described in Chapter 5 is employed because of its computational efficiency; it allows multiple iterations. The ADCIGs are computed before imaging, as described in Chapter 3.

The third step is to analyze the prestack partial images. The reflectors' geometry is determined by picking reflectors from the image cube. Figure 23 shows three orthogonal sections of the image cube obtained with the starting velocity. The asterisks superimposed onto the images indicate the interpreted reflectors. The RMO semblance spectra are computed and analyzed along the interpreted horizons. The twelve cubes displayed in Figure 24 are RMO spectra extracted at locations in physical space corresponding to the

Figure 22. Orthogonal slices that cut through the initial velocity cube. Figure is adapted from Clapp (2001).

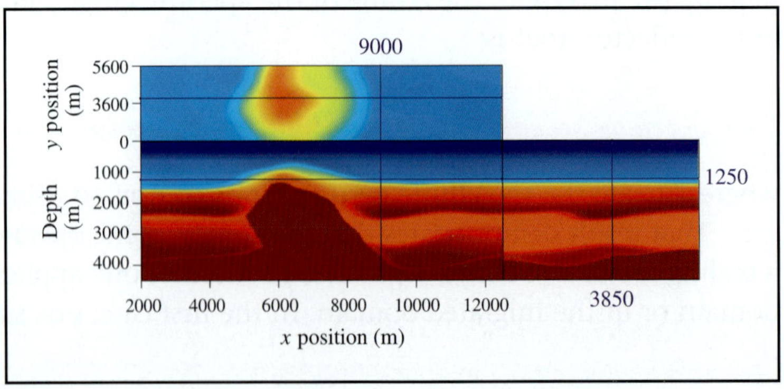

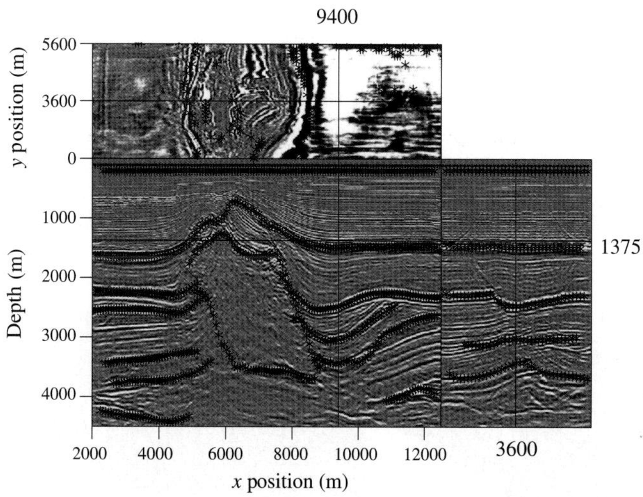

Figure 23. Orthogonal slices that cut through the migrated image obtained using the starting velocity. The asterisks superimposed onto the image indicate the interpreted reflectors. Figure is from Clapp (2001).

interpreted reflectors. The vertical axes of these cubes represent the RMO parameter, and the horizontal axes correspond to the image's horizontal coordinates. The RMO analysis for the shallow reflectors (top left) is sharp and clean, but as expected, the resolution decreases as the depth of the reflectors increases. The RMO analysis becomes difficult for the bottom-of-salt reflectors (middle column, second from the top), because the actual residual moveouts in the image are substantially different from the one-parameter function used for computing the RMO spectra.

The next step is to set up and solve the tomographic estimation problem. The reflector geometry and the aperture-angle information in the ADCIGs are used to determine the raypaths corresponding to the RMO values measured from the prestack images. The velocity covariance matrix used to constrain the optimization problem (equation 7.12) is estimated using the procedure described in the next section. The tomographic MVA operator is evaluated according to equation 7.13. Solution of the optimization problem yields the velocity function shown in Figure 25. This new velocity function is used for a new depth migration, and the process is repeated until a satisfactory result is obtained.

The velocity cube obtained after two iterations is displayed in Figure 26. By comparing this final velocity with the initial velocity (Figure 22), one can see differences in the interpretation of the salt boundaries as well as in the layers' velocities, particularly to the right of the salt body. The chalk layer and the sediments below it are significantly slower in the new velocity model than in the initial one.

Such changes in the velocity function substantially improve the migrated results, as one can appreciate by comparing the orthogonal slices shown in Figure 27 (initial velocity) and Figure 28 (final velocity). Those slices are cut through the image cube at the same locations as the slices cut through the velocity cubes shown in the previous figures. The most noticeable differences between the images are highlighted by ellipses. The sediments below the salt edge (ellipse A) are imaged coherently farther in the direction of the salt in Figure 28 than in Figure 27. Lowering the velocity in the chalk and in the layer below it leads to a substantial improvement in the image of the two faults visible in the crossline section (ellipse D) and of the synclinal structure highlighted by ellipse C.

- Automatic MVA applied to subsalt imaging

The example described in Figures 23 through 28 illustrates the application of conventional tomographic MVA on the basis of explicit horizon picking that usually requires substantial interpretative efforts by geophysicists. As computers become more powerful and cost-effective, the time required for the interpretative component of MVA is becoming increasingly predominant and prevents interpreters from performing as many MVA

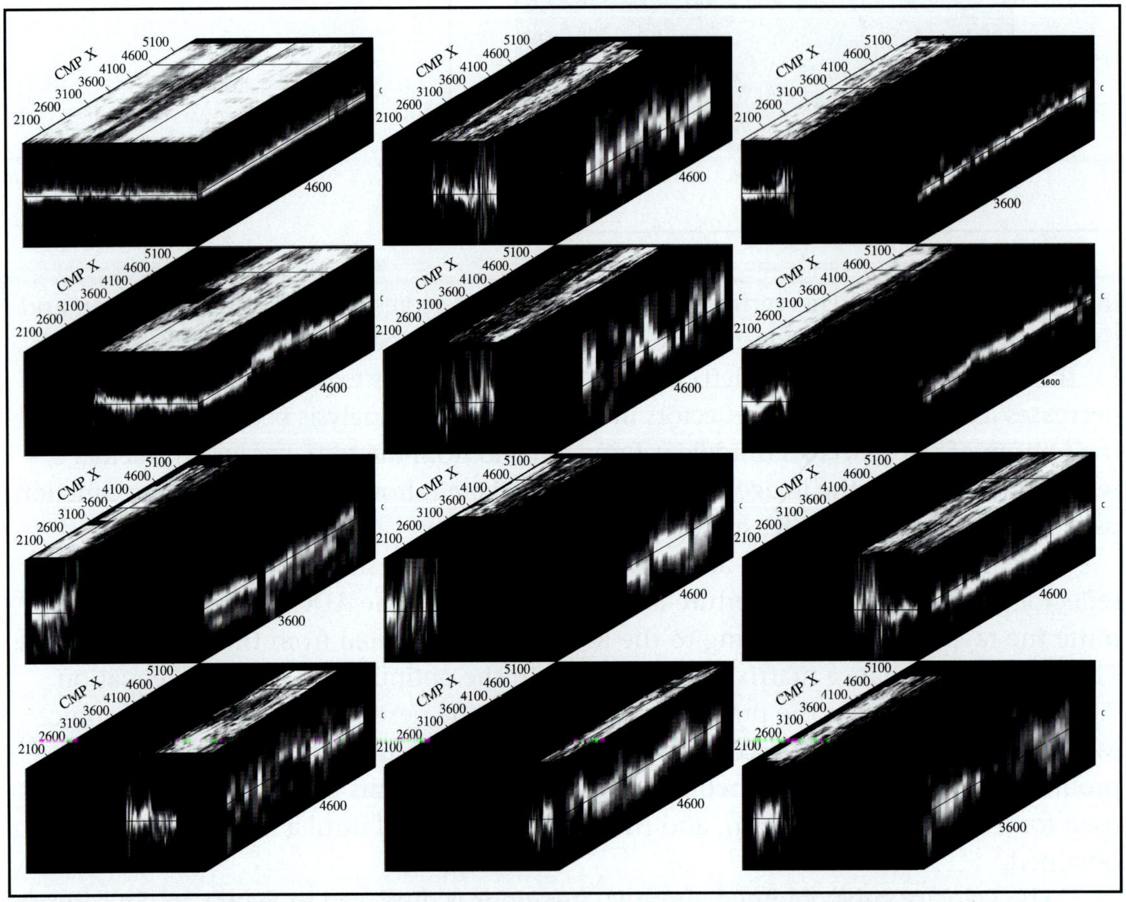

Figure 24. RMO semblance-analysis cubes for the interpreted reflectors. The vertical axes of these cubes represent the RMO parameter, and the horizontal axes correspond to the image's horizontal coordinates. Figure is from Clapp (2001).

Figure 25. Orthogonal slices that cut through the velocity cube after one iteration of the MVA tomographic updating. Figure is adapted from Clapp (2001).

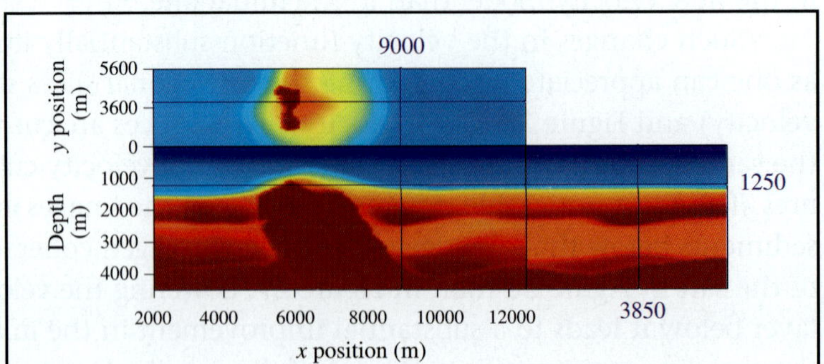

iterations as otherwise could be possible. Therefore, a significant advantage lies in automating as much of the MVA process as possible, and algorithms and workflows are being developed to support this goal (Hardy, 2003; Bevc et al., 2006). The following case study

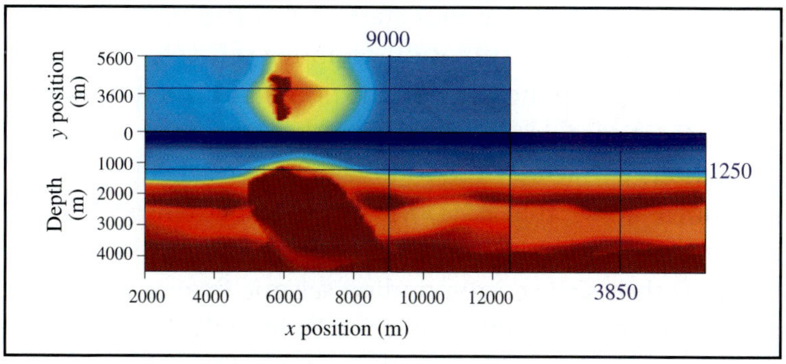

Figure 26. Orthogonal slices that cut through the velocity cube after two iterations of the MVA tomographic updating. Figure is adapted from Clapp (2001).

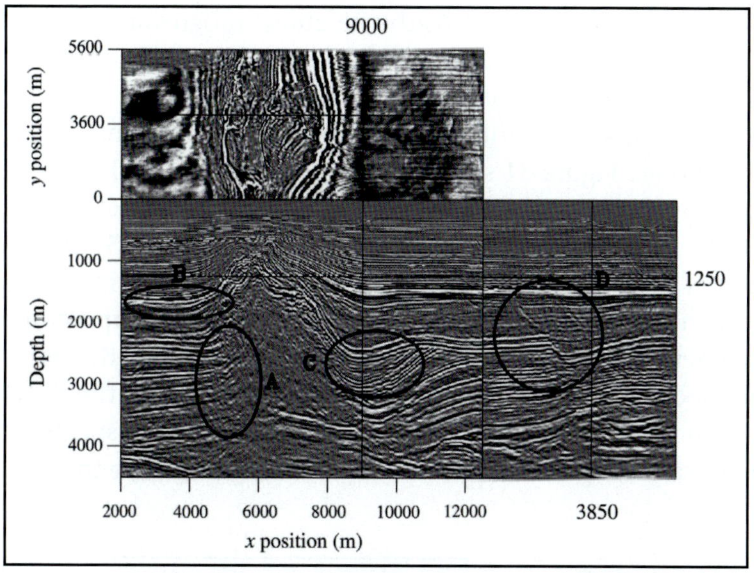

Figure 27. Orthogonal slices that cut through the migrated image obtained using the starting velocity. Figure is from Clapp (2001).

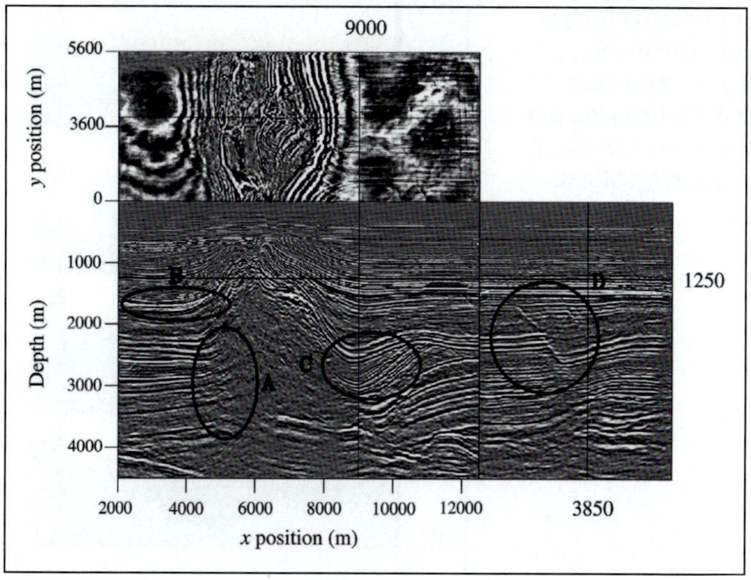

Figure 28. Orthogonal slices that cut through the migrated image obtained using the velocity function obtained after two iterations of the MVA tomographic updating. Figure is from Clapp (2001).

employs the techniques described by Bevc et al. (2006). Those authors obtained the results I show.

In addition to illustrating the application of automated MVA methods, the following example also illustrates a typical procedure applied when one is imaging subsalt targets. The basic procedure is based on the following four steps, but it can vary according to the particular challenges of the project: (1) estimation of above-salt velocity, (2) interpretation of the top of the salt body, (3) application of salt flood and interpretation of the bottom of the salt body, and (4) estimation of below-salt velocity.

Figure 29 displays the inline sections extracted from four velocity-model cubes defined before each of the steps outlined above; that is: (a) the initial velocity model, (b) the result from above-salt tomography, (c) the result from salt flooding, and (d) the result from picking the salt bottom. Figure 30 shows the corresponding seismic images obtained by prestack wavefield-continuation migration, assuming the velocity models shown in Figure 29. The red and yellow lines superimposed onto the images in panel (b) and panel (c) represent the picked top and bottom of the salt body. As the iterations progress from the top to bottom, the images are increasingly better focused at increasingly deeper levels.

Figures 31 through 33 show the kinds of information that can be extracted from the prestack images by automated methods; the information can be used for tomographic updating of the velocity model. Figure 31 shows the local dips of the reflectors computed along the inline (top) and crossline (bottom) directions. The top panel in Figure 32 displays a measure of the coherency of the local dips shown in Figure 31. These coherency measurements (together with the value of the semblance in the corresponding ρ-spectra) are then used to select the most reliable reflector segments to use for updating the velocity. Those reflector segments are represented by the dip bars shown in the bottom panel in Figure 32. At each location, the value of the parameter describing the RMO measured from the ADCIG (e.g., ρ) is picked automatically from precomputed ρ-spectra. Figure 33 shows the dip bars corresponding to the selected reflector segments for the

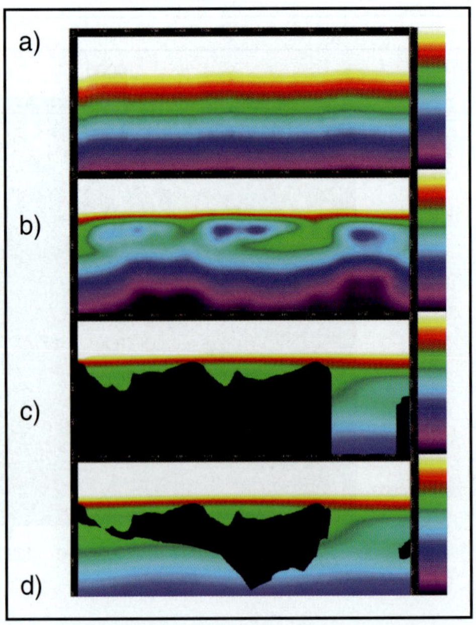

Figure 29. Inline sections extracted from velocity-model cubes corresponding to the migration results shown in Figure 30: (a) the initial velocity model, (b) the result of above-salt tomography, (c) the result after salt flooding, and (d) the result after picking the salt bottom. The salt body is shown in black in these figures. (This image was kindly provided by Moritz Fliender at 3DGeo.)

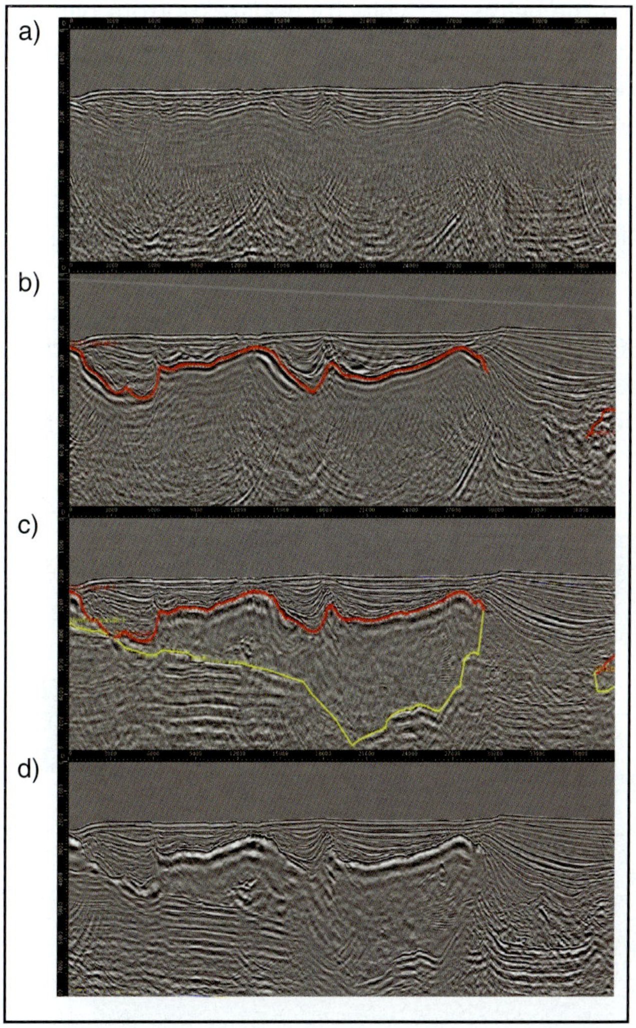

Figure 30. Inline sections extracted from prestack-migration cubes corresponding to the velocity models shown in Figure 29: (a) the initial velocity model, (b) the result of above-salt tomography, (c) the result after salt flooding, and (d) the result after picking the salt bottom. The picked top and bottom of the salt are indicated with the red and yellow lines, respectively. (This image was kindly provided by Moritz Fliender at 3DGeo.)

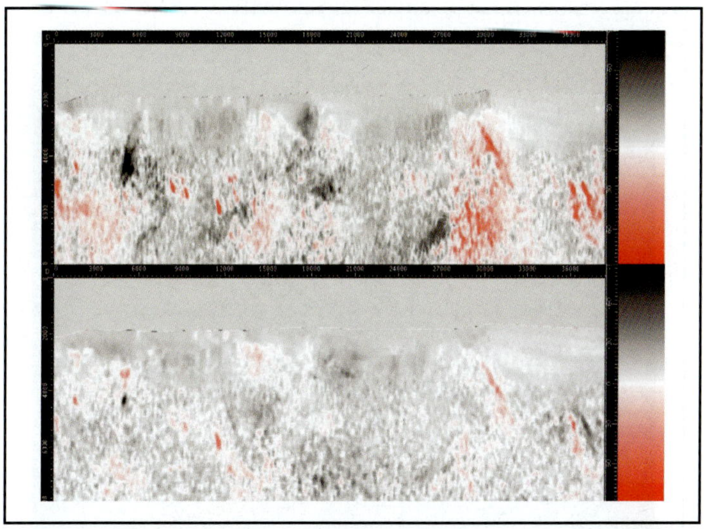

Figure 31. Inline dips (top) and crossline dips (bottom) automatically picked from the image cube obtained using the initial velocity model (Figure 29a). (This image was kindly provided by Moritz Fliender at 3DGeo.)

above-salt velocity update (top panel) and the below-salt velocity update (bottom panel). The dip bars are color coded according to the value of ρ picked from the ρ-spectra. This color-coded display of the dip bars conveys all the information that is inverted when one uses the tomographic MVA described above to estimate a velocity-model update; that is, the location of the dip bars, their inline and crossline dips, and the picked values of ρ. Figure 34 shows both the inline section extracted from the final velocity model obtained by inverting the picks shown in the bottom panel of Figure 33 and the corresponding prestack image.

In this last example, an interpreter picked the salt boundaries. The next and final step to achieve full automation of the MVA process would be to pick the salt boundaries automatically. Lomask et al. (2006) present an algorithm to pick the salt-sediment interface automatically by solving a large global optimization problem. Their algorithm is fair-

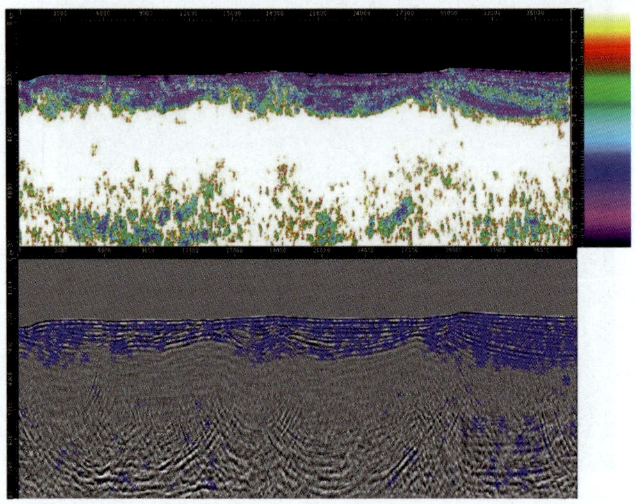

Figure 32. Top: Coherency of the dips picked from the image cube obtained using the initial velocity model (Figure 29a) and shown in Figure 31. Bottom: Dip bars automatically selected on the basis of the dip-coherency cube and the semblance values of the ρ-spectra. The dip bars represent the data points used in the above-salt tomographically derived velocity update. (This image was kindly provided by Moritz Fliender at 3DGeo.)

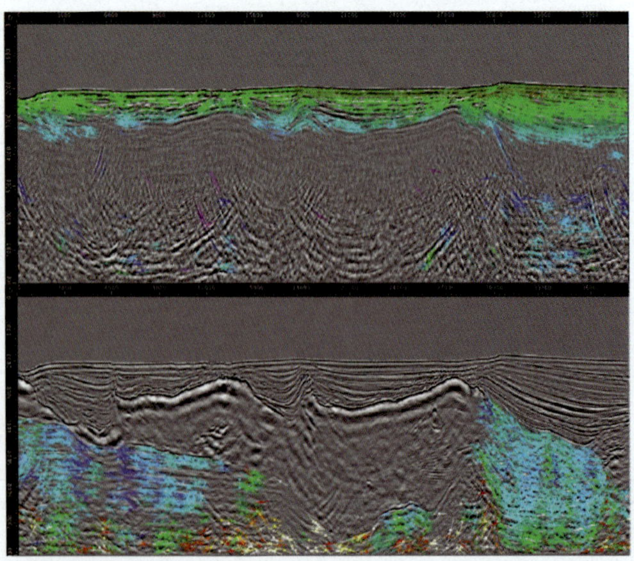

Figure 33. Dip bars corresponding to the selected data points for above-salt tomographic update (top panel) and below-salt tomographic update (bottom panel). The dip bars have been color coded with the ρ values automatically picked from the ρ-spectra. (This image was kindly provided by Moritz Fliender at 3DGeo.)

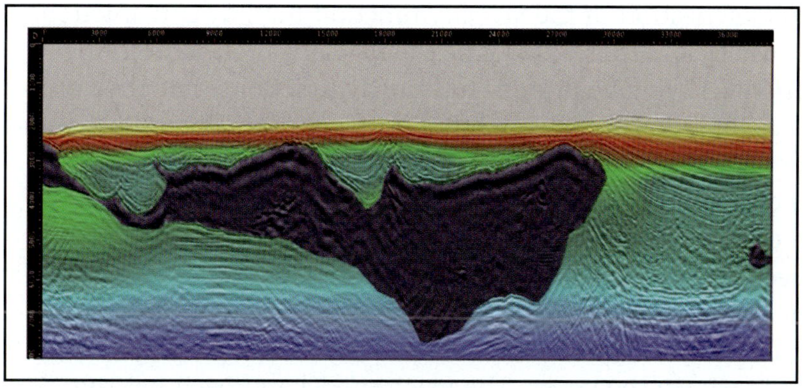

Figure 34. The image and velocity model resulting from the below-salt (and last) tomographic update of the velocity. (This image was kindly provided by Moritz Fliender at 3DGeo.)

ly robust even where the image of the salt-sediment interface is discontinuous. However, human editing of the automatic-picking results still is required in the most challenging situations. Further advancements in automated interpretation algorithms are needed to achieve the goal of a robust and completely automated MVA procedure.

Improving the convergence of tomographic MVA

Tomographic MVA is applied to the most challenging imaging projects, in which the difficulties caused by the nonunique and nonlinear nature of the velocity-estimation problem are magnified and may seriously impact the final success of the method. The nonuniqueness of the MVA solution causes poor local convergence during solution of the linearized problems encountered at each nonlinear iteration. The nonlinearity of the relationship between the migrated images and the velocity model creates multiple local minima in the objective function. Global convergence to the optimal solution is not assured, and it depends strongly on the starting model. The two problems are interrelated, but for simplicity, I will discuss them separately.

- Improving the local convergence of tomographic MVA

The velocity perturbations are not defined uniquely by the available data, because several velocity functions yield migrated CIGs that are equally flat, at least in a least-squares sense. That problem could be addressed either by making better use of the information in the seismic data or by integrating additional information into the inversion process. For example, to make better use of information contained in the seismic data, one could analyze the image's focusing in the physical space (Sava et al., 2005), in addition to the flatness of the CIGs. Alternatively, one could update the velocity function by using a wavefield-continuation operator that is capable of exploiting the frequency dispersion caused by complex wave propagation (Symes and Carazzone, 1991; Sava and Biondi, 2004a,b). The last chapter in Biondi (2006) presents a comprehensive discussion of several full-wavefield methods that have been proposed for performing MVA.

In this section, I discuss more conventional methods that use geologic information to better constrain the linearized optimization problem. Such geologic constraints are designed to guide the inversion process toward solutions that satisfy a geologic model of how the velocity should vary among layers and within each layer. Two main methods exist for guiding the velocity-estimation process toward geologically plausible solutions:

(1) Use a parsimonious parameterization of the velocity function that, by construction, allows only geologically plausible solutions. (2) Define the regularization term in the objective function of equation 7.12 to give priority to velocity functions that are consistent with the a priori geologic model. These two methods are not mutually exclusive and can be used in combination.

The parameterization of the velocity model by discrete layers of constant velocity is natural and has been employed since the earliest development of velocity-analysis techniques. Often, it has been described as the velocity macromodel strategy (van de Made et al., 1984; Wapenaar and Berkhout, 1985). The constant-velocity-layers parameterization can be generalized easily to nonconstant velocity layers, with a polynomial defining the velocity function within the layers. The macromodel approach has the advantage of being simple and requiring very few parameters to define the velocity function. The solution is tightly constrained and, because of the small number of parameters used, it requires few iterations to converge. In areas where the underlying assumptions of a strong correlation between lithology and velocity are appropriate (e.g., the North Sea), the macromodel approach has been successful. Its main disadvantage is its lack of flexibility. Because the parameterization is fixed a priori, it does not adapt automatically to situations in which optimal focusing of the data requires velocity variations that are more complex than the given parameterization allows. When unexpected velocity variations occur, the solution defined with a macromodel approach may completely miss them. In other words, the macromodel approach does not allow for surprises, whereas the real world is full of them.

Parameterization of the velocity model by cubic splines is a parsimonious velocity representation that, like the macromodel approach, has the advantage of requiring few parameters. However, the cubic-spline method is more flexible where the data require it (Biondi, 1990; Etgen, 1990; van Trier, 1990). The simplest way to use cubic splines is to use them globally — that is, to parameterize the whole velocity model as one single set of cubic splines. When that approach is clearly insufficient, such as when one represents velocity functions with known sharp discontinuities (e.g., salt and sediments), the cubic-spline approach can be combined with the macromodel approach and a set of spline functions for each macrolayer can be used (Perdrizet and Sinoquet, 2003).

When one uses a dense grid of cubic-spline nodes, the cubic-spline approach behaves similarly to a straightforward gridded representation of the velocity model. In this approach, the velocity model is defined on a dense Cartesian grid, and the task of stabilizing the inversion is left entirely to the regularization term ($\mathbf{C}_s$) in the objective function (equation 7.12). In Bayesian estimation theory, the matrix $\mathbf{C}_s$ is interpreted to be the a priori covariance matrix of the model parameters (Tarantola, 1987). Commonly, one assumes that the velocity model is smooth, and consequently one sets $\mathbf{C}_s^{-1}$ to be an isotropic roughening operator, like a Laplacian operator (Toldi, 1985). However, the choice of an isotropic roughening operator is driven more by the knowledge of the null space of the tomographic operator and by the goal of assuring stability to the inversion procedure than by geologic considerations.

Geologic information can be introduced in the regularization operator by making it directional — that is, by rewarding velocity functions that are smooth along some particular directions. Because velocity often is driven by a combination of compaction and sedimentation, the smoothing directions are chosen according to the geologic dips and/or the known compaction direction. The important advantage of directional regularization operators is that one lets geology (e.g., the structural image along with its geologic interpretation) drive the determination of $\mathbf{C}_S$. Therefore, the final model is likely to satisfy the geologist in addition to satisfying the geophysicist (e.g., to fit the seismic data). Because the geologic interpretation may vary with the nonlinear iterations as reflectors come in and out of focus, the estimate of the covariance matrix also changes with iterations.

Two methods have been proposed to introduce a directivity in the regularization term: (1) Penalize the dot product of the velocity gradient with the smoothing direction (Sinoquet, 1993) or (2) use nonstationary steering filters aligned along the smoothing direction (Clapp, 2001; Clapp et al., 2004). Nonstationarity of the directional regularization operators is essential when one images complex geologic structure with rapidly varying geologic dips. It also allows us to model constant-velocity layers (e.g., salt bodies or water columns), a task that is accomplished easily with the macromodel approach.

Our ability to guide the smoothing of a gridded velocity model according to the structural image makes the gridded-model approach share some of the attractive characteristics of the macromodel approach. However, the simplistic approach to smoothing regularization formalized in the objective function (equation 7.12) is likely to slow the convergence of the optimization problem. When one penalizes roughness in the velocity functions by using a roughening operator such as the inverse of the model covariance matrix, $\mathbf{C}_s^{-1}$, the resulting quadratic optimization problem is severely ill conditioned because the matrix $\mathbf{C}_s^{-1}$ itself is ill conditioned (Golub and Loan, 1983). Ill conditioning is caused by the widely variable scaling of the spatial wavenumbers performed by $\mathbf{C}_s^{-1}$. The low wavenumbers are strongly attenuated and, at the limit, the constant component is in the null space of $\mathbf{C}_s^{-1}$. An alternative and more intuitive description of this problem is that the regularization term is a short operator. Therefore, a conjugate-gradient method (Claerbout, 2004) requires many iterations to apply the regularization term (and its adjoint) a sufficient number of times to spread velocity perturbations that are localized in space.

The slow-convergence problem can be solved by preconditioning the optimization problem with a change of variable in the model parameters (Harlan, 1995). Clapp (2001) describes an efficient numerical algorithm for applying preconditioning when $\mathbf{C}_s^{-1}$ is defined in terms of steering filters. Meyerholtz et al. (1989) describe an analogous methodology based on convolutional quelling that has the same computational advantages that Clapp's steering-filter methodology has, but it is applied to isotropic smoothing instead of to directional smoothing. The possibility of preconditioning the regularization term enables the linearized problem to converge quickly when it is solved with a conjugate-gradient algorithm. Acceleration of the convergence decreases the computational cost of velocity updating in tomographic MVA by reducing the number of times the linearized tomographic operator $\mathbf{M}$ is applied.

The combination of directional regularization and preconditioning makes the gridded-velocity approach as capable of including structural-geology information and as computationally effective as the macromodel approach is. Because of its flexibility in adapting to expected and unexpected velocity variations, the gridded-velocity model is used in place of or in combination with the macromodel approach in an increasing number of imaging projects.

The gain in computational efficiency one achieves by preconditioning the regularization term is useful when velocity updating is performed with ray operators, as is the case for the method described in this section. However, for complex imaging projects, the relative cost of applying a ray-based tomographic operator is constantly decreasing compared with the cost of performing expensive wavefield-continuation migrations. Indeed, for these complex-imaging problems, one would like to update the velocity model by using full-wavefield operators (Biondi, 2006). When one uses wavefield operators to update the model, it becomes essential to reduce the number of iterations by preconditioning.

- Improving global convergence of tomographic MVA

The issue of assuring global convergence of the velocity-estimation problem is even more challenging than improvement of the local convergence. Solutions are less well developed, and they tend to be based more on heuristic considerations than on firm theoretical understanding. At present, the main discriminants between local minima in the objective function are geologic criteria that interpreters use to decide whether the solutions provided by the estimation process make geologic sense. However, many incentives exist for not relying exclusively on the subjectivity of human judgment. One unexplored area of seismic-imaging technology is the inclusion of quantitative geologic constraints to guide global convergence in the velocity-estimation process, in addition to the use of directional-smoothing operators described in this section.

Two main approaches exist for improving global convergence. The first is to apply a robust optimization algorithm for minimizing the objective function defined in equation 7.7. The hope is that a more robust optimization algorithm should avoid local minima and find the global minimum. The second approach is to modify the problem by defining a new objective function that is more quadratic in the global sense — that is, that has fewer local minima than the objective function defined in equation 7.7.

In theory, one always has the option of using a global optimization algorithm to solve a nonquadratic optimization problem — that is, one can use a genetic algorithm (Holland, 1975) or simulated annealing (Kirkpatrick et al., 1983). However, such algorithms require too many evaluations (thousands or more) of the forward-modeling operator to be practical in 3D velocity-estimation problems (Jin and Madariaga, 1994; Jervis et al., 1996). One heuristic approach that improves the chances of avoiding local minima is to start from a simple low-resolution model and let complexities enter the velocity function only gradually over several nonlinear iterations (Biondi, 1990). The rationale for this approach is that by adding model complexities when the current model is far from the true model, one risks locking the estimation process into the neighbor of a local minima and never being able to escape. That argument is based on the nature of the nonlinearities of the relationship between kinematics in the data and in the velocity model.

A modification of the conventional tomographic MVA problem that is successful for improving global convergence is to define the interval velocity model as a function of the two-way traveltime τ instead of as a function of depth (Kosloff et al., 1996; Biondi et al., 1998; Clapp, 2001; Alkhalifah, 2003). That modification does not require a change in the structure of the objective function; it requires only a change in the domain of definition of the velocity function and the reflectors' geometry. Tomographic velocity updating is based on the information measured at the reflectors, and mapping the reflectors in depth is a function of the current migration velocity. Consequently, locations of the velocity updates also are functions of the current velocity.

Ideally, one would like to remap the velocity updates in depth using the mapping function defined by the updated velocity. Such remapping introduces a nonlinear relation between the velocity function and the kinematic errors measured from the data (or the image). Thus, it cannot be captured by a linearized tomographic operator defined in the depth domain. In contrast, by defining both the reflectors and the velocity in the τ domain instead of in the depth domain, one substantially reduces the reflector geometry's dependence on the current velocity function, and one removes the need for the remapping procedure just described.

The vertical updating process described earlier in this chapter is a simple situation in which the advantage of performing the updates in the τ domain is evident. In that case, one updates the interval-velocity function in the τ domain, after one transforms the measured average-velocity errors into the τ domain from the depth domain. The interval-velocity function is mapped back to depth for a new migration only as the last step, using the mapping determined by the new velocity.

Velocity updating in the τ domain does not involve a change in the structure of the objective function for tomographic MVA. In contrast, the DSO method (Symes and Carazzone, 1991) defines an objective function that differs slightly from the one optimized by conventional MVA methods (see the earlier section titled "Objective function of tomographic migration velocity analysis" for a brief discussion of the DSO objective function). Because of the differential nature of its objective function, DSO may avoid some of the local minima caused by inaccurate interpretation of the migrated image. Chauris and Noble (1998) demonstrated that the DSO functional is convex in fairly simple cases. Mulder and ten Kroode (2002) showed that coherent noise (e.g., multiple reflections) can cause the presence of local minima in the DSO functional, but global convergence can be achieved if the effect of the multiple reflections is removed from the migrated CIGs.

- Example of improved convergence of tomographic MVA

In the following example, I illustrate two of the concepts introduced in this section for improving the convergence of tomographic MVA. I compare two results of one velocity-updating iteration on a 2D slice of the 3D North Sea data set shown earlier in this chapter. The two results were obtained (1) with the velocity model defined in the depth domain and using a conventional isotropic regularization operator (e.g., a Laplacian operator) and (2) with the velocity model defined in the τ domain and using a directional regularization operator (e.g., steering filters). In both cases, the regularization term was preconditioned.

Figure 35 shows the starting velocity model. It is a smooth velocity function, with the addition of a salt body interpreted from a preliminary migration of the data. Figure 36 shows the velocity updated using conventional MVA methods of defining the velocity function in depth and using an isotropic regularization operator. In contrast, Figure 37 shows the velocity updated using two of the methods discussed in this section for improving the convergence of tomographic MVA. A directional regularization operator was used in the objective function, and the velocity model was defined in the τ domain.

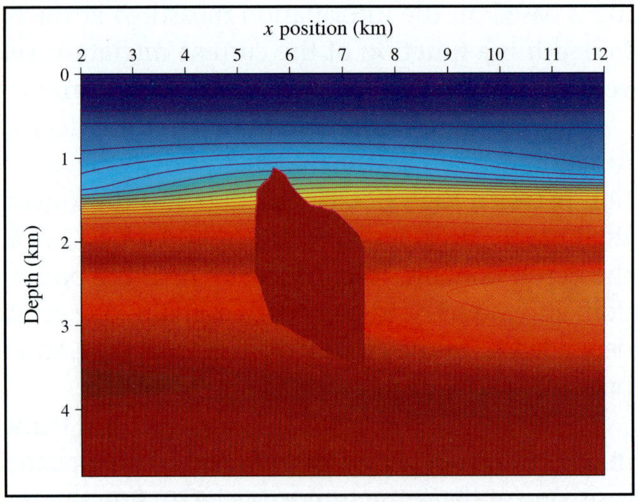

Figure 35. The initial velocity model. Figure is from Clapp (2001).

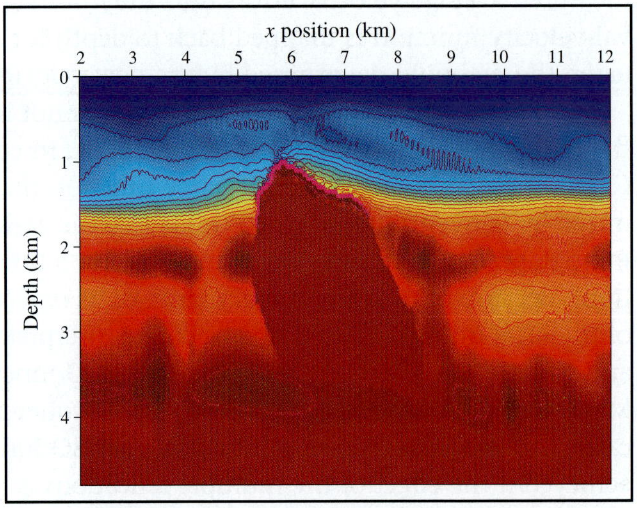

Figure 36. The velocity model after velocity updating in the depth domain with an isotropic (Laplacian) regularization operator. Figure is from Clapp (2001).

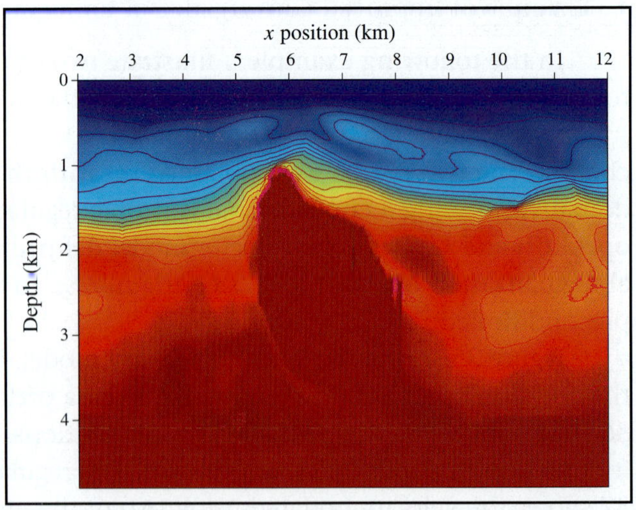

Figure 37. The velocity model after velocity updating in the τ domain with a directional (steering-filter) regularization operator. Figure is from Clapp (2001).

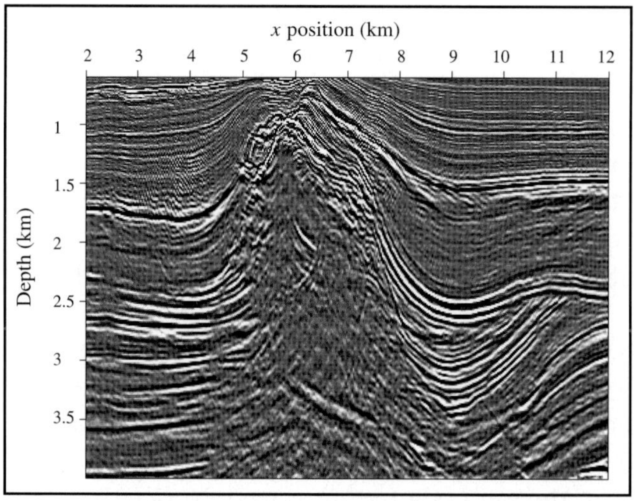

Figure 38. Depth migration using the initial velocity model shown in Figure 35. Figure is from Clapp (2001).

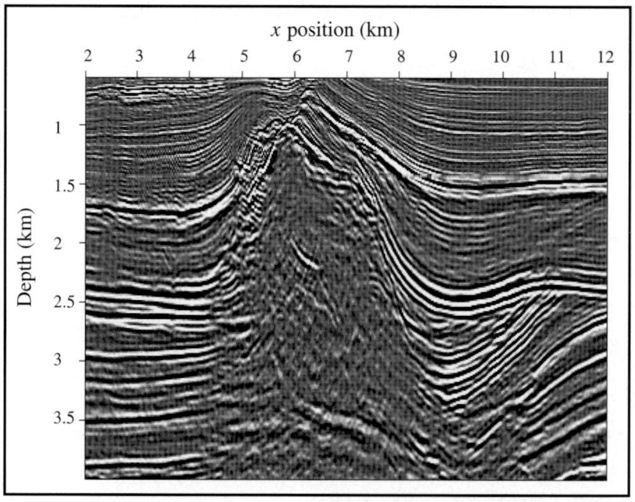

Figure 39. Depth migration using the velocity model shown in Figure 36, which was obtained by velocity updating in the depth domain with an isotropic (Laplacian) regularization operator. Figure is from Clapp (2001).

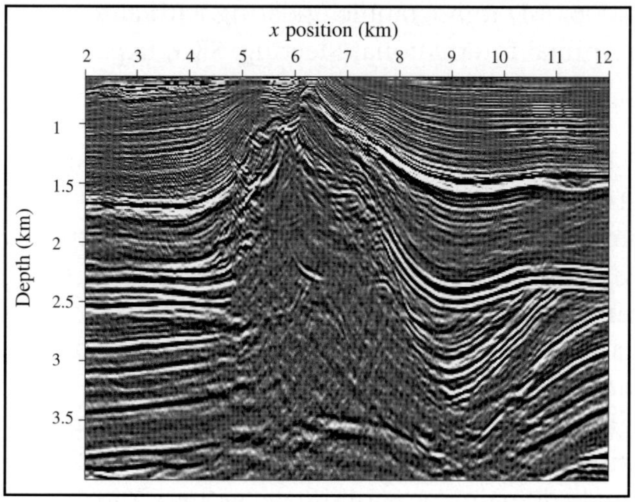

Figure 40. Depth migration using the velocity model shown in Figure 37, which was obtained by velocity updating in the τ domain with a directional (steering-filter) regularization operator. Figure is from Clapp (2001).

Notice that for comparison, all the figures are displayed in the depth domain, even when they were computed in the τ domain. The regularization directions then were determined in the τ domain by a dip analysis of the depth-migrated image obtained with the initial velocity model shown in Figure 35, after mapping the image into the τ domain (Clapp, 2001). This depth image is shown in Figure 38. The benefits of the directional regularization operator on velocity updating are evident when one compares the two velocity functions. In particular, both the chalk layer on top of the salt body and the synclinal structure to the right of the salt body are represented better in Figure 37 than in Figure 36.

Figures 39 and 40 show the migrated images obtained by using the velocity models shown in Figures 36 and 37, respectively. The bottom of the salt reflector is better imaged when one is using the methodologies described in this section (Figure 40), and the salt sediments under the edge of the salt are more coherent. Furthermore, the focusing of the synclinal structure to the right of the salt body has improved. Notice that 2D migration and MVA have inherent limitations in the area covered by this survey, because the subsurface geology has many 3D features.

References

Alkhalifah, T., 2003, Tau migration and velocity analysis: Theory and synthetic examples: Geophysics, **68**, 1331–1339.

al-Yahya, K. M., 1989, Velocity analysis by iterative profile migration: Geophysics, **54**, 718–729.

Audebert, F., J. P. Diet, P. Guillaume, I. F. Jones, and X. Zhang, 1997, CRP-scans: 3-D preSDM velocity analysis via zero-offset tomographic inversion: 67th Annual International Meeting, SEG, Expanded Abstracts, 1805–1808.

Audebert, F., J. P. Diet, and X. Zhang, 1996, CRP-scans from 3-D pre-stack depth migration: A powerful combination of CRP-gathers and velocity scans: 66th Annual International Meeting, SEG, Expanded Abstracts, 515–518.

Bevc, D., M. Fliender, J. VanderKwaak, 2006, 3-D tomographic updating with automatic volume-based picking: 76th Annual International Meeting, SEG, Expanded Abstracts, TOM 1.6.

Biondi, B., 1990, Velocity analysis using beam stacks: Ph.D. thesis, Stanford University.

———, 2003, Amplitude balancing of 3-D angle-domain common-image gathers: Stanford Exploration Project, Report 114, 45–56.

———, 2005, Angle-domain common image gathers for anisotropic migration: 75th Annual International Meeting, SEG, Expanded Abstracts, 1922–1925.

———, 2006, 3D seismic imaging: SEG Investigations in Geophysics, No. 14.

Biondi, B., and P. Sava, 1999, Wave-equation migration velocity analysis: 69th Annual International Meeting, SEG, Expanded Abstracts, 1723–1726.

Biondi, B., and W. W. Symes, 2004, Angle-domain common-image gathers for migration velocity analysis by wavefield-continuation imaging: Geophysics, **69**, 1283–1298.

Biondi, B., and T. Tisserant, 2004, 3-D angle-domain common-image gathers for migration velocity analysis: Geophysical Prospecting, **52**, 575–591.

Biondi, B. L., R. G. Clapp, S. B. Fomel, and T. A. Alkhalifah, 1998, Robust reflection tomography in the time domain: 68th Annual International Meeting, Society of Exploration Geophysicists, Expanded Abstracts, 1847–1850.

Brandsberg-Dahl, S., M. de Hoop, and B. Ursin, 1999, Velocity analysis in the common scattering-angle/azimuth domain: 69th Annual International Meeting, SEG, Expanded Abstracts, 1715–1718.

Chauris, H., and M. S. Noble, 1998, Testing the behavior of differential semblance for velocity estimation: 68th Annual International Meeting, SEG, Expanded Abstracts, 1305–1308.

Claerbout, J. F., 2004, Image estimation by example: Stanford Exploration Project, http://sepwww.stanford.edu/sep/prof/index.html, accessed May 26, 2004.

Clapp, R. G., 2001, Geologically constrained migration velocity analysis: Ph.D. thesis, Stanford University.

Clapp, R., and B. Biondi, 2000, Tau domain migration velocity analysis using angle CRP gathers and geologic constrains: 70th Annual International Meeting, SEG, Expanded Abstracts, 926–929.

Clapp, R. G., B. Biondi, and J. F. Claerbout, 2004, Incorporating geologic information in reflection tomography: Geophysics, **69**, 533–546.

Deregowski, S. M., 1990, Common-offset migrations and velocity analysis: First Break, **8**, 224–234.

Doherty, S. M., and J. F. Claerbout, 1976, Structure-independent velocity estimation: Geophysics, **41**, 850–881.

Etgen, J., 1990, Residual prestack migration and interval velocity estimation: Ph.D. thesis, Stanford University.

Fomel, S., 2003, Velocity continuation and the anatomy of residual prestack time migration: Geophysics, **68**, 1650–1661.

Gardner, G. H. F., W. S. French, and T. Matzuk, 1974, Elements of migration and velocity analysis: Geophysics, **39**, 811–825.

Golub, G. H., and C. F. V. Loan, 1983, Matrix computations: Johns Hopkins University Press.

Hardy, P., 2003, High resolution tomographic MVA with automation, EAGE/SEG Summer Research Workshop on "Processing and imaging of seismic data — using explicit or implicit velocity model information?", T27.

Harlan, W. S., 1995, Regularization by model reparameterization: http://www.billharlan.com/pub/papers/regularization.html, accessed June 22, 1997.

Holland, J., 1975, Adaptation in natural and artificial systems: University of Michigan Press.

Jervis, M., M. K. Sen, and P. L. Stoffa, 1996, Prestack migration velocity estimation using nonlinear methods: Geophysics, **61**, 138–150.

Jiao, J., P. L. Stoffa, M. K. Sen, and R. K. Seifoullaev, 2002, Residual migration-velocity analysis in the plane-wave domain: Geophysics, **67**, 1252–1269.

Jin, S., and R. Madariaga, 1994, Nonlinear velocity inversion by a two-step Monte Carlo method: Geophysics, **59**, 577–590.

Kirkpatrick, S., C. D. Gelatt, and M. P. Vecchi, 1983, Optimization by simulated annealing: Science, **220**, 671–680.

Kosloff, D., J. Sherwood, Z. Koren, E. MacHet, and Y. Falkovitz, 1996, Velocity and interface depth determination by tomography of depth migrated gathers: Geophysics, **61**, 1511–1523.

Lafond, C. F., and A. R. Levander, 1993, Migration moveout analysis and depth focusing: Geophysics, **58**, 91–100.

Liu, Z., and N. Bleistein, 1995, Migration velocity analysis: Theory and an iterative algorithm: Geophysics, **60**, 142–153.

Lomask, J., R.G. Clapp, and B. Biondi, 2006, Parallel implementation of image segmentation for tracking 3D salt boundaries: 68th Annual International Meeting, EAGE, F004.

MacKay, S., and R. Abma, 1993, Depth-focusing analysis using a wavefront-curvature criterion: Geophysics, **58**, 1148–1156.

Meng, Z., and N. Bleistein, 2001, On velocity/depth ambiguity in 3-D migration velocity analysis: Geophysics, **66**, 256–260.

Meyerholtz, K. A., G. L. Pavlis, and S. A. Szpakowski, 1989, Convolutional quelling in seismic tomography: Geophysics, **54**, 570–580.

Mosher, C., S. Jin, and D. Foster, 2001, Migration velocity analysis using common angle image gathers: 71st Annual International Meeting, SEG, Expanded Abstracts, 889–892.

Mulder, W. A., and ten Kroode, A. P. E., 2002, Automatic velocity analysis by differential semblance optimization: Geophysics, **67**, 1184–1191.

Ottolini, R., and J. F. Claerbout, 1984, The migration of common-midpoint slant stacks: Geophysics, **49**, 237–249.

Perdrizet, T., and D. Sinoquet, 2003, Adaptive model parameterization designed for reflection tomography: 73rd Annual International Meeting, SEG, Expanded Abstracts, 674–677.

Rocca, F., and L. Salvador, 1982, Residual migration: 52nd Annual International Meeting, SEG, Expanded Abstracts, 4–7.

Rothman, D. H., S. A. Levin, and F. Rocca, 1985, Residual migration — Applications and limitations: Geophysics, **50**, 110–126.

Sarkar, D. and I. Tsvankin, 2004b, Migration velocity analysis in factorized VTI media: Geophysics, **69**, 708–718.

Sattlegger, J. W., 1975, Migration velocity determination — Part I: Philosophy: Geophysics, **40**, 1–5.

Sava, P., and B. Biondi, 2004a, Wave-equation migration velocity analysis — I: Theory: Geophysical Prospecting, **52**, 593–606.

———, 2004b, Wave-equation migration velocity analysis — II: Subsalt imaging example: Geophysical Prospecting, **52**, 607–623.

Sava, P., B. Biondi, and J. Etgen, 2005, Diffraction-focusing migration velocity analysis: Geophysics, **70**, U19–27.

Sava, P. C., 2003, Prestack residual migration in frequency domain: Geophysics, **68**, 634–640.

Sinoquet, D., 1993, Modeling a priori information on the velocity field in reflection tomography: 63rd Annual International Meeting, SEG, Expanded Abstracts, 591–594.

Stolt, R. H., 1978, Migration by Fourier transform: Geophysics, **43**, 23–48.

———, 1996, A prestack residual time migration operator: Geophysics, **61**, 605–607.

Stork, C., 1992, Reflection tomography in the postmigrated domain: Geophysics, **57**, 680–692.

Symes, W. W., and J. J. Carazzone, 1991, Velocity inversion by differential semblance optimization: Geophysics, **56**, 654–663.

Tarantola, A., 1987, Inverse problem theory: Methods for data fitting and model parameter estimation: Elsevier.

Thomsen, L., 2002, Understanding seismic anisotropy in exploration and exploitation: SEG Distinguished Instructor Short Course.

Toldi, J., 1985, Velocity analysis without picking: Ph.D. thesis, Stanford University.

Tsvankin, I., 2001, Seismic signatures and analysis of reflection data in anisotropic media: Elsevier Science.

van de Made, P. M., P. van Riel, and A. J. Berkhout, 1984, Velocity and subsurface geometry inversion by a parameter estimation in complex inhomogeneous media: 54th Annual International Meeting, SEG, Expanded Abstracts, 373–376.

van Trier, J., 1990, Tomographic determination of structural velocities from depth migrated seismic data: Ph.D. thesis, Stanford University.

Wapenaar, C. P. A., and A. J. Berkhout, 1985, Velocity determination in layered systems with arbitrarily curved interfaces by means of wave-field extrapolation of common-midpoint data: Geophysics, **50**, 63–76.

Whitmore, N. D., and J. D. Garing, 1993, Interval velocity estimation using iterative prestack depth migration in the constant angle domain: The Leading Edge, **12**, 757–762.

Yilmaz, O., 2001, Seismic data analysis, *in* M. R. Cooper and S. M. Doherty, eds., Seismic data analysis, v. 2: SEG, 1001–2027.

Yilmaz, O., and R. E. Chambers, 1984, Migration velocity analysis by wave-field extrapolation: Geophysics, **49**, 1664–1674.

Chapter 8 Imaging and Aliasing

Introduction

The quality of 3D images is influenced strongly by the spatial sampling of the data and by whether the imaging operators properly take into account the data sampling. Strong aliasing artifacts degrade the images when the data are sampled poorly and the imaging operators are not implemented carefully. The sampling problem is more acute in 3D imaging than in 2D imaging, because the spatial axes of 3D data often are sampled sparsely and irregularly. In this chapter, I analyze the problems caused by regular data grids that are sampled too coarsely, although regularly. Chapter 9 discusses the issues related to irregularity of both data acquisition and reflector illumination.

Three types of aliasing are relevant to seismic imaging: data aliasing, image aliasing, and operator aliasing. Data aliasing and image aliasing are fairly straightforward to understand by using standard sampling theory, whereas operator aliasing is more specific to imaging operators and thus requires detailed analysis. From the earliest development of Kirchhoff migration methods (Schneider, 1978), it has been evident that aliasing artifacts appear in the image if one does not consider data and operator aliasing when one implements the summation operator (Gardner et al., 1974). Thus, methods for preventing operator aliasing during Kirchhoff migration are well established (Bevc and Claerbout, 1992; Gray, 1992; Lumley et al., 1994; Abma et al., 1998; Biondi, 2001). In the first part of this chapter, I discuss aliasing for Kirchhoff migration.

In contrast, operator aliasing becomes an issue for wavefield-continuation migration only when we are migrating prestack data. Therefore, our understanding of operator aliasing for wavefield-continuation migration is less well developed. In the second part of this chapter, I present the basic concepts underlying our understanding of operator aliasing in wavefield-continuation migrations. Some of those concepts are relatively well established (Zhang et al., 2003), whereas others are the subjects of active research (Artman et al., 2003). At the end of the chapter, I apply the general concepts governing operator aliasing in wavefield-continuation migration both to the analysis of aliasing in prestack migration and to the definition of a simple procedure to avoid or at least attenuate operator-aliasing artifacts.

Aliasing Fundamentals

The basic principles of sampling and aliasing are understood easily for plane waves. Plane waves are stationary signals in time and space. Thus, they can be analyzed in the frequency-wavenumber domain as well as in the original time-space domain. For a plane wave with slope p, the frequency ω and the wavenumber k are related by the expression

$$p = \frac{k}{\omega}. \tag{8.1}$$

Plane waves with a given frequency ω_d are well sampled if their slopes are limited within the range determined by the condition

$$|p| \le \frac{k_N}{\omega_d}, \tag{8.2}$$

where k_N is the Nyquist wavenumber given as a function of the spatial sampling Δx, as

$$|k_N| = \frac{\pi}{\Delta x}. \tag{8.3}$$

Aliasing occurs when a condition (equation 8.2) is violated because the dip range (dip bandwidth) is too wide or the waveform is too high in frequency or the spatial sampling is too coarse.

Figure 1 shows a classic wavenumber representation of aliasing. The drawing at the top of the figure (Figure 1a) shows the spectrum, for a single temporal frequency, of a family of plane waves sampled adequately in space. Because the signal is discretized in space, the original spectrum of the continuous signal (center) is replicated at intervals of twice the Nyquist wavenumbers (left and right). The signal is not aliased because its maximum wavenumber is lower than the Nyquist wavenumber. In contrast, aliasing occurs in Figure 1b because the original spectrum of the signal is stretched, either by an increase in temporal frequency or by a widening of the dip bandwidth of the plane waves. Consequently, plane waves that are steeply dipping and that correspondingly have a high wavenumber appear to have a negative wavenumber and to be dipping in the opposite direction. Figure 1c shows the aliasing caused by doubling the spatial sampling Δx and consequently halving the Nyquist wavenumber.

Figure 1. Spectra of a family of plane waves in which there is (a) no aliasing, (b) aliasing caused by an increase in frequency or dip bandwidth, and (c) aliasing caused by doubling the spatial sampling.

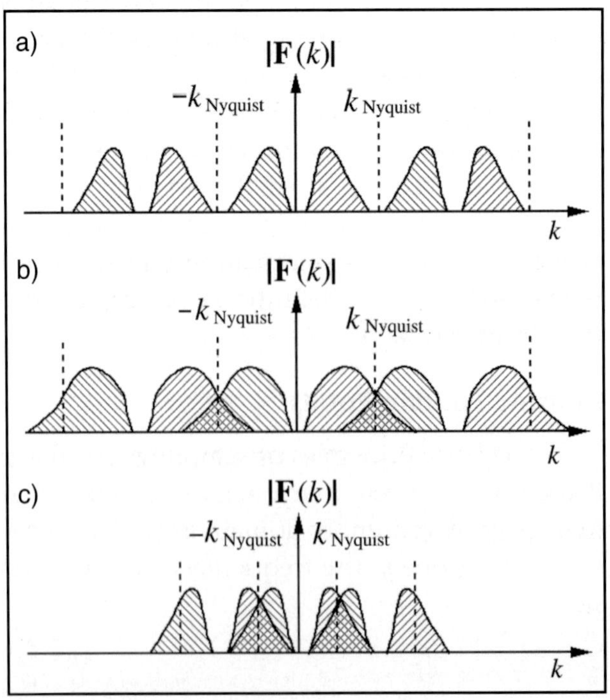

Figure 2 illustrates the aliasing bounds in the multidimensional case. For simplicity, the bounds on the data wavenumbers are set as a maximum of the absolute value of the wavenumber vector. In that case, all the data wavenumbers are contained in a circular region. The sampling along the crossline direction y is assumed to be coarser by a factor of two than that along the inline direction x. When the temporal frequency is low enough (Figure 2a), the data dips are not aliased. However, if the temporal frequency ω_d increases, the replicas of the circles overflow in the main band (Figure 2b). To avoid aliasing artifacts, only the wavenumber components within the hourglass-shaped area should be included in the image, or, even more conservatively, the imaging should be limited to the wavenumber components within the dark rectangular area in Figure 2.

The principles of aliasing embodied in Figures 1 and 2 can be applied directly to the analysis of data aliasing and image aliasing. Aliasing of imaging operators still is based on the same fundamental concepts, but it is less obvious. To understand operator aliasing, one needs to analyze how imaging operators connect the data space with the image space. Operator aliasing is a function of the local time derivatives of the operator's summation surface, often called operator dip, and of the frequency bandwidth and dip bandwidth of the data.

In this chapter, I will analyze methods used to avoid any of the three kinds of aliasing that cause artifacts in seismic images. The second section of this chapter presents the theory of aliasing and antialiasing for Kirchhoff imaging operators. The third section builds on the understanding we gain in the first part and shows us how to analyze aliasing in wavefield-continuation migration — particularly for prestack wavefield-continuation migration.

The antialiasing dilemma

Before plunging into the theory, I discuss an example of imaging artifacts caused by operator and data aliasing, and I present a fundamental dilemma that one faces when one images aliased data. Because prestack seismic data almost always are aliased, one constantly must determine the appropriate trade-off between imaging artifacts and image resolution. Although the following discussion is based on results obtained by Kirchhoff migration, prestack migration by wavefield-continuation methods presents similar challenges.

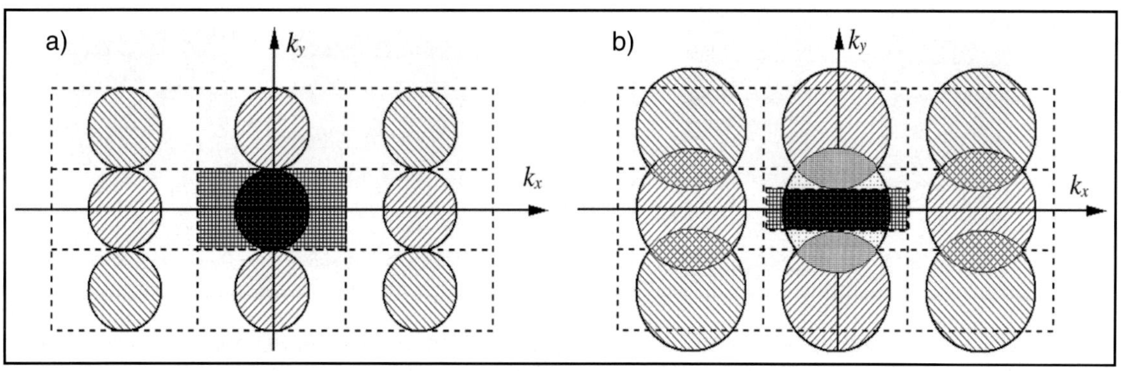

Figure 2. Wavenumber representation of operator aliasing in 2D. The signal (circle centered at the origin) is not aliased in (a) and is aliased in (b). The black-shaded areas are the admissible regions for the operator dip.

In Kirchhoff imaging, we usually avoid operator aliasing by suppressing some high-frequency components during the summation process. Unfortunately, this low-pass filtering decreases the image resolution when the data are aliased. Therefore, we face a dilemma when we image aliased data. On one hand, to avoid aliasing noise, we need to apply antialiasing filters; on the other hand, we do not want to lose image resolution. Figure 3 exemplifies the problem. It shows the results of 3D zero-offset time migration of a salt-dome flank in the Gulf of Mexico. The section in Figure 3a was 3D migrated without the use of any antialiasing filter, whereas the section in Figure 3b was obtained with a standard antialiased migration. Note that the image obtained without antialiasing is much noisier than the antialiased one, and it also has higher resolution. In the shallow part of the section shown in Figure 3a, the aliasing noise is so strong that we cannot appreciate the higher resolution of Figure 3a compared with Figure 3b.

However, when I zoom into the deeper parts of the sections and compare them (Figure 4), it is apparent that one loses resolution by applying antialiasing. In particular, the high-frequency dipping event at about CMP X = 700 m and time = 2.2 s is resolved poorly in the antialiased migration (Figure 4b). The antialiased migration misses a whole wavelet cycle of the sediment truncation against the salt flank. If one considers that hydrocarbon reservoirs often are located at the sediment-salt interfaces, one appreciates the potential advantages of improving the resolution of such events.

In this example, migration without antialiasing achieves higher resolution than antialiased migration does because it images data components that are aliased in the data space. In particular, the steeply dipping energy reflected from the salt flanks, visible in the data window shown in Figure 5, is aliased. Figure 6 shows the frequency-wavenumber spectrum of the data window in Figure 5. Figure 6 shows the central band of the spectrum and its two replicas on either side. The vertical white lines correspond to the Nyquist wavenumbers. The two replicated bands are caused by the spatial discretization, and their existence is at the root of the aliasing

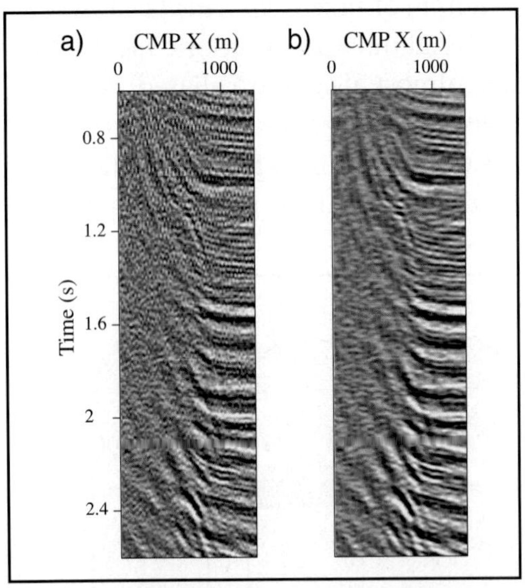

Figure 3. 3D migrations of salt-dome flanks in the Gulf of Mexico: (a) migration obtained without the use of any antialiasing filter, (b) migration obtained with application of a standard antialiasing filter.

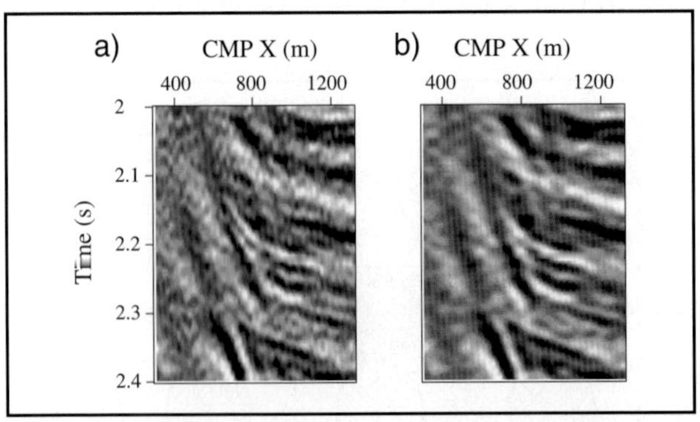

Figure 4. A zoom into the 3D migrations of salt-dome flanks in the Gulf of Mexico shown in Figure 3: (a) migration obtained without the use of any antialiasing filter, and (b) migration obtained with application of a standard antialiasing filter.

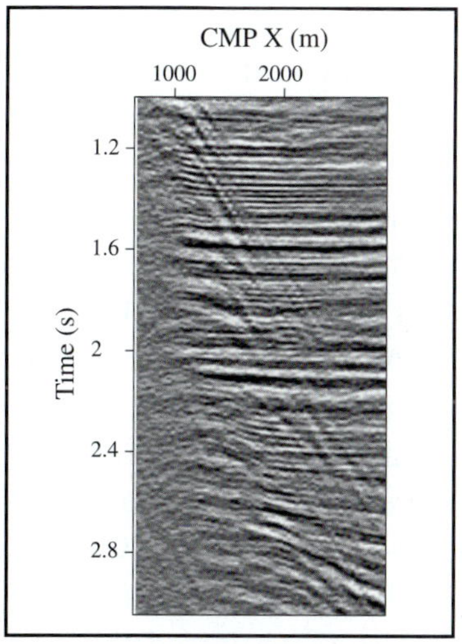

Figure 5. A data window containing aliased reflections from the salt flanks.

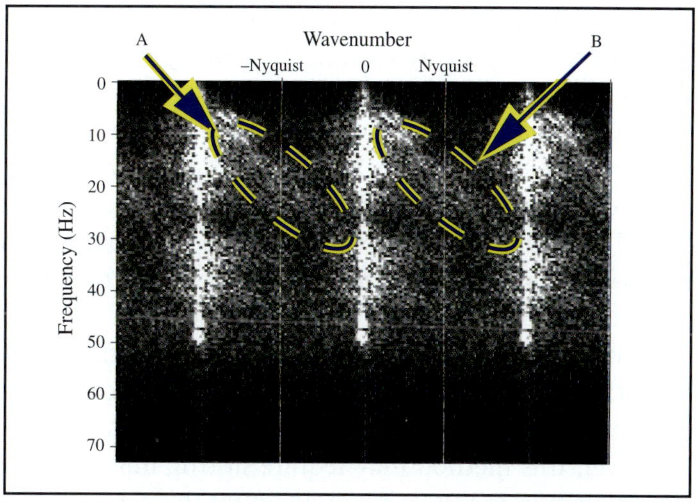

Figure 6. The frequency-wavenumber spectrum of the data window shown in Figure 5. Notice that the aliased events correspond to the clouds in the spectrum contoured with ellipses in the figure. The ellipse marked as B contours the energy that can be migrated correctly. The ellipse marked as A contours the energy that cannot be migrated correctly and should be avoided during the summation.

problem. The dipping events that appear aliased in the data window correspond to the clouds in the spectrum contoured with ellipses in Figure 6. The ellipse marked as **B** contours the energy that can be migrated correctly. The cloud starts in the main band but crosses the positive Nyquist line and trespasses on the replicated band on the positive side. The portion of the cloud that crosses the positive Nyquist line corresponds to the signal components that are lost when conventional antialiased migration is applied (Figures 3b and 4b) and that are responsible for the high resolution achieved by migration without antialiasing (Figures 3a and 4a). The ellipse marked as **A** contours the energy that cannot be migrated correctly and should be avoided during the summation.

Aliasing in Kirchhoff Migration

In the previous sections, I reviewed the basic concepts of aliasing, and I illustrated the effects of aliasing on the migrated images. In this section, I will apply those concepts to the problem of eliminating artifacts in the images created by the application of Kirchhoff migration operators. I will analyze in detail the aliasing of 3D prestack migration, but the same concepts can be applied to all other integral operators presented in this book, such as DMO (Hale, 1991) and AMO (Biondi et al., 1998) presented in Chapter 4.

Aliasing in image space

The simplest type of aliasing related to the imaging operator is image-space aliasing. It occurs when the spatial sampling of the image is too coarse to represent adequately the steeply dipping reflectors that the imaging operator attempts to build during the imaging process. In other words, image-space aliasing is caused by excessively coarse sampling of the image space, with consequent aliasing of the migration ellipsoid. With

Kirchhoff imaging, the choice of image sampling is arbitrary, and thus image aliasing can be avoided easily by increasing the density of the image. However, the cost of migration grows proportionally with image density. To avoid image aliasing for a given spatial sampling of the image, one must limit the frequency content of the image during the migration as a function of reflectors' dips. That can be accomplished by applying a dip-dependent temporal low-pass filtering of the input data during the summation process.

The triangle-filters method described in Claerbout (1995) is an efficient method of low-pass filtering time-domain data that have variable frequency. An alternative method is to precompute low-passed versions of the input traces and to select the appropriate input data during summation (Gray, 1992). This second method potentially is more accurate than triangle filtering, and it is more efficient computationally when each input trace is summed into many output traces, as happens for 3D migrations (Abma et al., 1998). However, this method may require storing many versions of the input data. To reduce the storage requirements without affecting the accuracy, I interpolate the low-passed traces linearly along the frequency axis during the summation.

The antialiasing constraints used to avoid image aliasing can be derived easily from basic sampling theory. For the case of time migration, when the coordinates of the image space are $\xi = (\tau_\xi, x_\xi, y_\xi)$, the pseudodepth frequency ω^ξ_τ must fulfill the following inequalities:

$$\begin{aligned} \omega^\xi_\tau &< \frac{\pi}{\Delta x_\xi \left| p^\xi_x \right|}, \\ \omega^\xi_\tau &< \frac{\pi}{\Delta y_\xi \left| p^\xi_y \right|}, \end{aligned} \tag{8.4}$$

where Δx_ξ and Δy_ξ are the image sampling rates of the x_ξ and y_ξ axes and p^ξ_x and p^ξ_y are the operators' dips in image space.

As discussed above, it is convenient to apply an antialiasing filter as a band-pass filter of the input traces, before the traces' contributions are summed into the output image. Therefore, I recast the constraints on the pseudodepth frequency ω^ξ_τ of equations 8.4 into constraints on the input data frequency ω_d. This distinction is important because stretching or compression of the time axis changes the frequency content of the seismic wavelet during the imaging process. In particular, during time migration, the wavelet always is stretched; neglecting this stretch would lead to antialiasing constraints that are too stringent. Notice that the seismic wavelet may be compressed, instead of stretched, by other imaging operators such as inverse DMO and AMO (Chapter 4). Thus, the pseudodepth frequency of the image and the temporal frequency of the data are linked by the wavelet-stretch factor $dt_D/d\tau_\xi$, as $\omega^\xi_\tau = \omega_d dt_D/d\tau_\xi$, where t_D is the time coordinate in the data space. Taking into account the wavelet-stretch factor, I can write the data-frequency constraints that control image aliasing, as a function of the image sampling rates Δx_ξ and Δy_ξ, the image dips p^ξ_x and p^ξ_y, and the wavelet-stretch factor $dt_D/d\tau_\xi$:

$$\begin{aligned} \omega_d &< \frac{\pi}{\Delta x_\xi \left| p^\xi_x \right| \frac{dt_D}{d\tau_\xi}}, \\ \omega_d &< \frac{\pi}{\Delta y_\xi \left| p^\xi_y \right| \frac{dt_D}{d\tau_\xi}}. \end{aligned} \tag{8.5}$$

To understand the effects of incorporating an antialiasing filter into a migration operator, I analyze the images generated when one single input trace is migrated into a cube. Figure 7 shows the result when one trace is migrated without using any antialiasing filter. The input trace was recorded at an offset of 2.4 km, and the velocity was assumed to be constant and equal to 1.5 km/s. The image sampling was 20 m in each direction ($\Delta x_\xi = \Delta y_\xi = 20$ m). Strong aliasing artifacts are visible in both the time slices and the vertical section.

Figure 8 shows the result produced when the same data trace is migrated with an appropriate antialiased operator. Notice that the aliasing artifacts disappear as the frequency content of the imaged reflectors progressively decreases — that is, as the dips increase. Figure 9 shows the result of migrating the same data trace when the effects of the wavelet stretch are not taken into account — that is, by setting $dt_D/d\tau_\xi = 1$. In that case, the antialiasing filter overcompensates for the image dip, and valuable resolution is lost at steep dips. Examining the time slice shown at the top of Figure 9, one notices that the loss of resolution is larger for regions of the migration ellipses with a steep dip along the crossline direction.

Figure 10 shows the effects of image-space aliasing on the migrated results from the salt-dome data set shown in Figure 3. Figure 10a shows the migration results obtained without the use of any antialiasing and with $\Delta x_\xi = 36$ m — that is, the original sampling of the zero-offset data. Shallow and steeply dipping reflectors are significantly aliased. The quality of the image improves when the image's spatial sampling is halved to Δx_ξ =18 m (Figure 10b). Note that the traces in Figure 10a are exactly the same as the odd traces in Figure 10b. Therefore, image aliasing does not add noise to the image; it merely makes the interpretation of the image more difficult and ambiguous. Application of the image-space antialiasing constraints expressed in equation 8.5 further improves the image

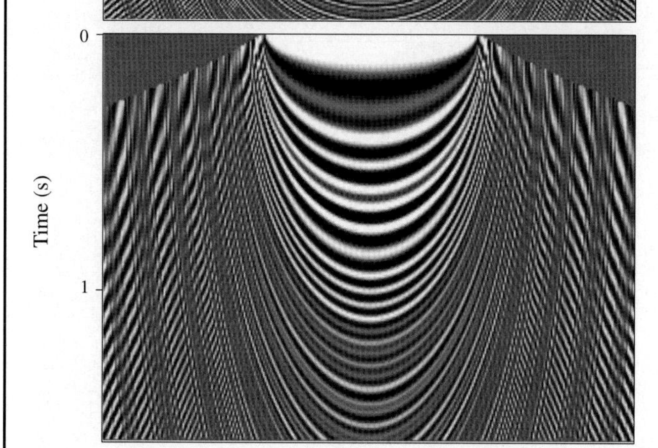

Figure 7. An image obtained by constant-velocity Kirchhoff migration without the use of any antialiasing ($\Delta x_\xi = \Delta y_\xi = 20$ m, $V = 1.5$ km/s, and offset = 2.4 km). The time slice is cut at time = 1 s, and the vertical section is cut at CMP Y = 1000 m.

quality, particularly for the shallower events (Figure 10c). Although the image in Figure 10c is less noisy than the images in Figure 10a and 10b, it still is contaminated by aliasing artifacts. These artifacts are caused by operator aliasing, or data-space aliasing. The next section analyzes the causes of operator aliasing and presents antialiasing constraints used to eliminate it.

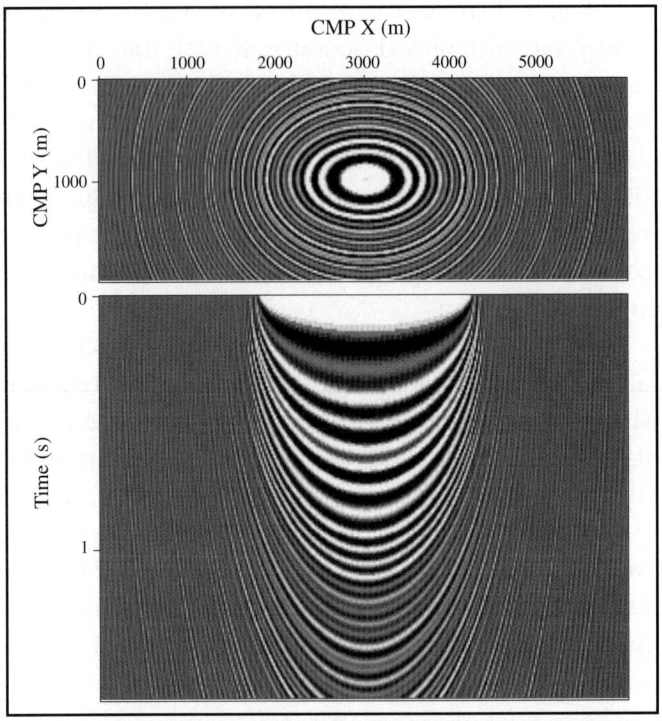

Figure 8. An image obtained by constant-velocity Kirchhoff migration using image-space antialiasing ($\Delta x_\xi = \Delta y_\xi =$ 20 m, $V =$ 1.5 km/s, and offset = 2.4 km). The time slice is cut at time = 1 s, and the vertical section is cut at CMP Y = 1000 m.

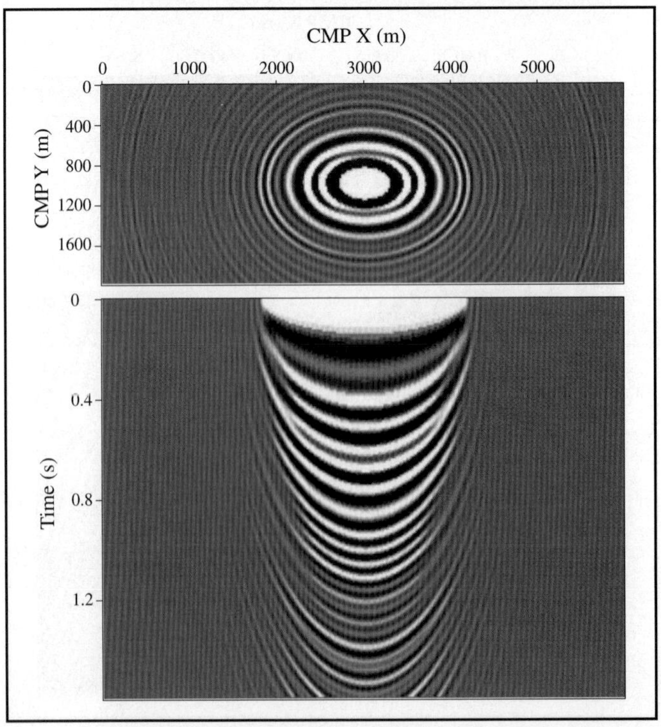

Figure 9. An image obtained when too-strong antialiasing is applied, because the effects of the wavelet stretch on the frequency content of the image are ignored ($\Delta x_\xi = \Delta y_\xi =$ 20 m, $V =$ 1.5 km/s, and offset = 2.4 km). The time slice is cut at time = 1 s, and the vertical section is cut at CMP Y = 1000 m.

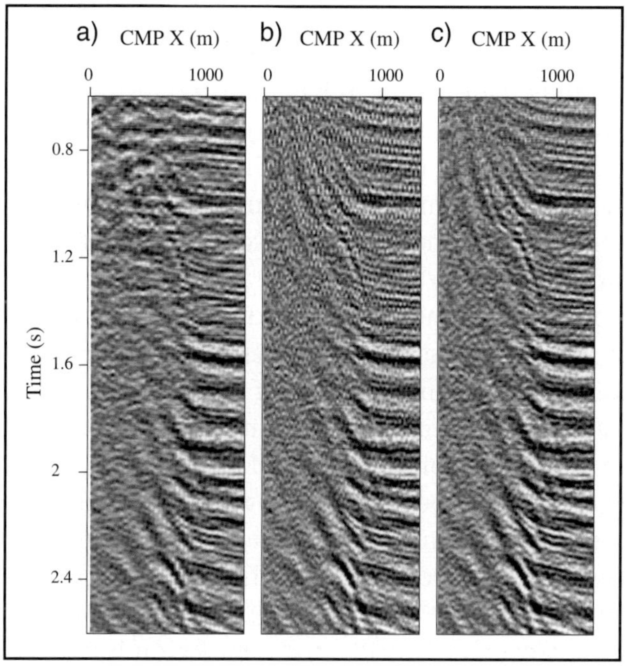

Figure 10. A migrated section from the Gulf of Mexico salt dome: (a) $\Delta x_\xi = 36$ m and no antialiasing, (b) $\Delta x_\xi = 18$ m and no antialiasing, and (c) $\Delta x_\xi = 18$ m and image-space antialiasing.

Operator aliasing

The previous section analyzed aliasing in the image space, which occurs when the spreading representation of the imaging operator is aliased. Data-space aliasing, the counterpart of image-space aliasing, occurs when the gathering representation of the imaging operator is aliased in the data space. Such data-space aliasing commonly is called operator aliasing. Operator aliasing is different from data aliasing and may occur even when the data are sampled sufficiently. When operator aliasing occurs, noise actually is added to the image, in contrast to the case of image-space aliasing, in which interpretability of the image is hampered by missing data, not by additive noise. In general, artifacts are generated when data that are not aligned with the summation path do not interfere destructively, as they should during the summation, and thus they stack coherently into the image. This phenomenon is the cause of the aliasing noise that degrades the image when operator aliasing occurs.

To avoid adding aliasing noise to the image, one could low-pass-filter the input data according to the operator dips. A reasonable strategy is to limit the migration operator to the data components in the main band of the data spectrum (e.g., to the data between the negative Nyquist and the positive Nyquist lines in the spectrum shown in Figure 6). Such a strategy leads to application of the following antialiasing constraints on the temporal frequency of the summation operator:

$$\begin{aligned} \omega_d &< \frac{\pi}{\Delta x_D \left| p_x^{\mathrm{op}} \right|}, \\ \omega_d &< \frac{\pi}{\Delta y_D \left| p_y^{\mathrm{op}} \right|}, \end{aligned} \tag{8.6}$$

where Δx_D and Δy_D are the sampling rates of the data axes and p_x^{op} and p_y^{op} are the operator dips in the data space, that is, the dips of the summation operator (i.e., the hyperboloid for migration). These or equivalent relationships have been presented by several authors (Bevc and Claerbout, 1992; Gray, 1992; Lumley et al., 1994; Abma et al., 1998).

The antialiasing conditions expressed in equation 8.6 can be either too restrictive or too lax. Although those conditions may be correct for several important cases, they do not take into account the fact that operator aliasing depends on the presence of conflicting and potentially aliased dips in the data. They implicitly assume that no dips are aliased in the data. When that implicit assumption is violated, as it was for the positive dip in the previous example, it can be harmful to apply the standard antialiasing conditions. To avoid problems with standard antialiasing conditions, Biondi (2001) introduced more general antialiasing conditions that take into account the existence of conflicting dips. Pica (1996) presented a model-based antialiasing method similar to the one proposed by Biondi (2001), although it yields different antialiasing conditions.

Figure 11 shows graphically the idea at the basis of more general antialiasing criteria. The shaded area is the widest possible area of the spectrum where there is no interference with aliased energy; that is, the data components in the shaded areas can be imaged safely without generating operator-aliasing noise. The boundaries of this pass region are determined as a function of the minimum data dips, $\mathbf{p}^{\min} = (p_x^{\min}, p_y^{\min})$, and maximum data dips, $\mathbf{p}^{\max} = (p_x^{\max}, p_y^{\max})$. On the negative-wavenumber side, the pass region is bounded by a line representing a replica of the maximum dip line. That is, it is bounded by a line that starts at $-2k_{\mathrm{nyq}}$ (where k_{nyq} is the Nyquist wavenumber) and that has a slope corresponding to the maximum dip. Similarly, on the positive-wavenumber side, the pass region is bounded by a line starting at $2k_{\mathrm{nyq}}$ and having a slope that corresponds to the minimum dip.

The boundaries for the pass region are expressed analytically by the inequalities

$$p_x^{\max} - \frac{2\pi}{\omega_d \Delta x_D} < p_x^{\mathrm{op}} < p_x^{\min} + \frac{2\pi}{\omega_d \Delta x_D}, \tag{8.7}$$

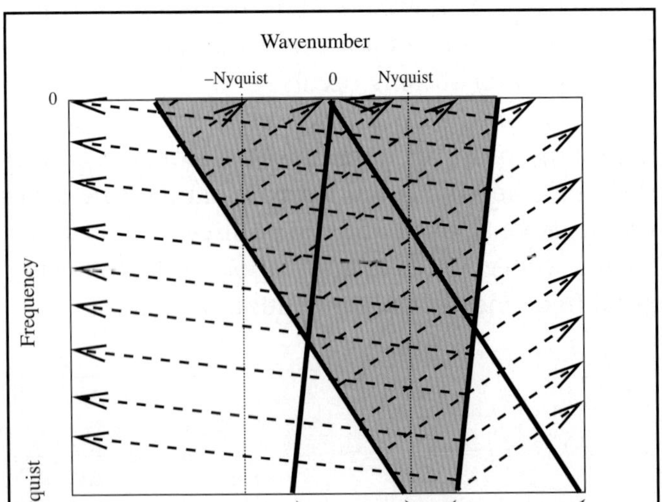

Figure 11. Graphic representation of the general antialiasing criteria (equations 8.7 and 8.8). The shaded area corresponds to the pass region. It is the widest possible area of the spectrum where there is no interference with aliased energy. On the negative-wavenumber side, the pass region is bounded by a line that starts at $-2k_{\mathrm{nyq}}$ and has a slope corresponding to the maximum dip $p^{\max}$. On the positive-wavenumber side, the pass region is bounded by a line starting at $2k_{\mathrm{nyq}}$ and with a slope corresponding to the minimum dip $p^{\min}$. The dashed lines cover the areas in which at least one of the inequalities in equation 8.7 is satisfied.

and

$$p_y^{\max} - \frac{2\pi}{\omega_d \Delta y_D} < p_y^{\text{op}} < p_y^{\min} + \frac{2\pi}{\omega_d \Delta y_D}. \tag{8.8}$$

The inequalities expressed in equations 8.7 and 8.8 can be recast easily as antialiasing constraints on the maximum frequency in the data, as

$$\omega_d < \frac{2\pi}{\Delta x_D \left(p_x^{\text{op}} - p_x^{\min}\right)}, \tag{8.9}$$

$$\omega_d < \frac{2\pi}{\Delta x_D \left(p_x^{\max} - p_x^{\text{op}}\right)}, \tag{8.10}$$

$$\omega_d < \frac{2\pi}{\Delta y_D \left(p_y^{\text{op}} - p_y^{\min}\right)}, \tag{8.11}$$

and

$$\omega_d < \frac{2\pi}{\Delta y_D \left(p_y^{\max} - p_y^{\text{op}}\right)}. \tag{8.12}$$

The constraints expressed in equations 8.9 through 8.12 can be used as alternatives to or in conjunction with the constraints expressed in equation 8.6, to design antialias summation operators.

By comparing equation 8.6 with equations 8.9 through 8.12, one can see that the two sets of constraints are equivalent when, for each frequency, the data-dip limits $p^{\min}$ and $p^{\max}$ are set respectively to $-k_{\text{nyq}}/\omega_d$ and k_{nyq}/ω_d. This is easy to verify by substituting $p_x^{\min} = -\pi/(\Delta x_D \omega_d)$ into equation 8.9 and $p_x^{\max} = \pi/(\Delta x_D \omega_d)$ into equation 8.10 and performing analogous substitutions into equations 8.11 and 8.12. This result implies that when one can assume that no spatial aliasing exists in the data, the new antialiasing conditions are equivalent to the standard conditions.

The dip limits of the data, $\mathbf{p}^{\min}$ and $\mathbf{p}^{\max}$, can vary in both time and space according to the local dips in the data. Therefore, the antialiasing filters applied to the data as a consequence of the constraints in equations 8.9 through 8.12 can be fairly complex and can depend on factors including local dips, time, and spatial coordinates. If no a priori knowledge of the local dips is available and if the summation is carried out along the midpoint axes, twice the inverse of propagation velocity is a reasonable bound on the absolute values of both $\mathbf{p}^{\min}$ and $\mathbf{p}^{\max}$. In contrast, in the case in which the summation is performed along the offset axes, as it is for CMP stacks, $\mathbf{p}^{\min}$ can be assumed safely to be positive and at worst equal to zero. Therefore, imaging operators that sum data over the offset axes, such as NMO followed by stack, are less prone to spatial aliasing than are imaging operators that sum data over midpoints, such as migration.

The most substantial benefits of applying the more general constraints expressed in equations 8.9 through 8.12 are achieved when asymmetric bounds on the dips in the data enable us to image the high-frequency components present in the data as aliased energy and still to minimize the aliasing artifacts. Those high-frequency components would be filtered out if the constraints in equation 8.6 were applied. Asymmetric bounds on the data dips are realistic in the important application of imaging steep salt-dome flanks, as in the Gulf of Mexico data set shown above. In that case, one can assume that the negative time dips in the data are small. According to equations 8.7 and 8.12, the increase in $p^{\min}$, that is, the decrease of its absolute value increases the maximum positive dip that can be imaged without operator aliasing. In practice, application of the generalized constraints in equations 8.9 through 8.12, with $\mathbf{p}^{\min} \neq \mathbf{p}^{\max}$, causes the migration operator to be asymmetric, with the dip bandwidth dependent on reflector direction.

Figures 12 and 13 show an example of the effects of asymmetric dip bounds on the migration operator. The image sampling is the same as that in Figure 8 ($\Delta x_\xi = \Delta y_\xi = 20$ m) for both images, but here the data sampling is assumed to be coarser than the image sampling by a factor of two; that is, $\Delta x_D = \Delta y_D = 40$ m. When the constraints in equation 8.6 are applied (see Figure 12), the operator has lower resolution than in Figure 8. However, if one assumes that $p_x^{\min} = 0$ s/m and applies the constraints in equations 8.9 through 8.12, the pass region for the operator is augmented by the band between $k_{\rm nyq}$ and $2k_{\rm nyq}$ (see Figure 11). Consequently, in Figure 13, the positive time dips are imaged with the same resolution as in Figure 8.

The constraints derived to avoid operator aliasing and image aliasing differ in general, and both sets of constraints must be respected if a high-quality image is to be assured. However, the two sets of constraints are equivalent in many practical situations. In those situations, only one set of constraints needs to be applied to assure that the image is both adequately sampled and free of operator-aliasing noise. Chapter 8 in Bion-

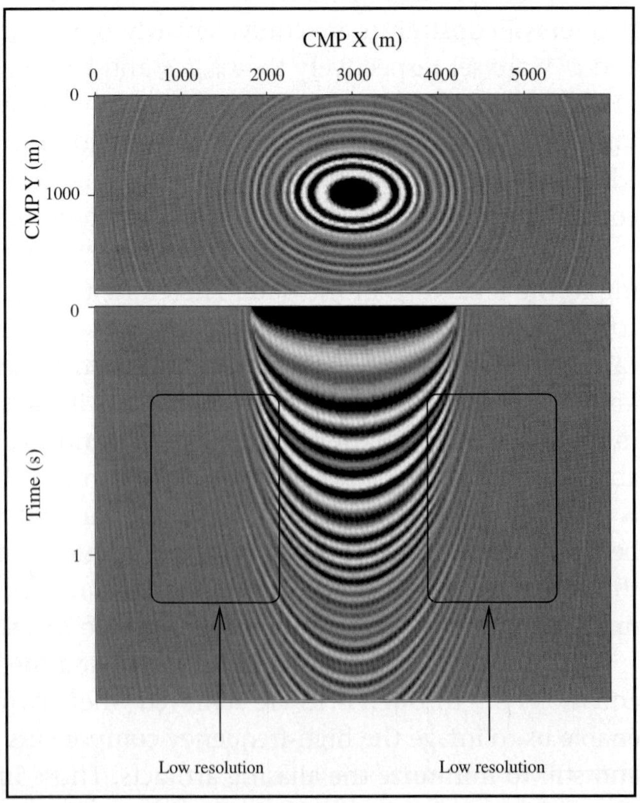

Figure 12. An image obtained by constant-velocity Kirchhoff migration using standard antialiasing ($\Delta x_\xi = \Delta y_\xi = 20$ m, $\Delta x_D = \Delta y_D = 40$ m, $V = 1.5$ km/s, and offset = 2.4 km). The time slice is cut at $t = 1$ s, and the vertical section is cut at CMP Y = 1000 m.

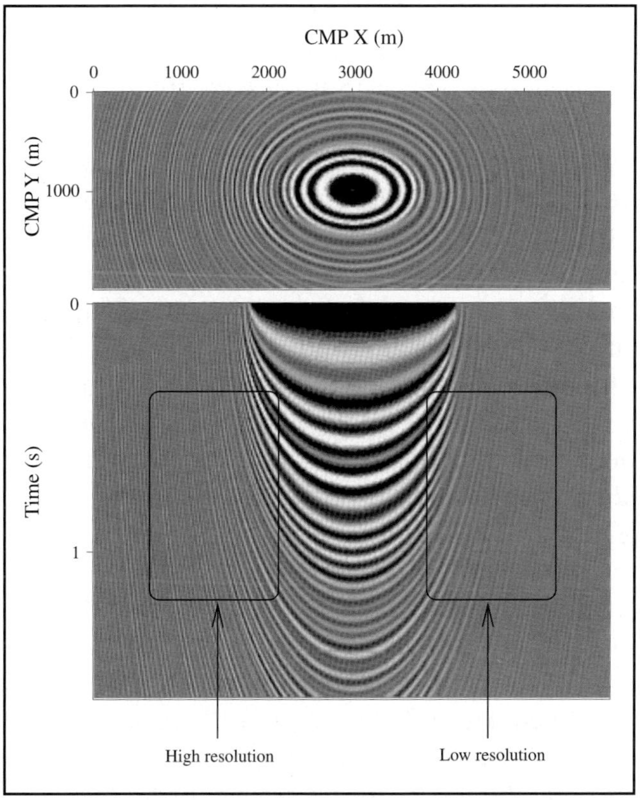

Figure 13. An image obtained by constant-velocity Kirchhoff migration with use of the directed antialiasing, assuming $p_x^{min} = 0$ s/m ($\Delta x_\xi = \Delta y_\xi =$ 20 m, $\Delta x_D = \Delta y_D = 40$ m, $V = 1.5$ km/s, and offset = 2.4 km). The time slice is cut at $t = 1$ s, and the vertical section is cut at CMP Y = 1000 m.

di (2006) discusses in greater detail the relation between operator antialiasing and image antialiasing.

The distinction between operator aliasing and image aliasing becomes important when a priori assumptions about the dips in the data permit one to set less stringent operator antialiasing constraints, so that one can image reflectors with high resolution but without operator-aliasing artifacts. In such cases, it is important for one to use a finer spatial sampling for the image than for the data, to avoid aliasing the image. The next section shows an example of that situation.

- High-resolution imaging of salt flanks

High-resolution imaging of salt-dome flanks is an important application of the general antialiasing method described in the previous section. In that case, often we can assume that in the proximity of the salt flanks, the data contain little or no energy dipping in the direction opposite that of the reflections from the flanks. According to the method I presented in the previous section, this assumption enables us to image the salt flanks with higher resolution than was otherwise possible. However, it is important to be aware that when we increase the image-frequency content by applying the constraints in equations 8.9 through 8.12, rather than those in equation 8.6, we run the risk of aliasing the image. As I discussed in the previous section, the conditions that avoid operator aliasing do not guarantee avoidance of image aliasing. Therefore, the constraints for avoiding image aliasing (equation 8.5) also must be taken into account, and the image sampling must be reduced so that both goals — avoiding image aliasing and preserving high-frequency components — can be achieved. Because of image-aliasing considerations, the images of the salt flanks from the Gulf of Mexico data that are shown in the following figures were sampled twice as densely ($\Delta x_\xi = \Delta y_\xi = 18$ m) as were the zero-offset data ($\Delta x_D = \Delta y_D = 36$ m).

The first step in achieving high resolution by applying the general antialiasing condition is to determine the appropriate values for the bounds on the data dips. For simplicity, I choose constant bounds for this example; that is, $p_x^{min} = -0.082$ s/km, $p_x^{max} = 0.48$ s/km, $p_y^{min} = -0.8$ s/km, $p_y^{max} = 0.8$ s/km. In more geologically complex cases, it may be advantageous to allow the data-dip bounds to vary in both time and space. Figure 14 shows the same data spectrum as in Figure 6. The dashed lines superimposed onto the spectrum cover the areas determined by the inequalities of equations 8.7 and 8.8, according to the chosen bounds for p_x^{min} and p_y^{max}. The areas that honor all antialiasing constraints and thus represent data components used by the imaging are covered by crossing dashed lines. A large swath of the aliased energy with positive time dips is used by the high-resolution imaging, whereas that energy would be discarded if standard antialiasing methods were used.

Figures 15 through 17 demonstrate the improvements in image resolution that the high-resolution antialiasing makes possible. Figure 15a shows the results of 3D poststack migration obtained without any antialiasing. Figure 15b shows the results obtained when the high-resolution antialiasing constraints in equations 8.9 through 8.12 are applied. Figure 15c shows the results produced when the standard antialiasing constraints in equation 8.6 are applied. The differences among the results are appreciated better by a comparison of windows zooming into smaller parts of the sections. Figure 16 shows such a comparison for the shallower part of the section. The image obtained without antialiasing (Figure 16a) is uninterpretable because of the strong aliasing noise. The image obtained with the high-resolution method shows better resolution of several dipping reflectors and of the steep salt flank (Figure 16b). Figure 17 demonstrates (for the same area shown in Figure 4) that the high-frequency sediment truncation against the salt flank (CMP X = 700 m and time = 2.2 s) is resolved well in the image obtained with the high-resolution method (Figure 17b), whereas it is resolved poorly in the image obtained with traditional methods (Figure 17c).

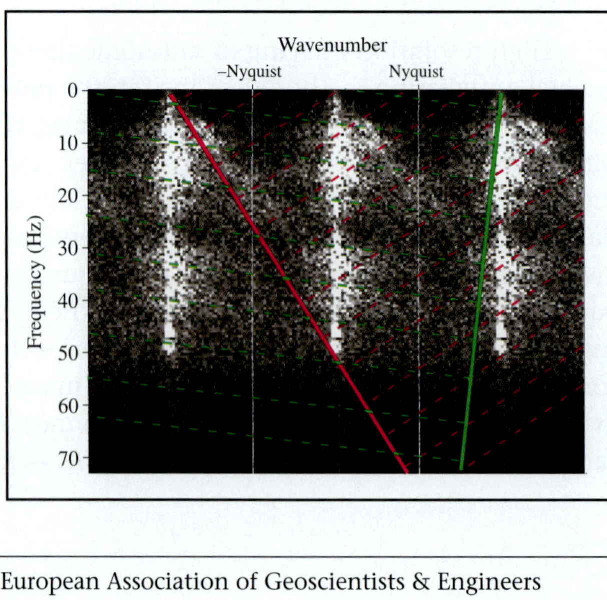

Figure 14. A frequency-wavenumber spectrum of a data window (the same one as in Figure 6). The dashed lines superimposed onto the spectrum cover the areas determined by the inequalities of equations 8.7 and 8.8. The areas that respect all the operator antialiasing constraints and thus represent data components used by the imaging are covered by dashed lines along both directions.

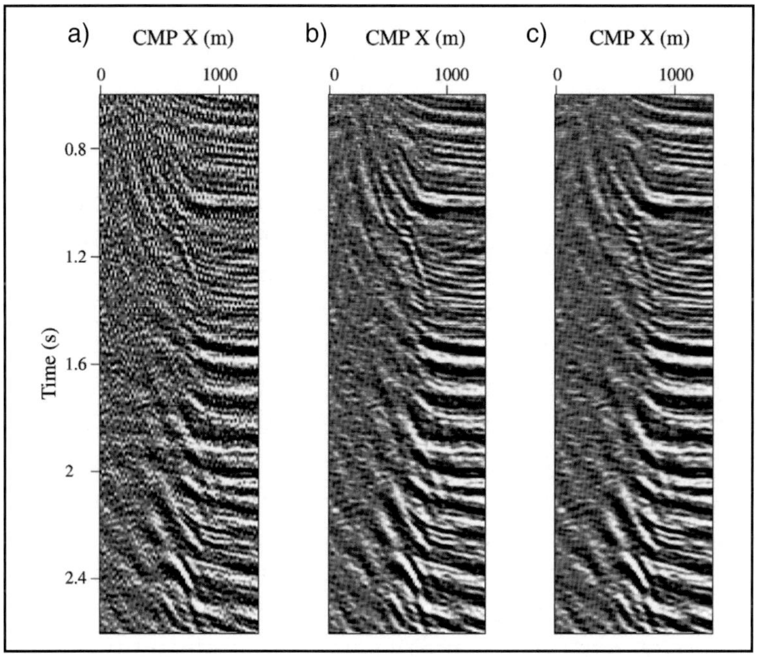

Figure 15. 3D migrations of a salt-dome flank in the Gulf of Mexico: (a) migration obtained without any antialiasing filter, (b) migration obtained with application of the high-resolution antialiasing filter, and (c) migration obtained with application of a standard antialiasing filter.

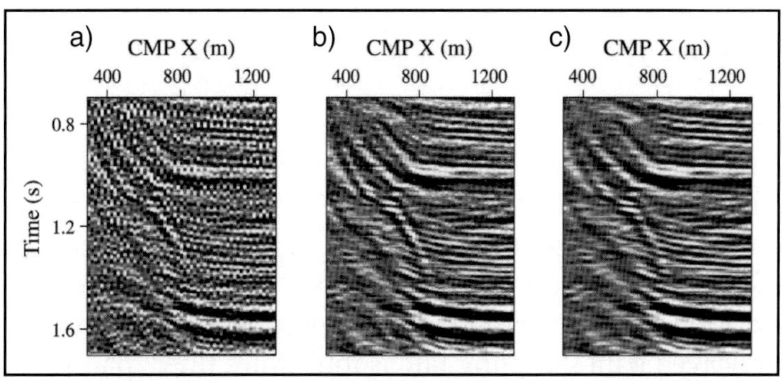

Figure 16. Magnification of the shallower part of the 3D migrations of a salt-dome flank in the Gulf of Mexico shown in Figure 15: (a) migration obtained without any antialiasing filter, (b) migration obtained with application of the high-resolution antialiasing filter, and (c) migration obtained with application of a standard antialiasing filter.

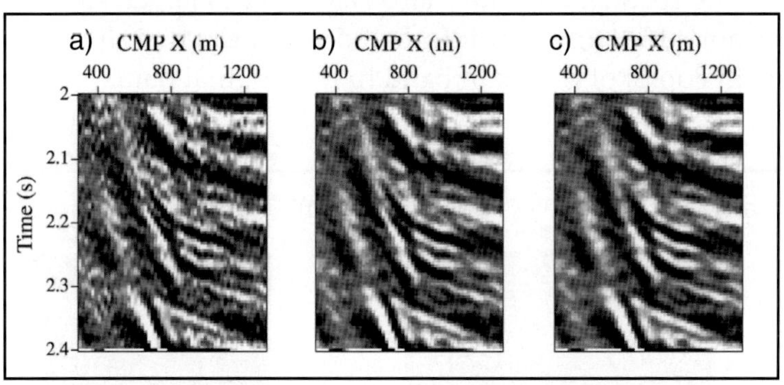

Figure 17. Magnification of the deeper part of the 3D migrations of a salt-dome flank in the Gulf of Mexico shown in Figure 15: (a) migration obtained without any antialiasing filter, (b) migration obtained with application of the high-resolution antialiasing filter, and (c) migration obtained with application of a standard antialiasing filter.

Aliasing in Wavefield-continuation Migration

Wavefield-continuation migration has long been considered immune to operator-aliasing artifacts. In contrast with Kirchhoff operators, which can sum the data on the side bands of the spectra in addition to the central band (see the earlier discussion on Kirchhoff operator aliasing), wavefield-continuation operators are limited intrinsically to propagating the wavefield in the central band of the discretely sampled data. Therefore, when data with dense and regular sampling are migrated, as usually is the case with zero-offset migration, no operator aliasing occurs. In contrast, operator aliasing often occurs when one is applying 3D prestack migration. The most common case of operator aliasing in wavefield-continuation migration occurs when the sources are sampled less densely than the receivers. This situation occurs routinely in both land and marine data sets. Aliasing along the source axes often is worsened by imaging-parameter choices. In particular, decimating the shots is a fairly common procedure used to reduce the computational cost of shot-profile migrations. Whereas the artifacts generated by this strategy might not be apparent in the stacked cube, they may deteriorate the partial-prestack images sufficiently to make them uninterpretable.

Wavefield-continuation migrations generate artifacts similar to those generated by Kirchhoff migrations when one interleaves the recorded traces with zero traces. Such interleaving is seldom done for zero-offset migration, but it is done, either implicitly or explicitly, for most prestack migration processes. An example of implicit interleaving is when one performs shot-profile migration on data that have coarser spatial sampling for the shots than for the receivers — a characteristic that is prevalent in 3D acquisition geometries. If one propagates both the source wavefield and the receiver wavefield on a grid sampled with the receiver sampling rate (finer sampling) and then evaluates the image using the standard imaging condition in equation 3.8, one effectively is interleaving the shot axes with as many zero traces as are needed to match the sampling rate of the receiver axes.

Basic Fourier-analysis theory teaches us that interleaving the data with zero traces is equivalent to replicating the central band in side bands, with the number of replicas being equal to the interleaving factor (Bracewell, 1999). Figures 18 and 19 illustrate the effects of interleaving applied to the Gulf of Mexico data set used in the previous section. Figure 18a shows the same data window shown in Figure 5 but with three zero traces inserted for each live trace. Figure 18b shows the frequency-wavenumber spectrum of the interleaved data. Because the interleaved data-sampling rate is four times finer than the sampling rate of the original data, the spatial bandwidth also is four times larger. However, by adding zero traces, one has not added any new information, and the high (positive and negative) wavenumbers contain copies of the main band in the original data.

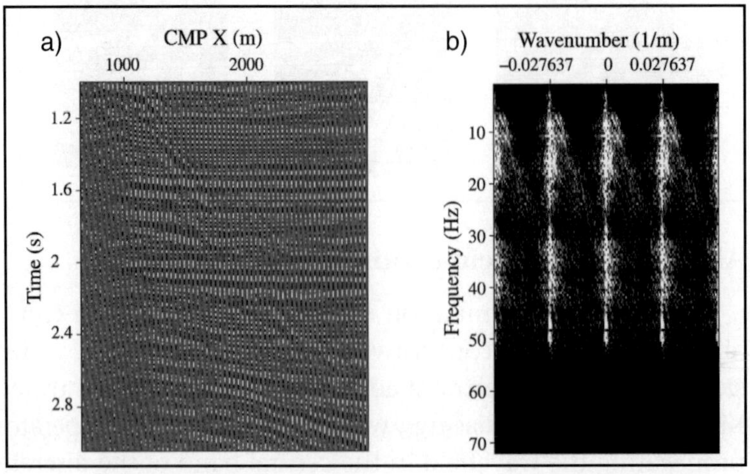

Figure 18. (a) The same data window of the Gulf of Mexico data that was shown in Figure 5, but here with three zero traces inserted for each live trace; (b) the frequency-wavenumber spectrum of the interleaved data.

When one migrates the interleaved data using a wavefield-continuation algorithm, the energy in the side copies of the main band is propagated incorrectly and generates aliasing artifacts in the image. Figure 19 shows the results of such an experiment. Figure 19a displays the image generated by performing zero-offset downward-continuation migration on the data obtained by interpolating the original traces onto a grid four times finer than the original sampling grid. Figure 19b displays the image generated by using the same migration algorithm and migration velocity, but applied on the data interleaved with zero traces shown in Figure 18. Addition of the spurious copies of the energy at high (positive and negative) wavenumbers creates the aliasing artifacts visible in Figure 19b.

As with Kirchhoff migration, adding the contributions of the side bands to the image improves the resolution of steeply dipping sediments that truncate against the salt body. Figure 19c shows the difference between the panels in Figure 19a and 19b. The difference panel is mostly incoherent noise, but a few coherent events correspond to the same dipping reflectors highlighted in the previous section; for example, the sediment truncation at about 3500 m deep is clearly visible in Figure 19c. Therefore, with wavefield-continuation migration, one faces the same dilemma that I discussed for Kirchhoff migration at the beginning of the chapter: how to set optimal trade-off parameters between aliasing artifacts and image resolution.

The aliasing artifacts that interleaving adds to the image in Figure 19b resemble the aliasing artifacts visible in Figure 15a, which was obtained by Kirchhoff migration without applying any antialiasing filter. Although the two images are similar, the amount of aliased energy is lower in the section obtained by wavefield-continuation migration than in the section obtained by Kirchhoff migration. The image obtained by wavefield continuation is equivalent to the image obtained by applying antialiased Kirchhoff migration using the standard antialiasing constraints and assuming the sampling rate to be equal to

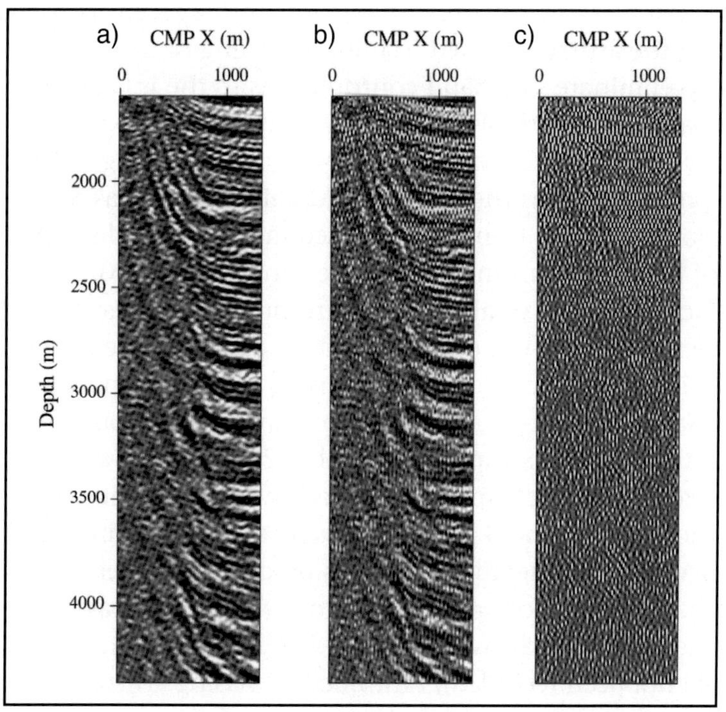

Figure 19. Images obtained by performing zero-offset downward-continuation migration (a) on the data obtained by interpolating the original data on a grid four times finer and (b) on the data obtained by interleaving the original data, with three zero traces for each live trace. (c) The difference between panel (a) and panel (b).

the sampling rate of the interleaved data. These antialiasing constraints are obtained by applying the relationships expressed in equations 8.9 through 8.12, with the assumption that the data-dip limits p^{min} and p^{max} are set, respectively, to $-k_{nyq}/4\omega_d$ and $k_{nyq}/4\omega_d$.

The previous example shows that the aliasing artifacts caused by interleaving zero traces with the data are similar to the artifacts caused by operator aliasing with Kirchhoff migration. Furthermore, if one interleaves data that are not aliased and then migrates the interleaved data using a wavefield-continuation migration, aliasing artifacts will be present in the image. Conversely, if one interleaves the original data with zero traces, one enables the migration algorithm to migrate wavefield components that were aliased in the original data, and thus one may increase the resolution of the image. In summary, aliasing caused by interleaving data with zero traces has characteristics similar to those of Kirchhoff-migration operator aliasing, and consequently I will refer to this kind of aliasing as wavefield-continuation operator aliasing. The next section analyzes operator aliasing for prestack migration and presents some simple solutions for reducing the effects of operator aliasing on the final image.

Aliasing of wavefield-continuation imaging operators

Operator aliasing seldom occurs in zero-offset wavefield migration because the data rarely are interleaved with zero traces. In contrast, operator aliasing is widespread in 3D prestack migration because of the sparse acquisition geometry. Effects of operator aliasing are particularly noticeable in the prestack partial images (e.g., angle-domain common-image gathers, or ADCIGs) created by wavefield-continuation methods. The effects on the stacked image are less visible because the artifacts tend to be incoherent over aperture angles, and thus the redundancy in the data contributes to attenuation of the artifacts in the stacked image.

As I discussed above, the most common case of operator aliasing in wavefield-continuation migration occurs when the sources are sampled less densely than the receivers. Two possible strategies exist for reducing the artifacts related to undersampling of the sources. A conservative strategy would be to eliminate side-band contributions to the image. That task can be achieved by spatially band-passing the wavefield either at the imaging point or directly at the source. Unfortunately, the conservative approach causes dipping reflectors to disappear from the image progressively as the source undersampling increases.

Figure 20 demonstrates this loss of energy from the migrated image of the dipping reflectors in the shallow section of the Marmousi model (Versteeg and Grau, 1990). Figure 20a shows the results of prestack migration of all the shots in the data; that is, $\Delta_s = \Delta_g = 25$ m. The other panels in the figure show the progressive degradation of the image as the shots are decimated progressively by a factor of two; that is, $\Delta_s = 50$ m for panel (b), $\Delta_s = 100$ m for panel (c), and $\Delta_s = 200$ m for panel (d). All these images were obtained by source-receiver migration using the nominal sampling interval of 25 m for the wavefield propagations. The decimated results were computed by interleaving the decimated shots with zero traces along the shot axis, followed by band-passing along the source axis before migration. Performing source-receiver migration after this process of interleaving and band-passing is equivalent to performing shot-profile migration with a band-passed source function. Notice that because the band-pass was performed after decimation (I assumed that the missing shots had not been recorded), I introduced aliasing artifacts in

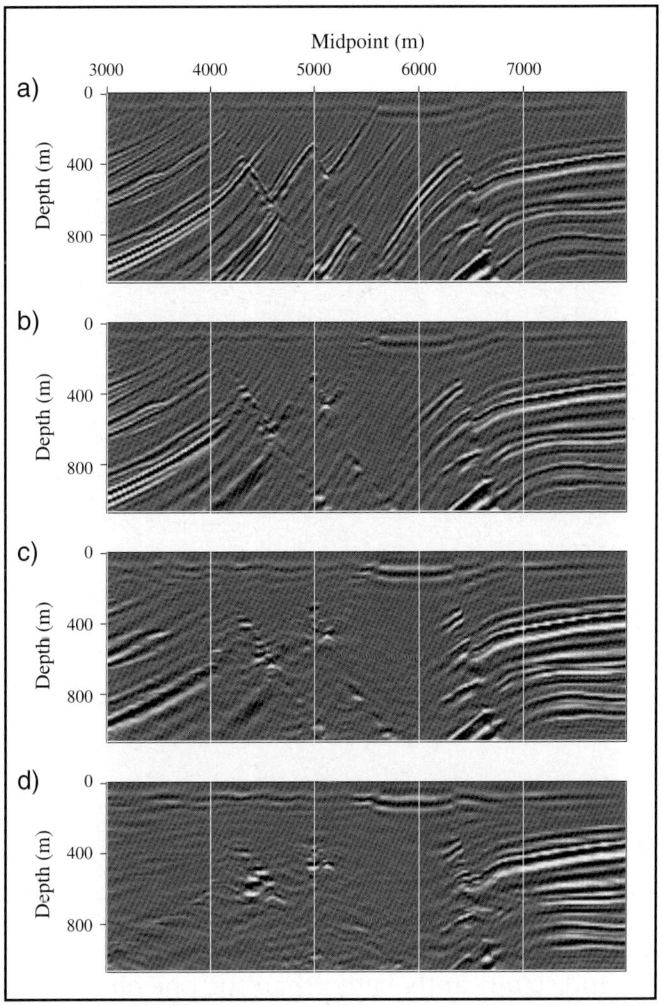

Figure 20. (a) A migrated section obtained by using all the shots in the data (Δ_s = 25 m). The other migrated sections are obtained by decimating the shots, followed by interleaving the decimated shots with zero traces along the shot axis and then band-passing along the source axis before migration. The shot intervals were: (b) Δ_s = 50 m, (c) Δ_s = 100 m, and (d) Δ_s = 200 m.

addition to losing the dipping events. The most notable aliasing artifacts are in Figure 20d, at $z \approx 600$ m and $x_m \approx 4500$ m.

As the migration examples shown in Figure 20 demonstrate, the conservative approach described above yields unsatisfactory results as soon as the shots are undersampled by more than a factor of two. An alternative strategy is to ignore the potential of creating aliasing artifacts and to migrate the decimated shots as if the shot-sampling interval were the same as the receiver-sampling interval. For source-receiver migration, such a strategy is implemented simply by interleaving the decimated shots with zero traces along the shot axis. For shot-profile migration, this strategy is implemented by propagating the source wavefield on a grid with the same sampling interval that was used for the receivers. Figure 21 shows the results of this strategy and of the antialiasing technique described below. In this figure, I show the ADCIGs at two midpoint locations (x_m = 4425 m and x_m = 7425 m), in addition to the section obtained by stacking the ADCIGs over angles.

Figure 21a shows the results obtained with all the shots (Δ_s = 25 m), Figure 21b shows the results obtained using one shot in every four (Δ_s = 100 m), and Figure 21d shows the results obtained using one shot in every eight (Δ_s = 200 m). As expected, both

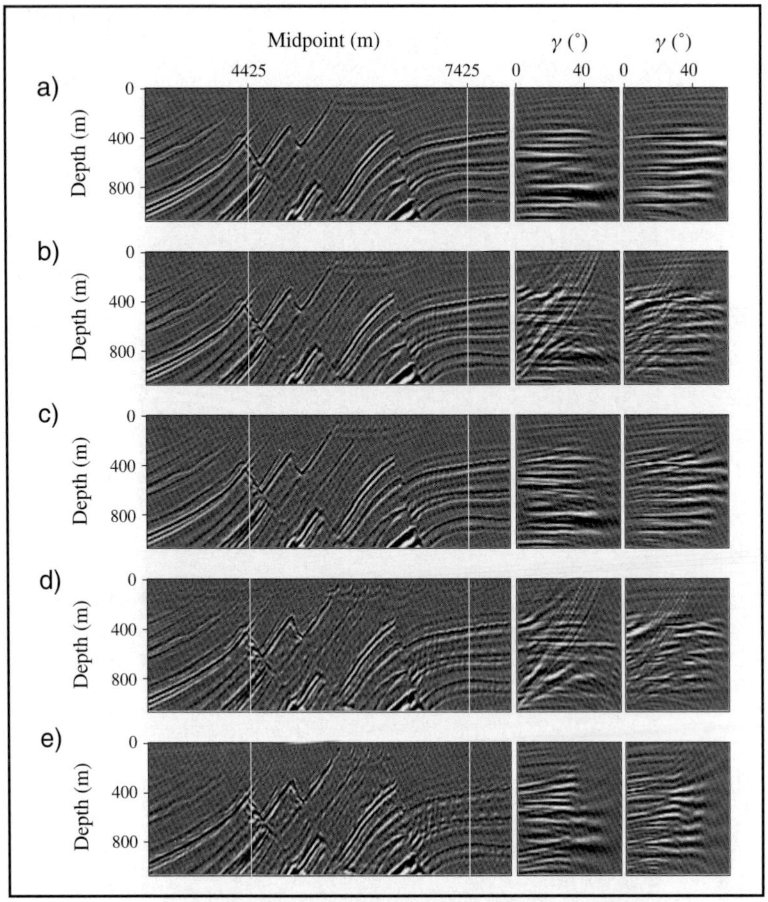

Figure 21. Migration results obtained by the correct velocity. (a) A migrated section and two ADCIGs obtained by using all the shots in the data (Δ_s = 25 m). The other migrated results were obtained by first decimating the shots and then interleaving the decimated shots with zero traces along the shot axis. The shot intervals were: (b) and (c) Δ_s = 100 m and (d) and (e) Δ_s = 200 m.

the stacked section and the ADCIGs are imaged well in Figure 21a. The stacked section in (b) is almost as good as the one in (a) and significantly better than the one obtained using the conservative approach with the same subset of shots (Figure 20d). The comparison between Figure 21b and Figure 20c demonstrates that when the subsampling is substantial, the side bands contain crucial energy, and elimination of their contributions to the image is damaging. Although the stacked section is almost free of aliasing artifacts, the ADCIG has strong coherent artifacts dipping along the axis of the aperture angle. Such artifacts would prevent any meaningful AVA analysis and would severely hamper velocity updating by use of the kinematic information contained in the ADCIGs (Chapter 7).

As the decimation factor is increased to eight, aliasing artifacts begin to appear also in the stacked section (Figure 21d), and the ADCIGs become too affected by aliasing artifacts to be of practical use. Many of the coherent events related to aliasing, which are visible in the stacked section shown in Figure 21d, are the same events visible in the stacked section shown in Figure 20d, which was obtained using the conservative approach.

To preserve as much of the useful energy as possible and to attenuate the strong aliasing artifacts, one can window the image along both the offset and the offset-wavenumber axes. Such an antialiasing strategy has the advantage of being a postprocessing task performed on the prestack image cube after migration. Therefore, the antialiasing parameters (i.e., the bounds of the pass region in subsurface-offset and offset-wavenumber domains) do not need to be determined a priori but can be determined easily by trial and error.

When one windows the prestack image cube in the offset domain around zero offset before one transforms to the angle domain, one removes a large amount of the aliased energy that creates the coherent dipping events in the ADCIGs. The implicit assumption behind windowing in the offset domain is that the energy corresponding to primary reflections has been focused around zero offset or that the migration velocity is approximately correct. By excluding the high offset wavenumbers, one effectively band-passes the image at the wide aperture angles and thus removes some of the high-frequency aliasing artifacts. For example, the strong aliasing artifact that appears around x_m = 4425 m in the stacked section obtained by taking one shot in every eight (Figure 21d) corresponds to the artifact that appears at wide-aperture angles (well beyond 40°) in the ADCIG. As is the case with windowing in the offset, also windowing in the offset wavenumber may create problems when one wants to use the ADCIGs for velocity estimation. Resolution of the measurements of the velocity errors from migrated ADCIGs depends on the availability of information at wide-aperture angles.

Figure 21c and 21e shows the result of applying this antialiasing methodology to the prestack images shown in Figure 21b and 21d, respectively. In both cases, the prestack image was windowed around zero offset; the offset range of the window was –200 m to 200 m. To create the ADCIGs shown in Figure 21c, only half of the available offset-wavenumber bandwidth was used, and only a quarter of the bandwidth was used in Figure 21e. The quality of the ADCIGs has improved significantly because of antialiasing, and the aliasing artifact present at x_m = 4425 m in the stacked section in Figure 21d has been attenuated strongly in Figure 21e. On the other hand, the stacked image of the flat shallow events around the ADCIG located at x_m = 7425 m has lost coherency in Figure 21e, compared with the same events in Figure 21d. That loss of coherency can be explained by analyzing the corresponding ADCIG in the fully sampled image (Figure 21a) and noticing that a large part of the energy contributing to the stacked image of the shallow events is reflected at fairly wide angles. However, the narrow angles have little energy because the minimum acquisition offset in the data is 200 m.

- Aliasing and velocity estimation

Velocity estimation is a primary use of prestack images such as the ADCIGs shown in Figure 21. Thus, it is important to analyze the effects of aliasing on the prestack images when the migration velocity is not correct and to verify under which conditions one can accurately extract velocity information from the migrated ADCIGs. Figure 22 shows the images obtained when the migration slowness is 5% higher than the correct one. Each of the panels in Figure 22 is obtained by applying the same procedure that was followed to compute the respective panels in Figure 21, except that in Figure 22, the migration slowness was scaled uniformly by a factor of 1.05.

As expected, the ADCIGs obtained with full shot sampling (Figure 22a) smile upward and provide good velocity information. Unfortunately, the images obtained when the velocity is too slow appear to be more susceptible to aliasing artifacts than are the images obtained when the velocity is correct. Undersampling the shots by a factor

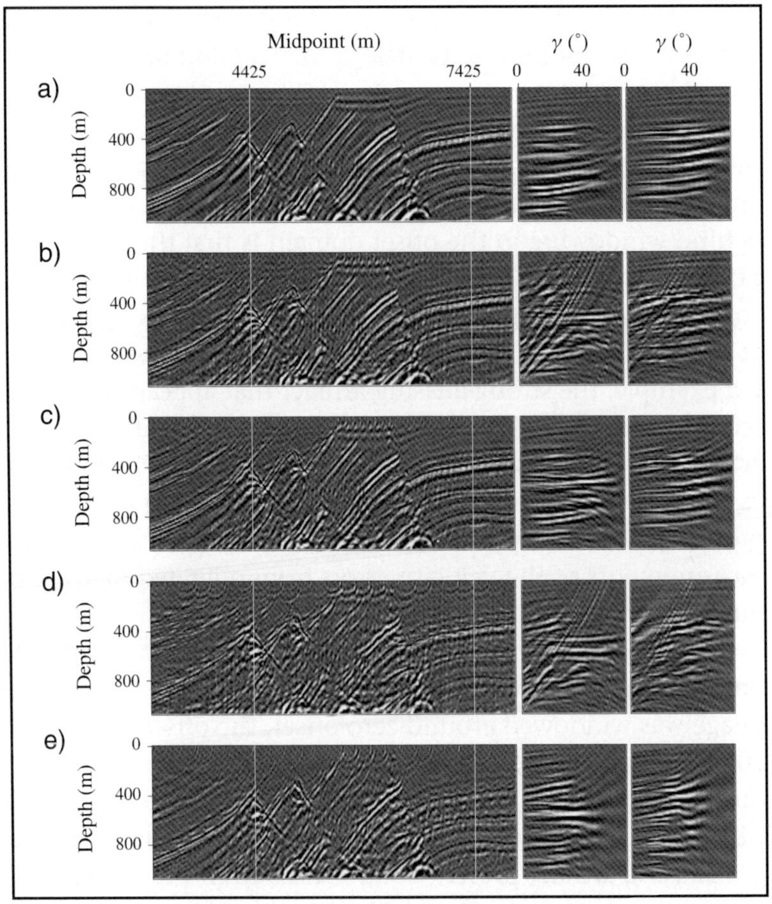

Figure 22. Migration results obtained with the slowness 5% higher than the correct one. (a) A migrated section and two ADCIGs obtained by using all the shots in the data (Δ_s = 25 m). The other migrated results were obtained by first decimating the shots and then interleaving the decimated shots with zero traces along the shot axis. The shot intervals were: (b) and (c) Δ_s = 100 m and (d) and (e) Δ_s = 200 m.

of four causes severe artifacts in the stacked section shown in Figure 22b, whereas the equivalent section shown in Figure 21b is almost free of artifacts. In particular, the aliasing artifacts that appear in Figure 22b are similar to the artifacts visible in the stacked section obtained with the correct velocity when the undersampling factor is doubled — that is, undersampled by a factor of eight instead of four (Figure 22d). Both the stacked section and the ADCIGs obtained by decimating the shots by a factor of four substantially improve after application of the antialiasing procedure described above (Figure 22c). Application of the antialiasing procedure has preserved the velocity information in the ADCIGs fairly well, with the exception of the aliased event that appears in the stacked section and that is flat along the angle axis in the ADCIG located at x_m = 7425 m. The flatness of the aliased event along the angle axis erroneously indicates a correct migration velocity for that event and consequently may mislead a velocity-updating procedure.

When shots are undersampled by a factor of eight (Figure 22d), the stacked image is affected so severely by aliasing artifacts that interpretation of the reflectors is impaired seriously. Application of the antialiasing procedure described above substantially improves the interpretability of the reflectors in the stacked image (Figure 22e) and thus could help the velocity-estimation process. Unfortunately, even after antialiasing, the ADCIGs are affected so severely by aliasing, it is unlikely that reliable velocity information can be extracted from prestack images.

Aliasing in image space

Zhang et al. (2003) described another type of aliasing artifact that can be created inadvertently when one is performing wavefield-continuation migrations, either by source-receiver migration or by shot-profile migration. Such aliasing artifacts are caused by undersampling the image space, and consequently the phenomenon is called image-space aliasing. This type of aliasing occurs even when the source and receiver spacings are the same. It is caused by the fact that the final image is built in midpoint-offset coordinates for both source-receiver migration (the imaging condition in equation 3.13) and shot-gather migration (the imaging condition in equation 3.8), even if the data are recorded and often propagated in source-receiver coordinates.

An alternative explanation of the occurrence of image-space aliasing is that the space-domain multiplication between source and receiver wavefields, which is required for evaluating the imaging condition, is equivalent to a convolution in the wavenumber domain. That convolution generates a signal with twice the bandwidth of the original one, hence the need for the image to be sampled twice as densely as the source and receiver wavefields are.

References

Abma, R., J. Sun, and N. Bernitsas, 1998, Anti-aliasing methods in Kirchhoff migration: Geophysics, **64**, 1783–1792.

Artman, B., B. Biondi, and J. Shragge, 2003, Operator aliasing in wavefield continuation migration: 73rd Annual International Meeting, SEG, Expanded Abstracts, 1091–1093.

Bevc, D., and J. Claerbout, 1992, Fast anti-aliased Kirchhoff migration and modeling: Stanford Exploration Project, Report 75, 91–96.

Biondi, B., 2001, Kirchhoff imaging beyond aliasing: Geophysics, **66**, 654–666.

Biondi, B., S. Fomel, and N. Chemingui, 1998, Azimuth moveout for 3-D prestack imaging: Geophysics, **63**, 574–588.

Bracewell, R. N., 1999, The Fourier transform and its applications, 3rd ed.: McGraw-Hill Science/Engineering/Math.

Claerbout, J. F., 1995, Basic earth imaging: Stanford Exploration Project, http://sep-www.stanford.edu/sep/prof/bei/toc_html/index.html, accessed December 5, 1995.

Gardner, G. H. F., W. S. French, and T. Matzuk, 1974, Elements of migration and velocity analysis: Geophysics, **39**, 811–825.

Gray, S. H., 1992, Frequency-selective design of the Kirchhoff migration operator: Geophysical Prospecting, **40**, 565–572.

Hale, D., 1991, A nonaliased integral method for dip moveout: Geophysics, **56**, 795–805.

Lumley, D. E., J. F. Claerbout, and D. Bevc, 1994, Anti-aliased Kirchhoff 3-D migration: 64th Annual International Meeting, SEG, Expanded Abstracts, 1282–1285.

Pica, A., 1996, Model-based anti-aliased Kirchhoff 3D PSDM: 58th Annual International Conference and Exhibition, EAEG, Extended Abstracts, X046.

Schneider, W. A., 1978, Integral formulation for migration in two dimensions and three dimensions: Geophysics, **43**, 49–76.

Versteeg, R., and G. Grau, 1990, Practical aspects of seismic data inversion, the Marmousi experience: 52nd Annual International Conference and Exhibition, EAGE, Extended Abstracts, 1–194.

Zhang, Y., J. Sun, and S. Gray, 2003, Aliasing in wavefield extrapolation prestack migration: Geophysics, **68**, 629–633.

Chapter 9 Imaging and Partial Subsurface Illumination

Introduction

In Chapter 8, I analyzed how the spatial sampling rate influences image quality. If data sampling is not sufficiently dense, the seismic image may lose resolution and/or it may be affected by artifacts. Unfortunately, however, density of spatial sampling is not the only problem encountered with realistic 3D acquisition geometries. An even more common problem is irregularity of the spatial sampling. Often, irregular sampling in space is a product of practical constraints, examples of which include cable feathering in marine acquisition and surface obstacles in land acquisition. In other cases (e.g., with button-patch geometries), irregular sampling geometry might be inherent in the survey design.

The main effect of irregular sampling geometries is either uneven illumination or incomplete illumination of the subsurface. Such partial illumination causes distortions in the image. In milder cases, distortions are limited to the image amplitudes, and they are clearly visible in depth or time slices. Those distortions often are called acquisition footprint. Figure 1 shows an example of acquisition footprints in a migrated depth slice taken from a marine data set. On the right-hand side, horizontal striping is clearly visible, superimposed over the image of a complex turbidite system with crossing channels. The horizontal striping is not linked to geology; it is along the direction of the sailing lines of the recording vessel.

When subsurface illumination is not only uneven but is also incomplete, the phase of the image is distorted, and strong artifacts are created. At the limit, when the acquisition geometry has holes, the data are aliased, at least locally. In such cases, a distinction between the effects of coarse sampling (which I called aliasing in Chapter 8) and the effects of irregular geometries obviously is artificial. However, it helps to analyze such effects separately and to develop independent methods for alleviating the problems.

Either uneven or incomplete illumination can be caused by complexity of the velocity function in the overburden, as well as by irregular acquisition geometries. Imaging under salt edges is an example of an important task that suffers from partial illumination of the reflectors. The problem often is caused by sharp velocity-model variations that prevent the seismic energy either from reaching the reflectors or from propagating back to the surface. Although the immediate causes of partial illumination differ in the two cases — irregular acquisition geometry versus complex overburden — the final manifestation is the same: The wavefield is not sampled sufficiently at depth for migration to image the reflectors without artifacts. The concepts and methods used to address the

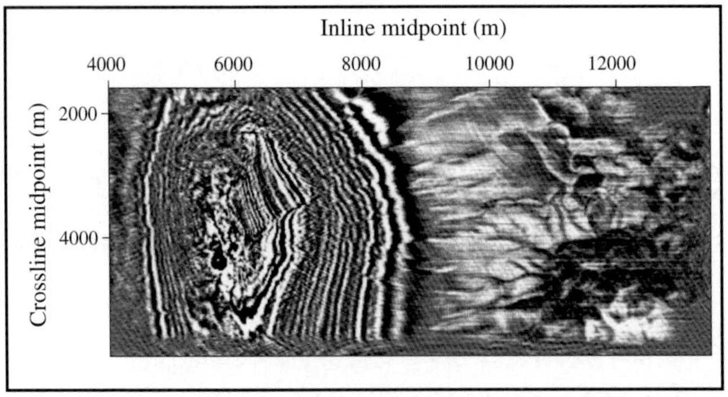

Figure 1. An example of acquisition footprint in a migrated depth slice. The horizontal stripes are related to the acquisition sail lines. Notice that the stripes bend where the reflectors start to dip in the vicinity of the salt ($x_m \approx$ 5500 m).

uneven-illumination problem are similar, regardless of its origin, and consequently I present them in a unified manner.

When illumination is uneven but without gaps, the image can be improved substantially by a simple normalization of the imaging operator or, as it often is called, by an operator equalization. In this chapter, I introduce the basic concepts of operator equalizations, using a simple imaging operator — interpolation followed by partial stacking — as a proxy for more complex imaging operators. In cases when uneven illumination of the reflectors relates mostly to irregular acquisition geometry and the velocity in the overburden is fairly simple, the DMO or AMO operators (Chapter 4) are normalized (Beasley and Mobley, 1988; Canning and Gardner, 1998; Chemingui, 1999). In more complex situations, in which the velocity in the overburden is sufficiently complex to distort the wavefield or even to cause illumination gaps, normalization should be applied in the image domain after full prestack migration (Bloor et al., 1999; Rickett, 2003).

Simple normalization of the imaging operators is not sufficient to remove imaging artifacts when illumination gaps are large. In such conditions, the data-modeling operator — which usually is defined as the adjoint of the imaging operator — should be inverted by a regularized inversion methodology. As is true for operator equalization, the methods proposed in the literature for inverting imaging operators can be divided into algorithms based on partial prestack migration (Ronen, 1987; Ronen and Liner, 2000; Chemingui and Biondi, 2002) and those based on full prestack migration. The latter methods use either a Kirchhoff operator (Nemeth et al., 1999; Duquet et al., 2000) or a wavefield-continuation operator (Kuehl and Sacchi, 2002; Prucha and Biondi, 2002). Iterative inversion is computationally expensive, especially when a full-prestack-migration operator is inverted. However, when there are large acquisition gaps or when the complexity of the overburden is responsible for incomplete illumination of the reflectors, the increasing cost-effectiveness of computing should help justify the computational cost of iterative regularized inversion. The method is therefore likely to be applied in challenging imaging problems in the future.

A radically different solution to the problem of incomplete illumination of the subsurface below complex overburden is to acquire data with a significantly wider azimuth than conventional marine acquisition can do by employing one single vessel. In Chapter 1, I discuss this innovative and expensive data-acquisition method as it has been employed for subsalt imaging. At the end of this chapter, I present synthetic-data and field-data imaging examples that indicate that wide-azimuth acquisition may substantially improve imaging results where poor subsurface illumination by conventional data is problematic.

Equalization of Imaging Operators

To explore the methods used to equalize imaging operators, I employ interpolation followed by partial stacking as a proxy of more complex imaging operators. Stacking is the operation of averaging seismic traces by summation. It is an effective way to reduce the size of data sets and to enhance reflections while attenuating noise. To avoid attenuating the signal along with the noise, the reflections need to be coherent among the traces that are being stacked. To increase trace coherency, one can apply simple normal moveout (NMO) before stacking, or a partial-prestack-migration operator such as DMO or AMO (Chapter 4).

Global stacking of all the traces recorded at the same midpoint location, regardless of their offset and azimuth, is the most common type of stacking. Partial stacking averages only those traces with their offset and azimuth within a given range. Partial stacking is useful if one wants to preserve differences among traces when those differences are functions of the trace offset and azimuth and thus one must avoid global averaging. AVO studies are a useful application of partial stacking. Partial stacking also is useful when simple transformations, such as NMO, are not sufficient to correct for the differences in time delays among traces with very different offsets and azimuths. Such a situation is common when velocity variations cause nonhyperbolic moveouts in the data. Because data redundancy is low in partial stacking, the results of partial stacking are more likely to be affected by artifacts related to irregular acquisition geometries than are the results of global stacking. Thus, in this section, I will focus my analysis on partial stacking, but the methods I present here obviously can be applied to global stacking operators too.

To start my analysis, I define a simple linear model that links the recorded traces (at arbitrary midpoint locations) to the stacked volume (defined on a regular grid). Each data trace is the result of interpolating the stacked traces and is equal to the weighted sum of the neighboring stacked traces. The interpolation weights are functions of the distance between the midpoint location of the model trace and the midpoint location of the data trace. The sum of all the weights corresponding to one data trace usually is equal to one. Because the weights are independent of time along the seismic traces, for notational simplicity, I collapse the time axis and consider each element d_i of the data space (recorded data) **d** and each element m_j of the model space **m** (stacked volume) as representing a whole trace. The relationship between data and model is linear and can be expressed as

$$d_i = \sum_j l_{ij} m_j , \tag{9.1}$$

subject to the constraint $\sum_j l_{ij} = 1$ on the interpolation weights l_{ij}. In matrix notation, equation 9.1 becomes

$$\mathbf{d} = \mathbf{L}\mathbf{m} . \tag{9.2}$$

Stacking is the summing of the data traces into the model traces, weighted by the interpolation weights. In operator notation, stacking can be represented as the application of the adjoint operator $\mathbf{L}'$ to the data traces (Claerbout, 2004); that is,

$$\mathbf{m} = \mathbf{L}'\mathbf{d}. \tag{9.3}$$

Application of simple stacking as an adjoint operator does not yield satisfactory results when the fold is distributed unevenly among midpoint bins. In the stack, the amplitudes of the bins with higher fold will be artificially higher than the amplitudes of the bins with lower fold. To compensate for this unevenness in the fold, it is common practice to divide the stacked traces by the inverse of the fold. This fold normalization also can be expressed in operator notation, when a diagonal operator $\mathbf{W}^m$ is added in front of the adjoint operator in computation of the stack:

$$\mathbf{m} = \mathbf{W}^m\mathbf{L}'\mathbf{d}. \tag{9.4}$$

The diagonal elements of $\mathbf{W}^m$, w_j^m, are given by the inverse of the fold, which can be computed simply by summing the elements in each column of $\mathbf{L}$; that is,

$$w_j^m = \left(\sum_i l_{ij}\right)^{-1}. \tag{9.5}$$

Figure 2 shows the fold as a function of the midpoint axes, computed for the data set shown in the examples presented in the upcoming section titled "Diagonal approximations of least-squares inverse." This is a marine data set recorded in the North Sea. Details of the data's geometry are presented in Chapter 1. The data were partially stacked over offsets ranging from 400 m to 600 m. The fold shown in Figure 2 was computed using a linear interpolator to define $\mathbf{L}$. Coverage gaps in the fold, like the one visible as a white area in Figure 2 around CMP X = 20,000 m, present a problem for fold normalization because they make the weights diverge to infinity in equation 9.5. To avoid instability, commonly one adds a small number ε_w to the actual fold, or one sets the weights

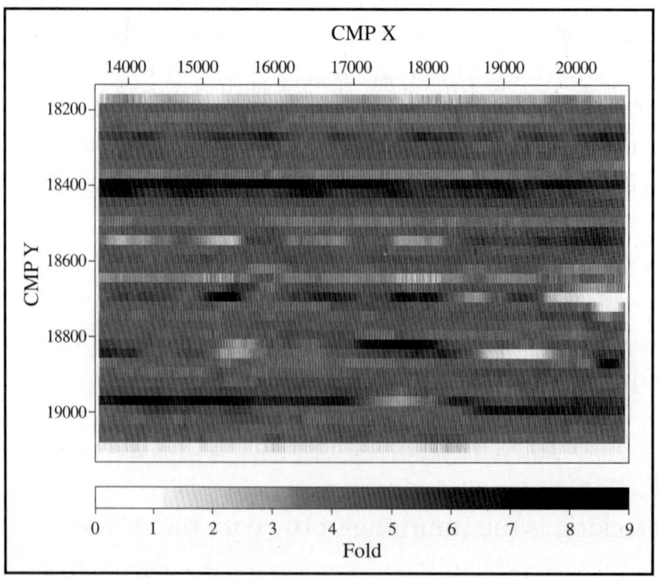

Figure 2. A fold of partial stacking by linear interpolation of a marine data set over offsets ranging from 400 m to 600 m.

to zero when the fold is smaller than ε_w, as in the following revised expression for the weights:

$$w_j^m = \begin{cases} \left(\sum_i l_{ij}\right)^{-1} & \text{if} \sum_i l_{ij} \geq \varepsilon_w, \text{ elsewhere.} \\ 0 & \end{cases} \tag{9.6}$$

I justified the fold normalization by a simple heuristic, and it may seem to be an ad hoc solution to the problem of normalizing the output of stacking. However, I can show that the weights used by fold normalization can be derived from applying the general theory of inverse least squares to the stacking normalization problem. In fact, fold normalization is equal to the exact least-squares inverse when nearest-neighborhood interpolation is used, whereas it is equal to an approximate inverse when more accurate interpolators are used.

In the next section, I show the relationship between fold normalization and inverse theory, and I derive simple principles that will be useful for analyzing operators that are more complex than stacking.

Least-squares inverse

The simple solution that I derived in the previous section to normalize stacking can be seen as a crude approximation to solution of a least-squares problem, in which the interpolation operator **L** is the forward-modeling operator. This least-squares problem can be expressed as the minimization of the objective function

$$J(\mathbf{m}) = \|\mathbf{Lm} - \mathbf{d}\|_2 \,. \tag{9.7}$$

In general, the operator **L** is not square and its inverse is not defined; therefore, I use its least-squares inverse (Strang, 1986). One formal solution for the least-squares inverse is expressed in terms of the operator **L** as

$$\mathbf{L}_m^{\dagger} = (\mathbf{L}'\mathbf{L})^{-1}\mathbf{L}' \,. \tag{9.8}$$

Applying the least-squares inverse is equivalent to applying the adjoint operator **L**′, and following that by a spatial filtering of the model space given by the inverse of **L**′**L**. The fold normalization can be seen as a particular approximation of the inverse of **L**′**L** with a diagonal operator. Biondi (2006) discussed in detail the properties of the least-squares inverse expressed in equation (9.8) and of an alternative solution.

Diagonal approximations of a least-squares inverse

When accurate interpolators (e.g., linear or sinc interpolators) are used for stacking, **L**′**L** is not diagonal, and computing its inverse is not straightforward. We also are interested in imaging operators that are more complex and expensive than stacking, for which even computing and storing the elements of **L**′**L** can be too expensive, and directly inverting the operators is impossible. In such cases, one has two choices. One can either (1) compute an analytical approximation to the inverse or (2) use an iterative method, such as the conjugate-gradient method, to compute a numerical approximation to the inverse (Claerbout, 2004). Even if one follows the second strategy, the availability of an analytical approximation to the inverse is useful, because the approximate inverse can be used as a preconditioner to accelerate convergence of the iterative inversion by the conjugate-gradient method (Golub and Loan, 1983).

I will discuss two methods for approximating the inverse of $\mathbf{L}'\mathbf{L}$. The first is based on the direct manipulation of the elements of $\mathbf{L}'\mathbf{L}$, such as extracting its diagonal or summing the elements in its columns (or rows). The second method is based on the idea that for capturing the most significant properties of $\mathbf{L}'\mathbf{L}$, one can sound it by measuring its effects when applied to a reference model ($\mathbf{m}_{\mathrm{ref}}$) (Claerbout and Nichols, 1994; Rickett, 2003). Then the approximation is evaluated as

$$\mathbf{L}'\mathbf{L} \approx \frac{\mathbf{diag}\left(\mathbf{L}'\mathbf{L}\mathbf{m}_{\mathrm{ref}}\right)}{\mathbf{diag}\left(\mathbf{m}_{\mathrm{ref}}\right)}. \tag{9.9}$$

This method has the important advantage of not requiring the explicit evaluation (and storage) of the elements of $\mathbf{L}'\mathbf{L}$, but instead, of requiring only its application to the reference model. Of course, the resulting approximation depends strongly on the choice of reference model. The closer the reference model is to the true model, the better the approximation is. In theory, if the reference model is equal to the true model, one will achieve optimal results.

Although these two methods seem unrelated, they yield equivalent results for specific choices of $\mathbf{m}_{\mathrm{ref}}$. Therefore, the second method can be used to analyze the assumptions that underlie the possible choices of approximations.

- Approximation of $\mathbf{L}'\mathbf{L}$ by the sum of its columns

I now analyze the properties of the approximate inverse that is defined by substitution of $\mathbf{L}'\mathbf{L}$ with a diagonal matrix $\mathbf{D}$, where each element of $\mathbf{D}$ is equal to the sum of the corresponding row of $\mathbf{L}'\mathbf{L}$. Notice that $\mathbf{L}'\mathbf{L}$ is symmetric and thus that summing over the columns would be equivalent to summing over the rows. This particular approximation for $\mathbf{L}'\mathbf{L}$ often is applied in numerical solution of finite-element problems. In that literature, it is known as mass-lumping approximation (Hughes, 1987).

It can be demonstrated that the diagonal elements of $\mathbf{D}$ are equal to the sum of the columns of the original operator $\mathbf{L}$; that is, they are equal to the fold computed for each corresponding model trace. That can be shown analytically by expressing the inverse of $\mathbf{D}$ as

$${}^{\Sigma}w_j^m = \left(\sum_j \sum_k l_{ki} l_{kj}\right)^{-1} = \left(\sum_k l_{ki} \sum_j l_{kj}\right)^{-1}. \tag{9.10}$$

The weights l_{ki} are interpolator weights that fulfill the constraint $\Sigma_j\, l_{kj} = 1$. It immediately follows that

$${}^{\Sigma}w_j^m = \left(\sum_k l_{ki}\right)^{-1}, \tag{9.11}$$

which is equivalent to equation 9.5. In summary, I just derived an approximation that is easy to compute for the least-squares inverse. That approximation is equal to the fold normalization defined by heuristic considerations. However, I can apply the definition of the approximate inverse when more complex imaging operators are involved for which

I lack a simple heuristic to define weights that improve the results obtained by applying the adjoint operator.

It also is easy to show that such an approximation is equivalent to the choice of a constant vector as $\mathbf{m}_{\text{ref}}$ in equation 9.9. Therefore, it biases the imaging process toward models that are constant. In the case of stacking, it privileges flat reflectors, which is consistent with the flat-reflector assumptions underlying the stacking process. In the case of imaging operators aimed at imaging complex structure, that bias toward flat reflectors may be less appropriate (Rickett, 2003).

- Approximation of $\mathbf{L}'\mathbf{L}$ by its diagonal

The matrix $\mathbf{L}'\mathbf{L}$ often is diagonally dominant. Thus, it is natural to approximate it with its diagonal (Nemeth et al., 1999; Duquet et al., 2000). For the stacking operator, evaluation of the diagonal elements is as simple as evaluation of the sum of the columns (or rows) $\mathbf{L}'\mathbf{L}$. Instead of being the sum of the elements in each column of $\mathbf{L}$ (equation 9.11), it is the sum of the squares of the elements in each column of $\mathbf{L}$; that is,

$$^{d}w_j^{m} = \left(\sum\nolimits_k l_{ki}^2\right)^{-1}. \tag{9.12}$$

In the case of nearest-neighborhood interpolation, $\mathbf{L}'\mathbf{L}$ is diagonal (and the elements l_{ki} are equal to either 0 or 1); thus, equation 9.11 and equation 9.12 are equivalent. In the case of linear interpolation (or more accurate interpolators), $^{\Sigma}w_i^{m} \leq {}^{d}w_i^{m}$ for every i, because $0 \leq l_{ki} \leq 1$. In this case, the values yielded by equation 9.11 and equation 9.12 can be significantly different. Figure 3 shows the values of the diagonal of $\mathbf{L}'\mathbf{L}$ for the North Sea data example. Compare Figure 3 with the fold values shown in Figure 2. The field shown in Figure 3 is rougher and closer to zero in most locations than the field shown in Figure 2.

No single reference model could be inserted in equation 9.9 to obtain the approximation equivalent to equation 9.12. However, the use of the diagonal of $\mathbf{L}'\mathbf{L}$ is equiv-

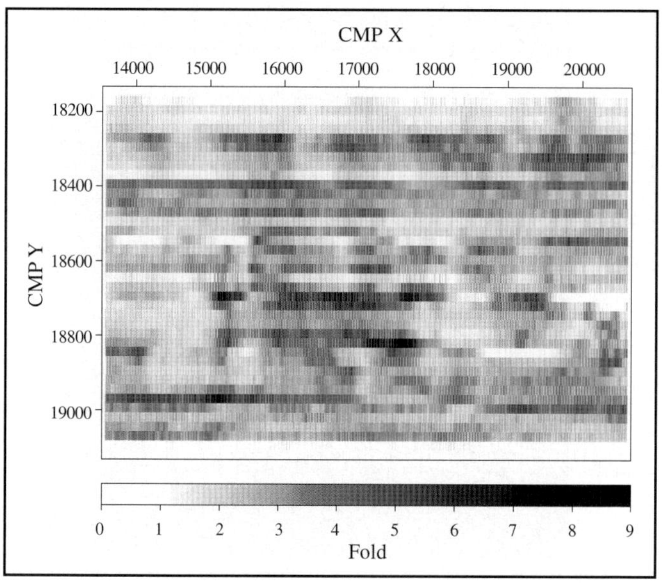

Figure 3. Values of the diagonal elements of $\mathbf{L}'\mathbf{L}$ when one performs partial stacking by linear interpolation of a marine data set over offsets ranging from 400 to 600 m. Compare with Figure 2.

alent to evaluating equation 9.9 with n_m reference models, each with a delta function at the corresponding model location. Therefore, choosing to approximate $\mathbf{L}'\mathbf{L}$ by its diagonal biases the imaging process toward models made of delta functions far apart from each other with respect to the wavelength of the seismic signal. This assumption is seldom realistic.

The theoretical analysis is confirmed by the results of applying the two kinds of approximate model-space inverses to the partial stacking of the North Sea data set. Figure 4 shows inline sections obtained by using linear interpolation to partially stack the data. The uppermost section of the figure was obtained without any normalization weights. For the one in the middle, ${}^{d}w_i^m$ were used, and for the one at the bottom, ${}^{\Sigma}w_i^m$ were used. For both approximate inverses, the inversion was stabilized by setting $\varepsilon_w = 0.1$ (equation 9.6). The gap in the coverage visible at the right end of the section is well compensated for by application of the ${}^{\Sigma}w_i^m$ weights, although one does this at the expense of a lower signal-to-noise ratio. However, the gap is compensated for very poorly by use of the ${}^{d}w_i^m$ weights, which introduce rapid fluctuations in the data amplitudes. Figure 5 shows the corresponding time slices. Again, the bottom image (weighted with ${}^{\Sigma}w_i^m$) is smoother and more continuous than the middle one (weighted with ${}^{d}w_i^m$).

Filling Illumination Gaps by Model Regularization and Preconditioning

In the previous section, we saw that data gaps are a challenge for simple operator equalization. Near the gaps, normalization weights may become large, and the signal-to-noise ratio of the results degrades substantially. Even worse, no data may be available to fill the gaps. Divergence of the normalization weights is a sign that the least-squares estimation problem defined by equation 9.7 is constrained poorly. To constrain the solution, one can add a regularization term to the data-fitting term of equation 9.7.

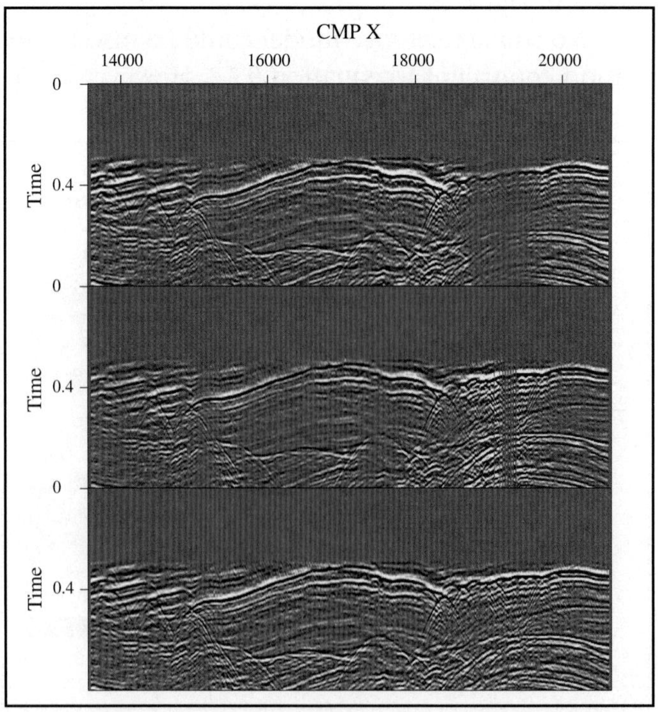

Figure 4. Inline sections obtained by partial stacking of the North Sea data set. Linear interpolation was used with application of no weights (top), ${}^{d}w_i^m$ weights (middle), and ${}^{\Sigma}w_i^m$ weights (bottom).

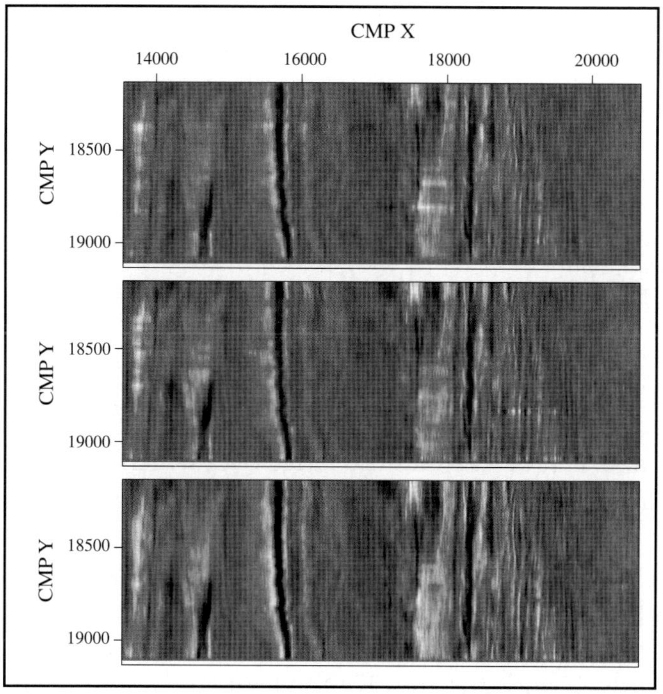

Figure 5. Time slices obtained by partial stacking of the North Sea data set. Linear interpolation was used with application of no weights (top), $^{d}w_{i}^{m}$ weights (middle), and $^{\Sigma}w_{i}^{m}$ weights (bottom).

One has several choices for a regularization term to constrain the data-interpolation problem. Often, regularization terms are based on a priori assumptions about the reflector's geometry or at least about the variability of reflectivity in spatial directions. When data coverage is quite poor, such statistical constraints often are unavoidable.

However, using the information from traces at the surrounding midpoints can cause a loss in resolution, because it may smooth true reflectivity changes. The challenge is to devise a method that maximizes the image resolution and minimizes artifacts.

If at all possible, one would like to constrain the estimation of data in the gaps by using methods that apply physical criteria instead of statistical criteria. Because of the nature of the reflection mechanism, one can assume that reflection amplitudes are a smooth function of the reflection angle and azimuth. That observation suggests the idea that smoothing the data over offset and azimuth could be performed without losing resolution. Ideally, such smoothing should be done over reflection angles (aperture and azimuth) at the imaging point, not over offset and azimuth at the surface (Bloor et al., 1999; Kuehl and Sacchi, 2002; Prucha and Biondi, 2002). However, smoothing at the reflectors would require a full migration of the data. The migration step would make the method dependent on accurate knowledge of the interval-velocity model. That reliance on the velocity model is inescapable when the gaps in the image are caused by complexities in the velocity model itself [e.g., with subsalt illumination (Bloor et al., 1999; Prucha and Biondi, 2002)], but it ought to be avoided when imaging problems are caused by irregularities in the acquisition geometries.

Regularization by imposing continuity over offsets and azimuths

In the context of least-squares inversion, smoothing along offset/azimuth in the model space (e.g., uniformly sampled offset/azimuth cubes) can be accomplished by introducing a model-regularization term that penalizes variations in seismic traces between the cubes. The simple least-squares problem of equation 9.7 then becomes

$$J_{\text{reg}}(\mathbf{m}) = \|\mathbf{Lm} - \mathbf{d}\|_2 + \varepsilon_D \mathbf{m}'(\mathbf{D}'_{\mathbf{h}}\mathbf{D}_{\mathbf{h}})'(\mathbf{D}'_{\mathbf{h}}\mathbf{D}_{\mathbf{h}})\mathbf{m}, \tag{9.13}$$

where the regularization operator $\mathbf{D}_{\mathbf{h}}$ is

$$\mathbf{D}_{\mathbf{h}} = \frac{1}{1-\rho_D}\begin{bmatrix} \mathbf{I}-\rho_D\mathbf{I} & 0 & \dots & 0 & 0 \\ -\rho_D\mathbf{I} & \mathbf{I} & \dots & 0 & 0 \\ 0 & -\rho_D\mathbf{I} & \dots & 0 & 0 \\ \dots & \dots & \dots & \mathbf{I} & 0 \\ 0 & 0 & \dots & -\rho_D\mathbf{I} & \mathbf{I} \end{bmatrix}. \tag{9.14}$$

The coefficient ρ_D must be between 0 and 1. It determines the range over which one smoothes the offset/azimuth cubes. The smaller one sets the value of ρ_D, the narrower the smoothing range is.

Regularization with a roughening operator such as $\mathbf{D}'_{\mathbf{h}}\mathbf{D}_{\mathbf{h}}$ has the computational drawback that it substantially worsens the conditioning of the problem and consequently slows the convergence of an iterative inversion (Fomel, 2001). Several iterations would be needed to smooth the solution sufficiently, if one were to solve the least-squares problem expressed in equation 9.13. Even worse, if one could not afford the cost of an iterative solution and attempted to approximate the least-squares inverse by a diagonal matrix, as I discussed in the previous section, the regularization term would have almost no effect on the solution. Fortunately, the problem is easy to precondition — $\mathbf{D}'_{\mathbf{h}}\mathbf{D}_{\mathbf{h}}$ is easy to invert because it already is factored into a lower block-diagonal operator $\mathbf{D}_{\mathbf{h}}$ and an upper block-diagonal operator $\mathbf{D}'_{\mathbf{h}}$. Each of these factors can be inverted easily by recursion. Chapter 9 in Biondi (2006) provides more details on the preconditioning of the regularization operator for geometry regularization.

Regularization by AMO

The main drawback of the method just described is that using the inverse of the simple roughening operator expressed in equation 9.14 to smooth over offset/azimuth cubes may cause a loss of resolution when geologic dips are present. Reflections from dipping interfaces are not flattened by NMO with the same velocity as are reflections from flat interfaces. However, one can generalize the method easily by substituting the identity matrix in the lower diagonal of $\mathbf{D}_{\mathbf{h}}$ with an appropriate operator that correctly transforms a common-offset/azimuth cube into an equivalent cube with a different offset and azimuth. This can be accomplished by using the AMO operator (Chapter 4) or by using an offset-continuation operator (Fomel, 2001). The roughening operator that includes AMO is expressed as (Biondi and Vlad, 2002; R.G. Clapp, 2006):

$$\mathbf{D}_{\mathbf{Th}} = \frac{1}{1-\rho_D}\begin{bmatrix} \mathbf{I}-\rho_D\mathbf{I} & 0 & \dots & 0 & 0 \\ -\rho_D\mathbf{T}_{\mathbf{h}_{1,2}} & \mathbf{I} & \dots & 0 & 0 \\ 0 & -\rho_D\mathbf{T}_{\mathbf{h}_{2,3}} & \dots & 0 & 0 \\ \dots & \dots & \dots & \mathbf{I} & 0 \\ 0 & 0 & \dots & -\rho_D\mathbf{T}_{\mathbf{h}_{n-1,n}} & \mathbf{I} \end{bmatrix}, \tag{9.15}$$

where $\mathbf{T}_{\mathbf{h}_{i,i+1}}$ is the AMO operator that transforms the offset/azimuth cube i into the offset/azimuth cube $i + 1$. The $\mathbf{D'}_{\mathbf{Th}}\mathbf{D}_{\mathbf{Th}}$ operator also can be inverted easily by recursion, and thus the least-squares problem obtained by substituting $\mathbf{D}_{\mathbf{Th}}$ for $\mathbf{D}_{\mathbf{h}}$ in equation 9.13 also can be preconditioned and normalized easily using the same techniques described in Biondi (2006).

The following two figures illustrate the improvements that can be achieved in the final image by the application of AMO regularization (R.G. Clapp, 2006). The images are obtained by common-azimuth migration (Chapter 5) of the same field-data shown in Figure 1 in this chapter, and in Figures 22 through 28 in Chapter 7. The common-azimuth prestack cube used as input for generating the depth slice shown in Figure 6a was created by iteratively solving a preconditioned inversion problem minimizing an objective function like the one expressed in equation 9.13, with the regularization operator including an AMO operator like the $\mathbf{D}_{\mathbf{Th}}$ operator expressed in equation 9.15. In contrast, simple interpolation and partial stacking were employed to produce the prestack-data cube that was used as input for the migration that produced the depth slice shown in Figure 6b. The ellipses superimposed onto the images highlight the areas where inclusion of regularized inversion in the processing flow improved the final image the most. The ellipses in the lower-middle area of the slices contour a region of the sediments in which several faults radiate from the salt body that pierces through a chalk layer and the sediments in the middle of the slice. The fault-plane reflections from these faults are well imaged in Figure 6a, whereas they are missing in Figure 6b. The ellipses in the upper-right corner of the slices contour a complex turbidite system with crossing channels; this system is imaged much better in all its details in Figure 6a than in Figure 6b.

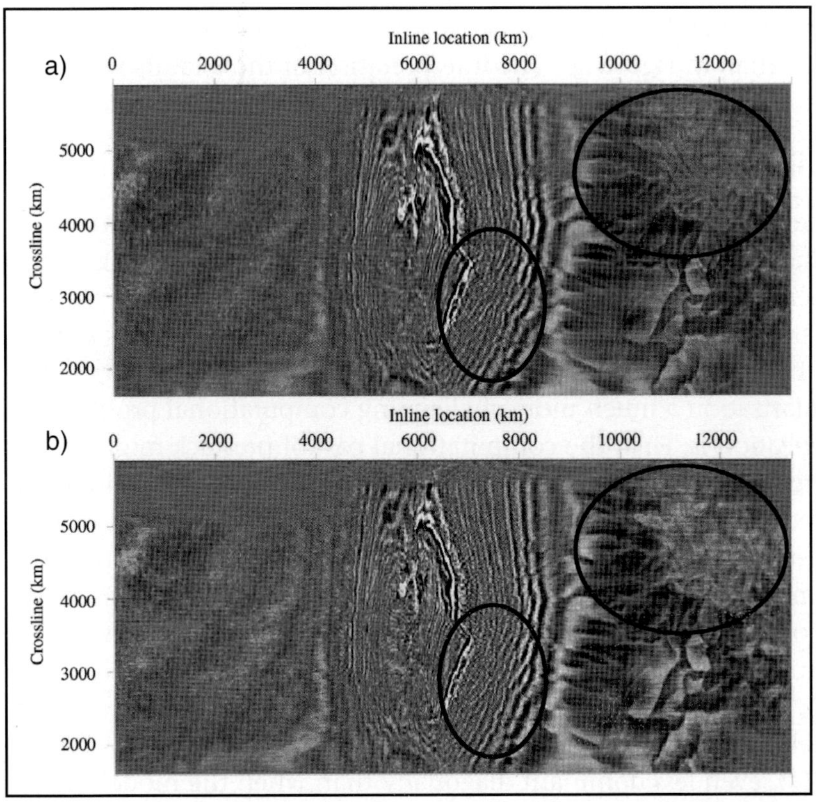

Figure 6. Effects of data-geometry regularization by least-squares inversion of the AMO operator before wave-equation migration on a depth slice (z = 870 m) of the final image: (a) the processing flow including least-squares AMO regularization and (b) the processing flow including a conventional interpolation and partial stacking. The ellipses superimposed onto the images highlight the areas where the inclusion of AMO regularization contributed to the most noticeable improvements in the image.

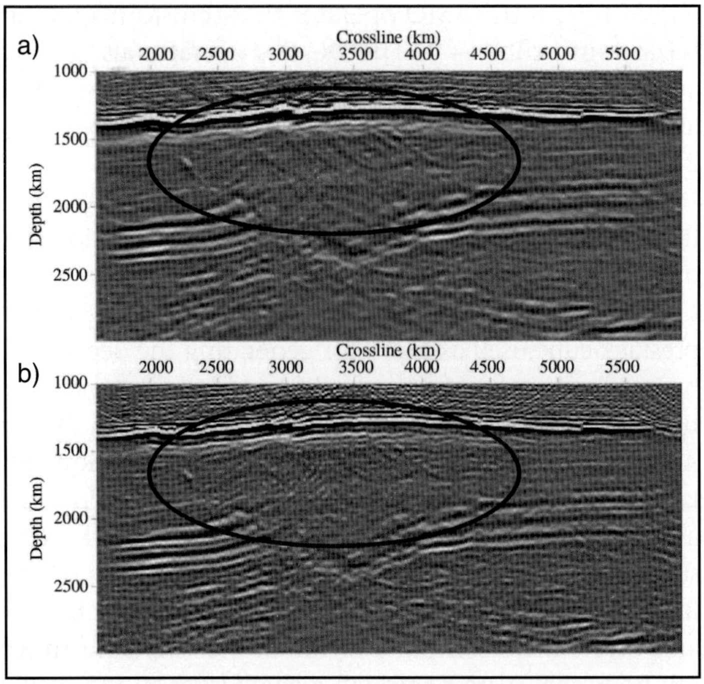

Figure 7. Effects of data-geometry regularization by least-squares inversion of the AMO operator before wave-equation migration on a crossline section (CMP Y = 8060 m) of the final image: (a) the processing flow including least-squares AMO regularization and (b) the processing flow including a conventional interpolation and partial stacking. The ellipses superimposed onto the images highlight the area where the inclusion of AMO regularization contributed to the most noticeable improvements in the image.

Figure 7 shows a comparison similar to the one in Figure 6. Here, the two images display two crossline sections cut through the common-azimuth migration results at CMPY = 8060. As in Figure 6, the top panel was created by including a regularized inversion in the processing flow before migration, whereas the bottom panel was created when conventional interpolation and partial stacking were employed. The ellipses superimposed onto the images highlight a system of faults in the chalk layer that is more clearly imaged in Figure 7a than in Figure 7b. The interpretation of these faults would be fairly straightforward from Figure 7a, whereas it would be challenging from Figure 7b.

Regularized Inversion of Prestack Migration

The data-domain methods I presented in the previous sections are appropriate when incomplete illumination of the reflectors is caused by irregular acquisition geometries. Unfortunately, when the challenge is posed by a complex velocity function in the overburden that distorts the wavefield, one must switch to image-domain regularization methods, which are based on full prestack migration. Two main factors conspire to make image-domain regularization a much more challenging computational problem than data-domain regularization is. First, the computational cost of prestack migration is substantially larger (by orders of magnitude) than the computational cost of the imaging operators (e.g., interpolation, AMO) used in the data-domain methods. Second, the noniterative normalization methods, which often are successful in the data domain, are less effective in eliminating imaging artifacts when applied in the image domain. This second characteristic is a direct consequence of the fact that the zone of influence of prestack migration (i.e., the migration aperture) is much wider than the corresponding aperture of data-domain operators. When the model space of the imaging operator **L** is the image space, gaps in the wavefield, caused either by irregular acquisition or by transmission effects, make the matrix **L′L** even less dominant diagonally than when the model space of **L** is still in the data space. Consequently, the diagonal approximations of the inverse of **L′L**, which are the basis of the noniterative regularization methods discussed in the previous sections, become less effective. Therefore, one must perform iterative inversion of the computationally expensive prestack migration operator. The compounding of these two problems makes the computational cost of image-domain regularization methods exorbitant, and they might be difficult to justify with today's computers (2007). However, expected improvement in the cost-effectiveness of computer technology likely will make these methods affordable a few years from now.

Valenciano et al. (2006) proposed a target-oriented approach that is intended to reduce the computational cost of regularized inversion for improving imaging under complex overburden. The target-oriented approach has the potential to make regularized inversion a practical tool for reservoir characterization, when the target of the imaging project (e.g. the hydrocarbon reservoir) must be imaged as accurately as possible and is only a small subset of the entire image.

Imaging under salt edges is an application of image-domain regularization methods that is likely to have a practical impact in the future. Figure 8 shows a migrated inline section (left panel) and an ADCIG (right panel) obtained from 2D prestack migration of a 3D data set acquired in the deep waters of the Gulf of Mexico. The vertical lines superimposed on the two panels mark the midpoint and aperture angle, respectively, where the other panel cuts through the prestack-image cube. The inline section shows a clear illumination gap below the nose of the salt and a less obvious but nonetheless troublesome gap farther right, under the salt edge ($x_m \approx 24$ km). Within the gap and on both of its sides, migration artifacts caused by the discontinuities of the wavefield affect the image. The ADCIG has a gap near the middle of the angular range. The artifacts are even more evident in the ADCIG than in the inline section; they appear as diffractions cutting across the illumination gap. A simple noniterative normalization might equalize the aperture across the illumination gaps, but it cannot remove the imaging artifacts.

Figure 9 shows an example of incomplete illumination under a salt edge in a synthetic data set (Paffenholz et al., 2002). The illumination problem here is similar to the

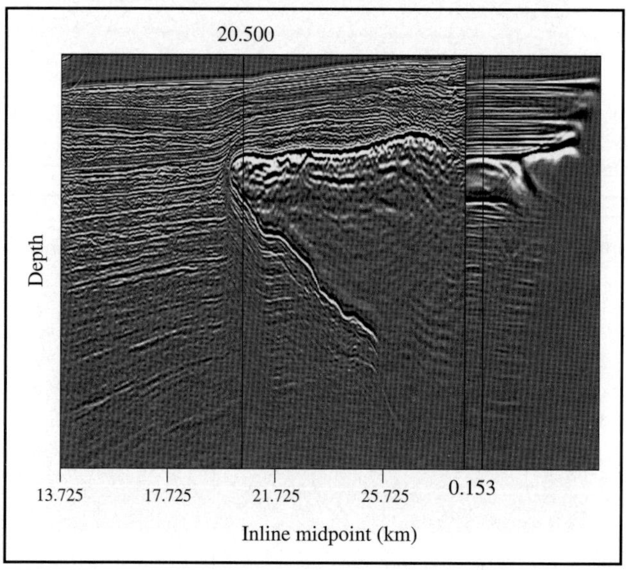

Figure 8. An inline section (left panel) and an ADCIG (right panel) extracted from the prestack-image cube of a 3D data set acquired in the deep waters of the Gulf of Mexico. Notice the illumination gaps and the imaging artifacts in both the inline section and the ADCIG. Figure is from M. L. Clapp (2005).

real Gulf of Mexico example shown in Figure 8. As does Figure 8, Figure 9 shows an inline section (left panel) and an ADCIG (right panel). However, in contrast to Figure 8, the inline section shown in Figure 9 is the result of the stack of all the aperture angles.

The advantage of analyzing a synthetic data set is that the velocity function is perfectly known. The imaging problems cannot be attributed to an inaccurate velocity estimate, as they could be in a real data case. Furthermore, one can gain valuable insight by modeling wave propagation accurately through the known velocity model with both ray-tracing and full-waveform modeling. Figure 10 shows the results of such modeling for a point source located in the poorly illuminated area of the subsurface, at the inline location of the ADCIG shown in Figure 9. The figure shows a snapshot of the wavefield at constant time; the corresponding rays and kinematic wavefront are superimposed onto the wavefield. In agreement with the imaging results shown in Figure 9, the energy corresponding to the middle of the aperture-angle range never reaches the surface. The left propagation lag of these events is deflected downward by the bottom of the salt nose.

Prucha and Biondi (2002) proposed to address the illumination problem under salt edges by a regularized inversion of wavefield-continuation migration. They proposed to minimize the objective function

$$J_{\text{reg}}(\mathbf{m}) = \|\mathbf{Lm} - \mathbf{d}\|_2 + \varepsilon_D \mathbf{m}'\mathbf{A}'\mathbf{Am}, \tag{9.16}$$

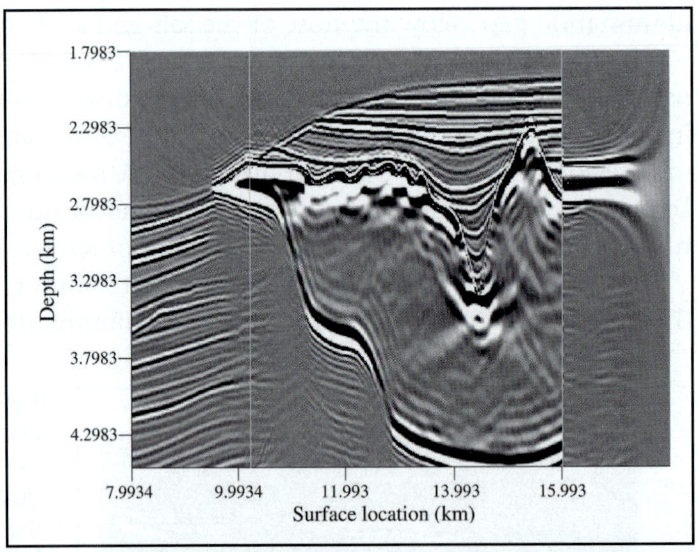

Figure 9. A migrated inline section (left panel) and an ADCIG (right panel) for a 2D synthetic data set. The illumination gaps and the imaging artifacts are similar to the real-data case shown in Figure 8.

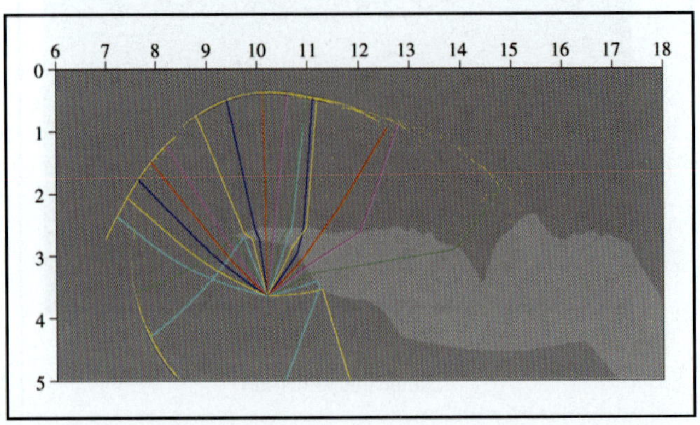

Figure 10. Waveform and ray-tracing modeling of seismic propagation from a point source located in the poorly illuminated area of the subsurface, at the inline location of the ADCIG shown in Figure 9. Notice that the energy corresponding to the middle of the aperture-angle range never reaches the surface.

where the modeling operator, **L**, is the adjoint of full-prestack wavefield-continuation migration and the regularization operator, **A**, is a roughening operator along the reflection-angle axes of the prestack image (see Chapter 3). Conceptually, this objective function is similar to the data-domain objective function expressed in equation 9.13. The regularization along the reflection angles is motivated similarly by the continuity of reflection coefficients as a function of aperture angle and azimuth, at least for precritical reflections. Because its origin is based on the physics of reflections, Prucha and Biondi dubbed this regularization method geophysical regularization. Similarly to the regularized inversion applied to velocity updating discussed in Chapter 7, the objective function in equation 9.16 needs a model-vector preconditioning for accelerating the convergence of iterative solutions (Biondi, 2006).

Figure 11 compares the prestack images obtained by simple migration (a) with the images obtained after 10 iterations of a preconditioned conjugate-gradient algorithm (b) applied to the minimization of the objective function expressed in equation 9.16, for the same inline location used for the images shown in Figure 8. The inversion was successful in removing most of the imaging artifacts from both the inline section and the ADCIG. In the inline section, several reflectors show better continuity at the edges of the illumination gaps. The ellipses superimposed onto the images highlight the areas where the inversion had the largest effects.

The improved lateral continuity of the reflectors in the illumination gaps is even more evident in the stacked sections shown in Figure 12. As in Figure 11, panel (a) in Figure 12 shows the results of simple migration and panel (b) shows the results after 10 iterations of the regularized inversion. Ellipses highlight the areas of most noticeable improvements.

The previous figures demonstrate that iterative inversion can improve the image of poorly illuminated reflectors. The illumination gaps have not been completely healed,

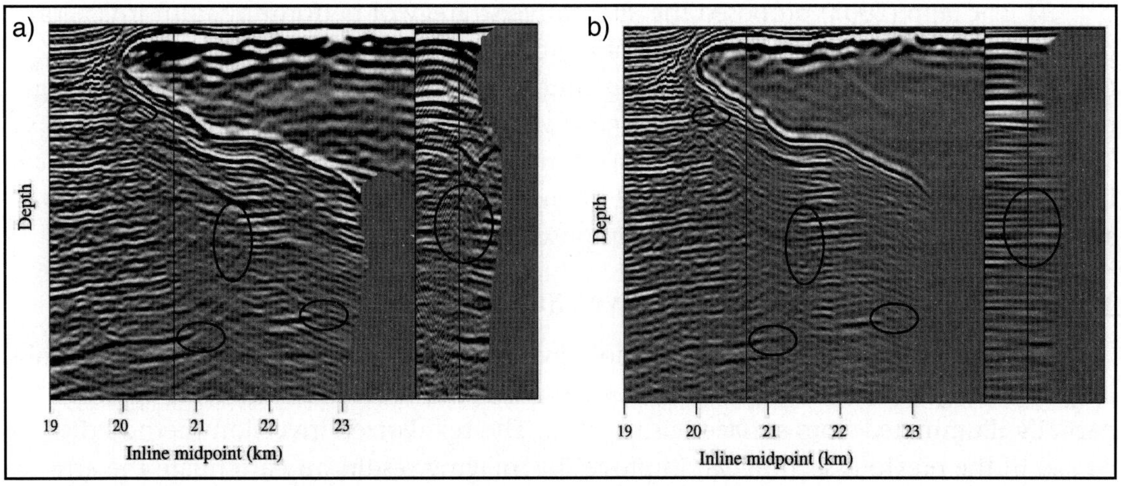

Figure 11. Inline sections (left panel) and ADCIGs (right panel) extracted from the prestack-image cubes obtained (a) by migration and (b) by 10 iterations of the regularized inversion. Notice the substantial reduction of imaging artifacts and increased continuity of the imaged events along both the midpoint and aperture-angle axes, achieved by the regularized inversion. Figure is from M. L. Clapp (2005).

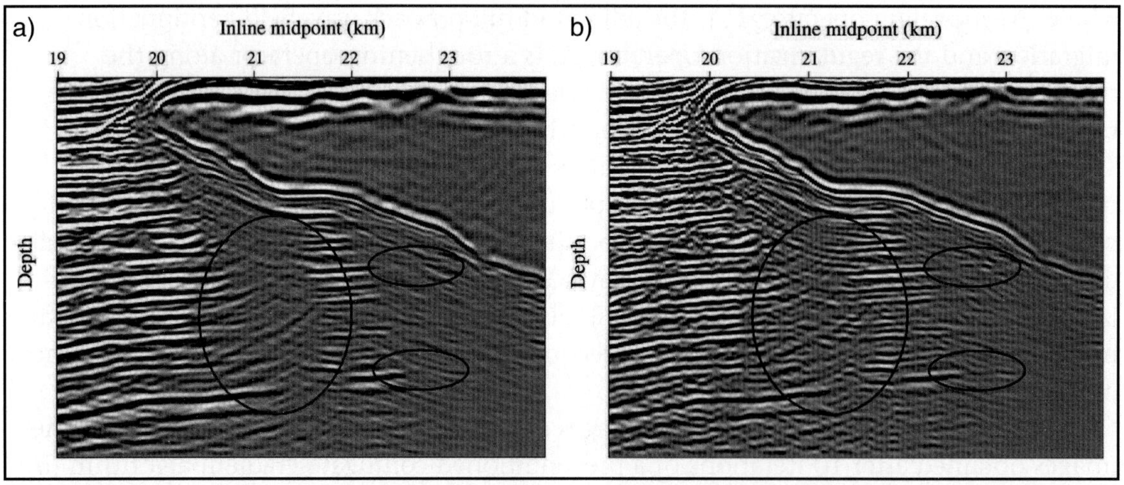

Figure 12. Inline stacked sections obtained (a) by migration and (b) by 10 iterations of the regularized inversion. The reflectors in the illumination gaps are more continuous in the inversion results than in the migration results. Figure is from M. L. Clapp (2005).

but the improved reflector continuities may lead to a more accurate geologic interpretation of the subsalt structure. The illumination gaps might have been healed even further, if more than 10 iterations had been performed. On the other hand, the computational cost of 10 conjugate-gradient iterations is already high — roughly equivalent to the computational cost of 20 prestack migrations. Prucha and Biondi (2002) proposed to add an additional regularization term — dubbed *geologic regularization* — that rewards interpreted geologic dips. The authors showed that imposition of these additional constraints might cause the illumination gaps to fill in faster than does the use of only geophysical regularization. However, their method has the drawback of biasing the inversion with a priori knowledge based on a geologic interpretation, so it is not fully objective.

M. L. Clapp (2003) proposed the alternative strategy of performing joint inversion of two or more 3D data sets acquired along different directions. The illumination gaps depend strongly on the prevailing data azimuth, and thus two data sets acquired along different directions are likely to complement each other in the poorly illuminated areas. This strategy is viable in practice, because many subsalt areas of commercial interest have been covered by several data sets, or, as I discuss in the next section, wide-azimuth data sets are now acquired to image under complex salt bodies.

Wide-azimuth Acquisition for Improved Subsurface Illumination

The imaging examples shown in the previous section clearly demonstrate the challenges of imaging under complex overburden when the subsurface reflectors are only partially illuminated from surface seismic data. The regularized inversion method discussed in the previous section can improve the imaging results and attenuate the artifacts in the final image. However, it obviously cannot create a good image from reflectors that have not been illuminated at all from the available data. An alternative solution to the poor imaging of subsalt reflectors is to acquire data with a wider azimuthal range than is usually done in conventional one-vessel marine acquisition. In Chapter 1, I describe recent advances in wide-azimuth acquisition using multiple boats instead of only one. In this section, I show (1) the imaging results obtained from a comprehensive synthetic-data study of the effects that the range of azimuths present in the input data exerts on the imaging of a realistic subsalt structure (Regone, 2006), (2) a comparison of the image obtained from a wide-azimuth field-data set with the image obtained from a conventionally acquired narrow-azimuth data set (Michell et al., 2006), and (3) comparison of the image obtained from a rich-azimuth field data set with the image obtained from a conventionally acquired narrow-azimuth data set (Howard and Moldeveanu, 2006).

The examples I show clearly demonstrate that broadening the acquisition azimuth significantly improves image quality and in many cases justifies the substantial increase in data-acquisition cost. Furthermore, the additional cost of wide-azimuth acquisition is likely to decrease as the technology is applied more broadly beyond field-development projects and extended to exploration-oriented projects. Widespread acquisition of wide-azimuth data, and the high costs incurred to acquire those data, also are likely to result in the practical application of imaging methods based on regularized inversion, such as the one presented in the previous section. This is because computational costs are bound to decrease faster than data-acquisition costs, and incentives will arise to extract as much information as possible from expensively acquired data.

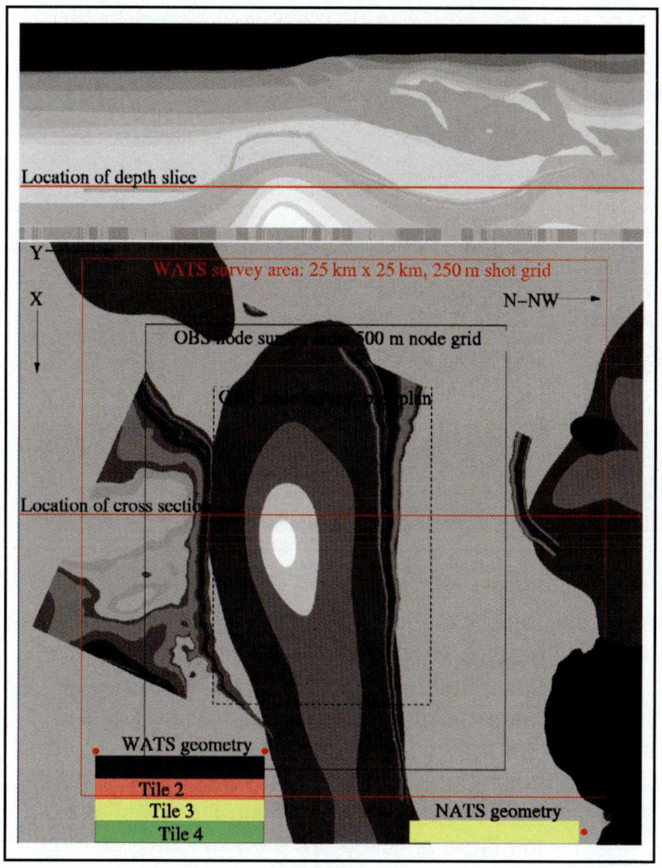

Figure 13. A density model assumed to model the synthetic data sets used to analyze the effect of acquisition geometry on subsurface illumination and imaging. The top panel shows an inline section through the model and the bottom panel shows a depth slice. The red lines superimposed onto each image indicate the location of the other section. The footprints of four boat passes (see Figure 5 in Chapter 1) are represented at the bottom-left of the figure and labeled Tile 1 through Tile 4. Figure is from Regone (2006).

Synthetic-data study of the effects of acquisition azimuths on subsalt imaging

Figures 13 through 18 show the results of a synthetic-data study conducted by Carl Regone at BP (2006) on the effects of acquisition azimuths on the quality of subsalt images. Figure 13 displays the density model assumed to model the synthetic data sets used in this study. The top panel shows an inline section through the model and the bottom panel shows a depth slice. The red lines superimposed onto each image indicate the location of the other section. The wide-azimuth towed-streamer acquisition geometry (WATS) described in Chapter 1 requires a maximum of four boat passes by the recording vessels; the footprints of four boat passes, each pass employing three boats (see Figure 5 in Chapter 1), are represented at the bottom-left of the figure and labeled Tile 1 through Tile 4. Notice that the right flank of the target anticline is in the shadow of a complex salt body, whereas the left flank is in the clear.

Figure 14 shows the inline section and depth slice of the image obtained from the data set modeled assuming conventional narrow-azimuth marine-acquisition geometry. The subsalt image is affected by strongly coherent noise, and the reflectors delimiting the right flank of the large anticline are difficult to interpret. In comparison, Figure 15

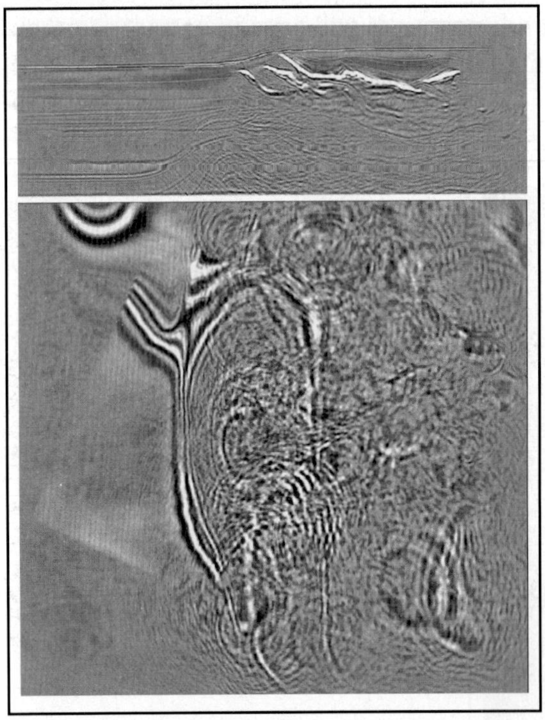

Figure 14. An inline section and a depth slice of the image obtained from the data set modeled assuming conventional narrow-azimuth marine-acquisition geometry. The subsalt image is affected by strongly coherent noise, and the reflectors delimiting the right flank of the large anticline are difficult to interpret. Figure is from Regone (2006).

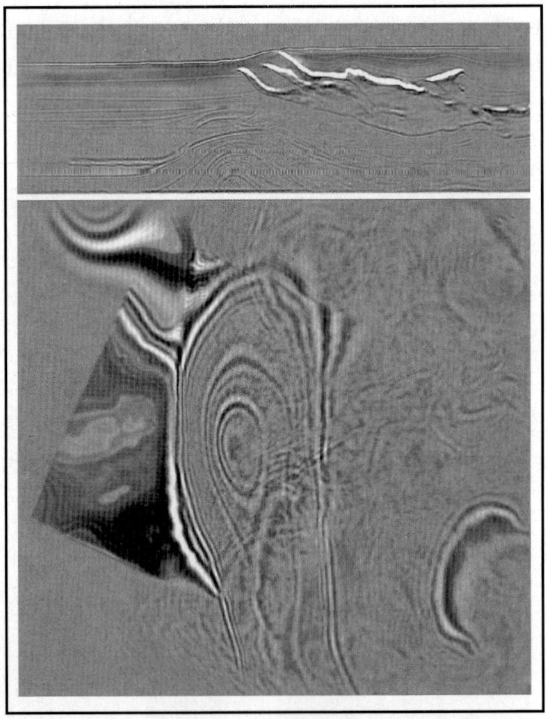

Figure 15. An inline section and a depth slice of the image obtained from the data set modeled assuming wide-azimuth acquisition using the data acquired in four boat passes; that is, all four tiles. The subsalt structure is well imaged and most of the strongly coherent noise present in Figure 14 has been eliminated. The reflectors delimiting the right flank of the large anticline are easy to interpret. Figure is from Regone (2006).

shows the inline section and depth slice of the image obtained from the data set modeled assuming wide-azimuth acquisition using the data acquired in four passes with three boats; that is, all four tiles in Figure 13. The subsalt structure is well imaged here and most of the strongly coherent noise present in Figure 14 has been eliminated. The reflectors delimiting the right flank of the large anticline are easy to interpret.

Figures 16 and 17 demonstrate that the high-quality results obtained by wide-azimuth data and shown in Figure 15 are sensitive to the number of boat passes and thus to the data-acquisition efforts. Figure 16 shows the inline section and depth slice of the image obtained from the data set modeled assuming wide-azimuth acquisition using the data acquired in two boat passes; that is, two tiles. The subsalt structure is well imaged, but some of the coherent noise present in Figure 14 starts to reappear in Figure 16 (compared with Figure 15, in which most of the noise has been eliminated). The reflectors delimiting the right flank of the large anticline are interpretable. Finally, Figure 17 shows the corresponding sections obtained from the data set modeled assuming wide-azimuth acquisition using the data acquired in only one boat pass; that is, one tile. The subsalt image is better here than in Figure 14, but the coherent noise significantly hampers the interpretation.

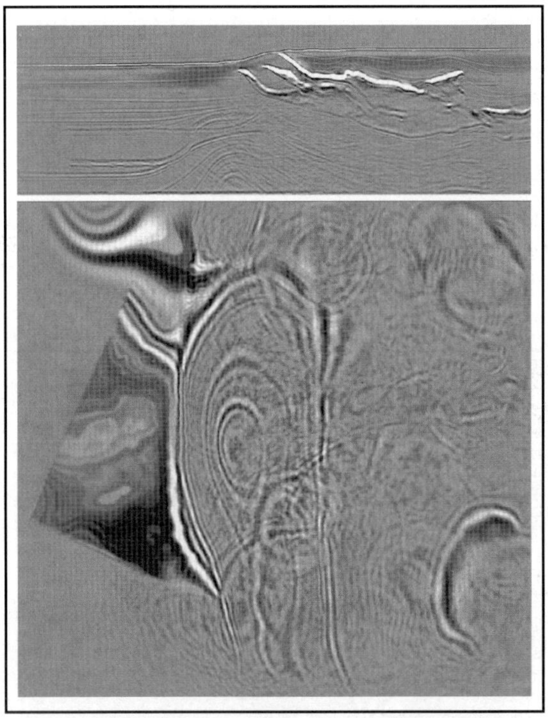

Figure 16. An inline section and a depth slice of the image obtained from the data set modeled assuming wide-azimuth acquisition using the data acquired in two boat passes; that is, two tiles. The subsalt structure is well imaged but some of the coherent noise present in Figure 14 starts to reappear here, compared with Figure 15. The reflectors delimiting the right flank of the large anticline are interpretable. Figure is from Regone (2006).

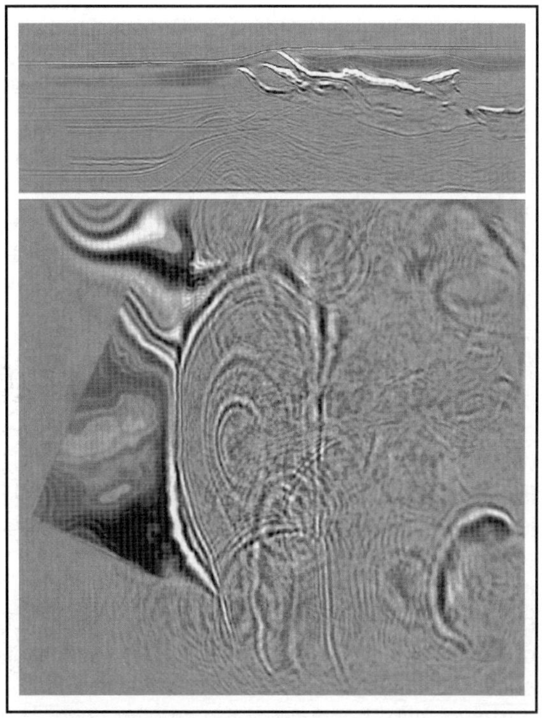

Figure 17. An inline section and a depth slice of the image obtained from the data set modeled assuming wide-azimuth acquisition using the data acquired in only one boat pass; that is, one tile. The subsalt structure is imaged better here than in Figure 14, but the coherent noise significantly hampers the interpretation. Figure is from Regone (2006).

For development projects, production infrastructure often impedes surface-streamer acquisition. Also, acquiring wide-azimuth data using surface streamers can be expensive and potentially risky. In such cases, ocean-bottom acquisition might be advantageous (see Chapter 1), and it allows the recording of wide-azimuth data that produce high-quality images. Figure 18 shows the inline section and depth slice of the image obtained from the data set modeled assuming an OBS acquisition geometry with a grid similar to the one represented in Figure 10 of Chapter 1. The subsalt structure is well imaged and most of the strongly coherent noise present in Figure 14 has been eliminated here. The reflectors delimiting the right flank of the large anticline are easy to interpret. The quality of the image is comparable to the one shown in Figure 15; however, the spatial extent is narrower here because of the smaller coverage that is economically affordable with OBS recording.

Field-data examples of wide-azimuth and rich-azimuth imaging for subsalt exploration and development

The encouraging results provided by the synthetic-data study presented above motivated BP to acquire wide-azimuth data above a subsalt reservoir in the Gulf of Mexico. The reservoir needed to be characterized accurately in order to be developed efficiently. Figure 19 shows the field-data comparison of images obtained from narrow-azimuth data and from wide-azimuth data. The image in (a) was obtained from the combination of two conventional narrow-azimuth surveys previously acquired in an attempt to satisfactorily image the reservoir. The image in (b) was obtained from a wide-azimuth survey acquired using four boat passes (four tiles) above the same reservoir. The wide-azimuth image of the subsalt reflectors clearly is superior to the narrow-azimuth image. The salt body itself also is better delineated, thus improving the velocity-model-building process.

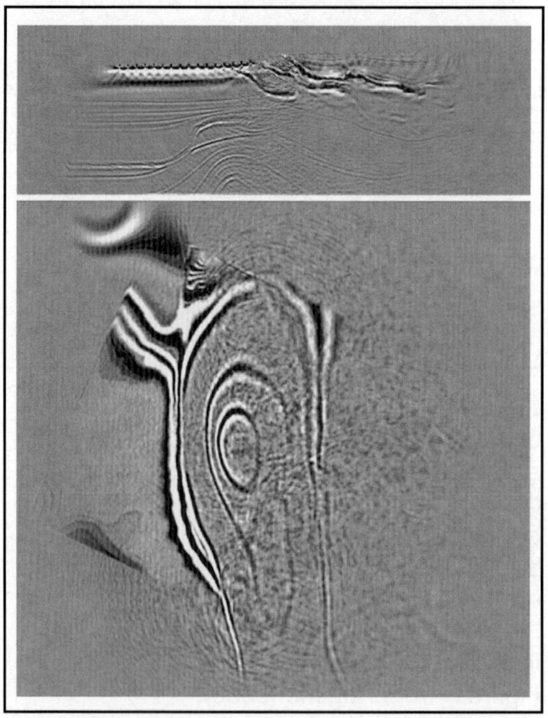

Figure 18. An inline section and a depth slice of the image obtained from the data set modeled assuming an OBS acquisition geometry with a grid similar to the one represented in Figure 10 of Chapter 1. The subsalt structure is well imaged and most of the strongly coherent noise present in Figure 14 has been eliminated. The reflectors delimiting the right flank of the large anticline are easy to interpret. The quality of the image is comparable to the one shown in Figure 15, but the spatial extent is narrower here because of the smaller coverage that is economically affordable with OBS recording. Figure is from Regone (2006).

Figure 20 shows the results obtained by imaging rich-azimuth (RAZ) data acquired using the acquisition geometry described in Chapter 1. Regarding the WATS example shown above, the goal of acquisition was to optimize the development plan for a large oil-reservoir in the subsalt of the Gulf of Mexico (Howard and Moldoveanu, 2006). Image (a) was obtained from a conventional narrow-azimuth survey, whereas image (b) was obtained from the rich-azimuth survey acquired using the acquisition geometry described in Chapter 1. The rich-azimuth image of the subsalt reflectors (Figure 20b)

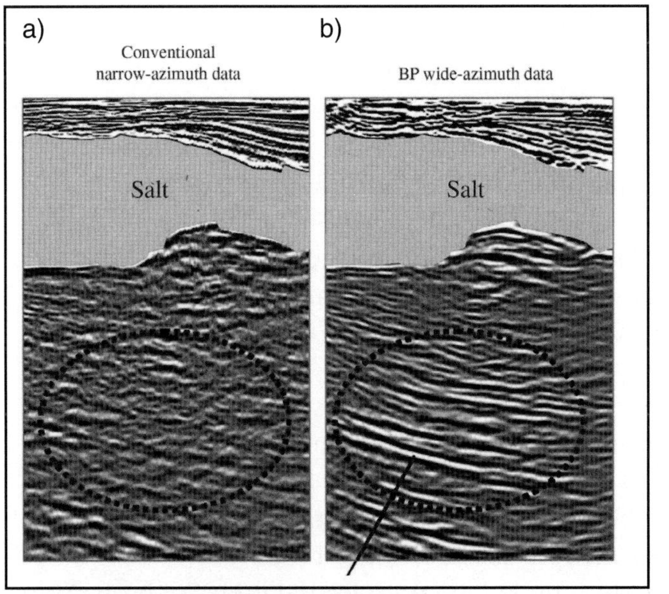

Figure 19. Field-data comparison of an image obtained from narrow-azimuth data and wide-azimuth data. (a) The image obtained from the combination of two conventional narrow-azimuth surveys acquired above a subsalt reservoir in the Gulf of Mexico. (b) The image obtained from a wide-azimuth survey acquired using four boat passes (four tiles) above the same reservoir. The wide-azimuth image of the subsalt reflectors clearly is superior to the narrow-azimuth image. The salt body itself also is delineated better, thus improving the velocity-model-building process. (This figure is courtesy of BP, www.bp.com, Technology Webcast, December 2005.)

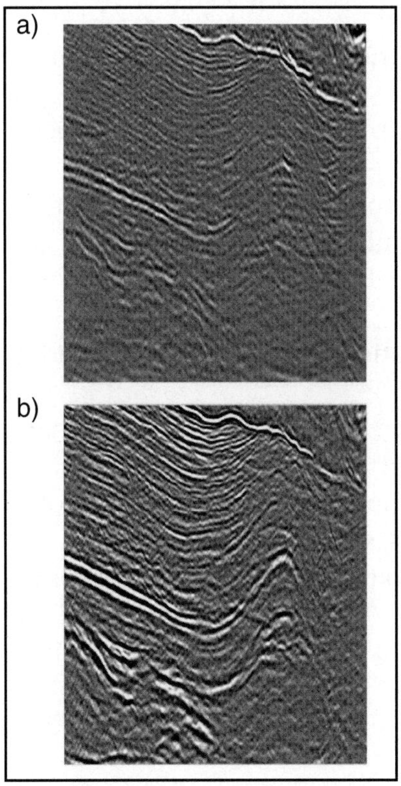

Figure 20. A field-data comparison of images obtained from narrow-azimuth data and rich-azimuth data. Image (a) was obtained from a conventional narrow-azimuth survey acquired above a subsalt reservoir in the Gulf of Mexico. Image (b) was obtained from a rich-azimuth survey acquired using the acquisition geometry described in Chapter 1. The rich-azimuth image of the subsalt reflectors clearly is superior to the narrow-azimuth image. The anticline under the salt body and the sediment terminations against the salt flank are imaged better by the rich-azimuth data. (This figure is courtesy of Mike Howard with BHP Billiton and Nick Moldoveanu with WesternGeco.)

clearly is superior to the narrow-azimuth image (Figure 20a). The anticline under the salt body and the sediment terminations against the salt flank are better defined by the rich-azimuth image.

The previous two field-data examples indicate that poorly imaged reflectors can be imaged substantially better from wide-azimuth data than from conventional narrow-azimuth data. However, accurate imaging of wide-azimuth data sets presents new challenges for the existing migration and velocity-estimation methods. Further algorithmic development is necessary to extract all the information contained in this promising— but expensive to acquire— new kind of marine survey.

References

Beasley, C. J., and E. Mobley, 1988, Amplitude and antialiasing treatment in $(x - t)$ domain DMO: 58th Annual International Meeting, SEG, Expanded Abstracts, 1113–1116.

Biondi, B., 2006, 3D seismic imaging: SEG Investigations in Geophysics, No. 14.

Biondi, B., and I. Vlad, 2002, Amplitude preserving prestack imaging of irregularly sampled 3-D data: 72nd Annual International Meeting, SEG, Expanded Abstracts, 2170–2173.

Bloor, R., U. Albertin, H. Jaramillo, and D. Yingst, 1999, Equalised prestack depth migration: 69th Annual International Meeting, SEG, Expanded Abstracts, 1358–1361.

Canning, A., and G. H. F. Gardner, 1998, Reducing 3-D acquisition footprint for 3-D DMO and 3-D prestack migration: Geophysics, **63**, 1177–1183.

Chemingui, N., 1999, Imaging irregularly sampled 3D prestack data: Ph.D. thesis, Stanford University.

Chemingui, N., and B. Biondi, 2002, Seismic data reconstruction by inversion to common offset: Geophysics, **67**, 1575– 1585.

Claerbout, J., and D. Nichols, 1994, Spectral preconditioning: Stanford Exploration Project, Report 82, 183–186.

Claerbout, J. F., 2004, Image estimation by example: Stanford Exploration Project, http://sepwww.stanford.edu/sep/prof/bei/toc_html/index.html, accessed May 26, 2004.

Clapp, M. L., 2003, Directions in 3-D imaging: Strike, dip, both?: Stanford Exploration Project, Report 113, 363–368.

———, 2005, Imaging under salt: Illumination compensation by regularized inversion: Ph.D. thesis, Stanford University.

Clapp, R. G., 2006, AMO inversion to a common azimuth: 76th Annual International Meeting, SEG, Expanded Abstracts, SI 3.4.

Duquet, B., K. J. Marfurt, and J. A. Dellinger, 2000, Kirchhoff modeling, inversion for reflectivity, and subsurface illumination: Geophysics, **65**, 1195–1209.

Fomel, S., 2001, Three-dimensional seismic data regularization: Ph.D. thesis, Stanford University.

Golub, G. H., and C. F. V. Loan, 1983, Matrix computations: John Hopkins University Press.

Howard, M. S., and N. Moldeveanu, 2006, Marine survey design for rich-azimuth seismic using surface streamers: 76th Annual International Meeting, SEG, Expanded Abstracts, SS 2.

Hughes, T. J. R., 1987, The finite element method: Linear static and dynamic finite element analysis: Prentice-Hall.

Kuehl, H., and M. Sacchi, 2002, Robust AVP estimation using least-squares wave-equation migration: 72nd Annual International Meeting, SEG, Expanded Abstracts, 281–284.

Michell, S., E. Shoshitaishvili, D. Chergotis, J. Sharp, and J. Etgen, 2006, Wide-azimuth streamer imaging of Mad Dog: Have we solved the subsalt imaging problem?: 76th Annual International Meeting, SEG, Expanded Abstracts, SS2.5.

Nemeth, T., C. Wu, and G. T. Schuster, 1999, Least-squares migration of incomplete reflection data: Geophysics, **64**, 208–221.

Paffenholtz, J., B. McLain, J. Zaske, and P. Keliher, 2002, Subsalt multiple attenuation and imaging: Observations from the Sigsbee2B synthetic dataset: 72nd Annual International Meeting, SEG, Expanded Abstracts, 2122–2125.

Prucha, M. L., and B. L. Biondi, 2002, Subsalt event regularization with steering filters: 72nd Annual International Meeting, SEG, Expanded Abstracts, 1176–1179.

Regone, C., 2006, Using 3D finite-difference modeling to design wide azimuth surveys for improved subsalt: 76th Annual International Meeting, SEG, Expanded Abstracts, SS 2.3.

Rickett, J., 2003, Illumination-based normalization for wave-equation depth migration: Geophysics, **68**, 1371–1379.

Ronen, J., 1987, Wave equation trace interpolation: Geophysics, **52**, 973–984.

Ronen, S., and C. L. Liner, 2000, Least-squares DMO and migration: Geophysics, **65**, 1364–1371.

Strang, G., 1986, Linear algebra and its applications: Harcourt Brace Jovanovich College Publishers.

Valenciano, A. A., B. Biondi, and A. Guitton, 2006, Target-oriented wave-equation inversion: Geophysics, **71**, A35-A38.

Index

J

K

L

M

N

O

S

W

Z